Post-Optimization Lead and Copper Control Monitoring Strategies

The mission of the Awwa Research Foundation (AwwaRF) is to advance the science of water to improve the quality of life. Funded primarily through annual subscription payments from over 1,000 utilities, consulting firms, and manufacturers in North America and abroad, AwwaRF sponsors research on all aspects of drinking water, including supply and resources, treatment, monitoring and analysis, distribution, management, and health effects.

From its headquarters in Denver, Colorado, the AwwaRF staff directs and supports the efforts of over 800 volunteers, who are the heart of the research program. These volunteers, serving on various boards and committees, use their expertise to select and monitor research studies to benefit the entire drinking water community.

Research findings are disseminated through a number of technology transfer activities, including research reports, conferences, videotape summaries, and periodicals.

Post-Optimization Lead and Copper Control Monitoring Strategies

Prepared by:
Gregory Kirmeyer and **Brian Murphy**
Economic and Engineering Services, Inc.
10900 NE 4th St, Ste 1110, Bellevue, WA 98004

Anne Sandvig
Sandvig Consulting
525 Belair Dr., Custer, SD 57730

Gregory Korshin and **Birol Shaha**
Department of Civil and Environmental Engineering, University of Washington
Box 352700, Seattle, WA 98195-2700

Massimiliano Fabbricino
Department of Hydraulic and Environmental Engineering Girolamo Ippolito
University of Naples Federico II
Via Claudio 21, 80125 Naples, Italy

and

Gary Burlingame
Bureau of Laboratory Services, Philadelphia Water Department
1500 East Hunting Park Avenue, Philadelphia, PA 19124-4941

Sponsored by:
Awwa Research Foundation
6666 West Quincy Avenue, Denver, CO 80235-3098

Published by:

Awwa Research Foundation
Advancing the Science of Water®

IWA Publishing

DISCLAIMER

This study was funded by the Awwa Research Foundation (AwwaRF). AwwaRF assumes no responsibility for the content of the research study reported in this publication or for the opinions or statements of fact expressed in the report. The mention of trade names for commercial products does not represent or imply the approval or endorsement of AwwaRF. This report is presented solely for informational purposes.

Published by IWA Publishing, Alliance House, 12 Caxton Street, London SW1H 0QS, UK

Telephone: +44 (0) 20 7654 5500; Fax: +44 (0) 20 7654 5555; Email: publications@iwap.co.uk;
Web: **www.iwapublishing.com**

AwwaRF report number 90996F. Originally published by AwwaRF for its subscribers in 2003.
IWA Publishing version published 2004.

ISBN 1 84339 8923

CONTENTS

TABLES

FIGURES

FOREWORD

The Awwa Research Foundation is a nonprofit corporation that is dedicated to the implementation of a research effort to help utilities respond to regulatory requirements and traditional high-priority concerns of the industry. The research agenda is developed through a process of consultation with subscribers and drinking water professionals. Under the umbrella of a Strategic Research Plan, the Research Advisory Council prioritizes the suggested projects based upon current and future needs, applicability, and past work; the recommendations are forwarded to the Board of Trustees for final selection. The foundation also sponsors research projects through the unsolicited proposal process; the Collaborative Research, Research Applications, and Tailored Collaboration programs; and various joint research efforts with organizations such as the U.S. Environmental Protection Agency, the U.S. Bureau of Reclamation, and the Association of California Water Agencies.

This publication is a result of one of these sponsored studies, and it is hoped that its findings will be applied in communities throughout the world. The following report serves not only as a means of communicating the results of the water industry's centralized research program but also as a tool to enlist the further support of the nonmember utilities and individuals.

Projects are managed closely from their inception to the final report by the foundation's staff and large cadre of volunteers who willingly contribute their time and expertise. The foundation serves a planning and management function and awards contracts to other institutions such as water utilities, universities, and engineering firms. The funding for this research effort comes primarily from the Subscription Program, through which water utilities subscribe to the research program and make an annual payment proportionate to the volume of water they deliver and consultants and manufacturers subscribe based on their annual billings. The program offers a cost-effective and fair method for funding research in the public interest.

A broad spectrum of water supply issues are addressed by the foundation's research agenda: resources, treatment and operations, distribution and storage, water quality and analysis, toxicology, economics, and management. The ultimate purpose of the coordinated effort is to assist water suppliers in providing the highest possible quality of water economically and reliably.

It was the goal of this research effort to provide corrosion control optimized utilities with tools that better address their monitoring needs in a manner that would be acceptable to regulators. In pursuit of this goal, the research team has developed a statistically valid alternative in-home tap monitoring protocol to more accurately track lead and copper levels following corrosion control optimization. This project also presents initial developmental achievements toward a new tool for predictive testing for lead and copper corrosion.

The true benefits are realized when the results are implemented at the utility level. The foundation's trustees are pleased to offer this publication as a contribution toward that end.

Edmund G. Archuleta, P.E.
Chair, Board of Trustees
Awwa Research Foundation

James F. Manwaring, P.E.
Executive Director
Awwa Research Foundation

ACKNOWLEDGMENTS

The authors would like to acknowledge the gracious financial support of the Awwa Research Foundation (AwwaRF), Pinellas County Water Department, and the San Francisco Water Department. Without their support, this project would not have been possible.

The authors wish to extend their thanks and appreciation to Jennifer Warner, AwwaRF Project Manager and the following AwwaRF Project Advisory Committee members for their expertise and constructive guidance and contributions throughout the project:

Dr. Darren Lytle, United States Environmental Protection Agency, Cincinnati, OH
Angela O'Brien, Zone 7 Water Agency, Livermore, CA
Dr. Loay Hidmi, Bitar Consultants, Amman, Jordan

The authors also extend a special thanks to Dr. Alexander Velichenko, Ph.D., Ukranian State Chemical Technology University, for his technical expertise; Dr. Steve Reiber, HDR, Inc. and Dr. John Ferguson, University of Washington for their in-kind contributions to the project and participation as members of the project's Technical Advisory Panel; and Lori Radder, Economic and Engineering Services, Inc. for writing and editorial support.

Several utilities from across the country offered significant time, staff, expertise, and facilities, for the gathering of data presented in this report in the form of case and field studies. The authors gratefully acknowledge their efforts:

Philadelphia Water Department, PA
Hastings Utilities, NE
Detroit Water and Sewerage Department, MI
North Clackamas County Water Commission, OR
Seattle Public Utilities, WA
San Francisco Water Utilities, CA
Pinellas County Water District, FL
City of Fort Worth, TX
Portland Water Bureau, OR
Skagit County Public Utility Dist. #1, WA
Metropolitan Utilities District, NE
Hillsborough County Water Dept., FL

The authors also wish to thank the state regulatory agencies who provided information on project surveys.

Finally, the authors wish to thank Druann O'Connor, Administrative Assistant of Economic and Engineering Services, Inc. for her efforts in bringing this work together into its final form.

EXECUTIVE SUMMARY

INTRODUCTION

Awwa Research Foundation (AwwaRF) project #2679, *Post-Optimization Lead and Copper Control Monitoring Strategies* was designed to explore the concerns that have been expressed by utilities that were previously optimized for corrosion control. The goal of the project was to develop alternative methods, which are acceptable to regulatory primacy agencies, of demonstrating that corrosion control is being maintained when treatment techniques and/or source waters are altered.

The intention of the alternative methodologies developed and described in this report was to provide primacy agencies and utilities with tools to verify that existing optimized corrosion control has been maintained; particularly, after changes in treatment or source waters. These methodologies are in no way intended to supplant Lead and Copper Rule (LCR) compliance monitoring requirements. The specific objectives of the project were to:

- Develop a statistically valid alternative home tap sampling protocol to track the metal uptake associated with lead and copper corrosion.
- Develop an on-line, accelerated testing methodology and associated equipment to track and predict impacts on lead and copper corrosion during treatment/source water changes.
- Perform case study investigations describing utility experience in achieving optimized corrosion control and the monitoring challenges following optimization.
- Assess regulator interpretation of the LCR *Minor Revisions* and their potential acceptance of alternative in-home tap sampling and predictive tools.

What is the impetus for alternative monitoring approaches?

The Lead and Copper Rule *Minor Revisions* (Minor Revisions) (January 12, 2000) state that…

> *"Any water system deemed to have optimized corrosion control…shall notify the State in writing [no later than 60 days after making the change]…of any change in treatment or the addition of a new source. The State may require any such system to conduct additional monitoring or to take other action the State deems appropriate to ensure that such systems maintain minimal levels of corrosion in the distribution system."*

The intent of the *Minor Revisions* is that systems continue to operate and maintain optimal corrosion control for both lead and copper while providing flexibility in the method of verification. The *Minor Revisions* left the determination of the method for verification of optimization at the discretion of the primacy agencies. Without alternative means to accomplish verification of continued corrosion control optimization, primacy agencies' only available option would be to require that water systems conduct additional customer in-home tap compliance sampling that may not be timely, cost-effective or sufficiently predictive and reliable.

Why is there a need for new monitoring approaches?

There were several identified reasons for the need to investigate alternative monitoring approaches. Those factors were variability in data, shrinking sample pools, increasing costs associated with monitoring programs, and the differing needs of utilities verifying optimization and corrosion control versus those trying to achieve optimization in the first place. As utilities achieve corrosion control, and lead and copper concentrations are reduced below the action levels, the variability in the data becomes a greater issue. As sample pools shrink due to home remodeling and lack of customer interest, the remaining sites will have less reliable results unless the variability can be reduced concurrently. This variability makes it difficult for a utility to accurately evaluate its corrosion control practices. In order to maintain a statistically representative sample pool, utilities report being faced with the conflict of trying to maintain a large sample pool in the face of fewer and fewer appropriate sites.

The alternative tap monitoring protocol recommended in this report simultaneously addresses control of data variability and attrition of sample pools by recommending the sampling of fewer sites more often. The alternative protocol provides utilities and regulators with a tool to continue to collect statistically sound data in the face of fewer qualified sample locations. Cost estimates by participating utilities also indicate that the alternative in-home tap monitoring protocol can be conducted at a significantly reduced cost.

The development of a predictive technology begun in this project may, with further development, go to the next step by eliminating the need for customer participation and in-home monitoring and reducing the cost of monitoring even further. This objective, however, was not realized by this project.

What was the approach to the project?

The project was conducted through six distinct project elements. These project elements were intended to characterize the current situation, develop potential methods for monitoring lead and copper levels, and evaluate the current state of the developed monitoring methods. Each project element is described below.

Element 1 - Regulator Survey

The regulatory agency survey was conducted as a means of gathering information from state primacy agencies for several reasons: to identify barriers to alternative compliance strategies for previously optimized utilities, to identify primacy agency expectations, and to assess the willingness of primacy agencies to accept alternative compliance methods such as those presented in this project.

The project survey represented responses from 33 of 50 state primacy agencies. Of the thirty-three respondents, twenty-eight reported having addressed the LCR *Minor Revisions* sufficiently to be able to provide information pertaining to their interpretation. The results of the survey and summaries of specific questions are included in the report.

Element 2 – Participating Utility Case Studies

The objective of the project's case studies was to document utility experiences with the process of attaining corrosion control optimization and the issues that have been addressed as treatment and source changes were considered and implemented. Eleven case studies are included in this report.

Element 3 – Development of Alternative In-home Tap Monitoring Protocol

Developing statistically valid alternative in-home tap monitoring guidelines was a core objective of the project. Monitoring guidelines were developed to address the issues of sample variability, shrinking sample pools, and utility resources. The monitoring guidelines provide utilities with a tool to more accurately track corrosion control changes with lead and copper levels.

The alternative protocol was developed around key statistical factors used to evaluate the monitoring data. These factors included: summary statistics; typical variabilities; and intra-site vs. inter-site variability.

Information on these key factors was summarized from historical literature and case studies of both regulatory monitoring results and alternative distribution system monitoring completed by utilities before and after corrosion control treatment implementation. Based on the historical data, draft alternative monitoring guidelines and corresponding statistical evaluation methods were developed. These guidelines include estimates of the number of sites and the frequency of sampling for a variety of desired accuracy and confidence levels.

Element 4 – Development of On-line Corrosion Monitoring Protocol

This project element investigated the potential of a centralized, on-line device to monitor changes in lead and copper corrosion rates. The project developed, tested and validated a novel monitoring and analytical protocol intended to provide utilities with a tool to accurately quantify effects of corrosion control and to detect and quantify minor differences in corrosion rates and metal release with higher accuracy, greater time resolution and lower cost.

Implementation and verification of the On-Line Corrosion Cell (OLCC) required development of the device and generation of an adequate amount of experimental data. The OLCC used Corrosion Potential-Stagnation/Flow (CPSF) techniques to evaluate and measure corrosion rates and corrosion potential change.

Development of the OLCC and the CPSF method included preliminary modeling of the relevant processes and the expansion of theoretical calculations to predict the behavior of corrosion potential and rate of metal release. The method development included consideration of the corrosion potential dependence on intrinsic corrosion conditions such as cell design, concentration of oxygen and chlorine, pH, and other water quality parameters.

Element 5 – Participating Utility Field Tests

Field test evaluations were conducted for the alternative in-home tap monitoring protocol, OLCC monitoring protocols, automated sampler operation and comparative pipe loop operations. Four utilities participated in the testing of in-home tap monitoring protocols; two utilities installed and tested the OLCC device; and two utilities provided data from long-term operation of pipe loops. As the sole participant in the automated sampler tests, the Philadelphia Water Department installed and operated multiple devices. The results from the various field test activities were used to refine the alternative tap monitoring protocol and verify the bench-scale findings of the on-line corrosion cell. Field test reports for each of the activities are included in the report.

Element 6 – Proposed Monitoring Strategies

The report concludes with a synthesis of the findings from case and field studies, data evaluations, and survey responses. The result is proposed alternative lead and copper corrosion control monitoring strategies for systems previously optimized for corrosion control. Included are proposed in-home tap sampling protocol and statistical evaluation methodologies and applicability of on-line corrosion cell predictive technology (including design and operation parameters). This section of the report discusses the applicability and drawbacks of each approach.

A report summary is provided that discusses the applicability of the project's findings, additional research opportunities identified during the course of the project, and the identified next steps in assisting water systems in addressing this on-going and increasingly challenging regulatory requirement.

What is the recommended alternative in-home monitoring protocol and what are the benefits?

Previous studies have shown that lead levels in tap water can vary significantly, both from similar sites that are in close proximity in the distribution system and are served by water of similar chemical characteristics, as well as from repeat samples taken from the same site under identical conditions.

While the existing regulatory protocol provides compliance information, it may not be the best protocol for assessing post-optimization conditions; i.e. changes in lead and copper levels after a water quality change in a previously optimized system.

The alternative tap monitoring protocol presented in this report is intended to provide primacy agencies and utilities with a tool to more accurately track lead and copper uptake following optimization. The project identified and ranked variability factors, along with the actual variability measured from historical monitoring studies, and used that information as a starting point in developing a draft alternative protocol. The actual variability measured from historical studies provided initial estimates of the recommended number of sites and the frequency of sampling. Table ES.1 summarizes the alternative tap monitoring protocol.

Table ES.1
Recommended alternative tap monitoring protocol summary

Metal	Lead	Copper
Number of sites	12-43	7-24 (per age category)
Number of samples/site	One per sampling period	
Evaluation period	Quarterly	
Site selection criteria	• Locate sites in areas of the distribution system that exhibit consistent water quality. • Select sites with similar physical conditions with respect to type and age of lead and/or copper source. • Select sites with consistent physical and age characteristics within each specific sample pool.	
	For Lead Sites: • Single family residences with lead solder installed prior to 1982 or lead service lines.	For Copper Sites: • Single family residences with copper piping. • Three age categories: >20 years ≥10 but ≤20 years < 10 years
Data collection and analysis criteria	• 1-Liter total. Split sampling with initial 125 mL followed by 875 mL. • Analyze total lead and copper levels • Minimum 6-hour stagnation time. All samples should have similar stagnation times	

What is an on-line corrosion cell and how can it be used?

An On-line Corrosion Cell (OLCC) is a device suitable for on-line monitoring of corrosion and metal release processes in drinking water. The device in this project was designed using the theories of the corrosion potential stagnation/flow (CPSF) method. The OLCC is designed to enable utility personnel to measure the corrosion potential (E_{corr}) during stagnation episodes, measure metal concentrations in water samples withdrawn from the corrosion cell, and correlate the data.

The OLCC that was deployed for field tests was the narrow rectangular channel (NRC) corrosion cell (Figure ES.1). Additional details of the cell design are shown in Figure ES.2. The cell's design allowed easy assembly or disassembly of the cell at the time of its commissioning or decommissioning.

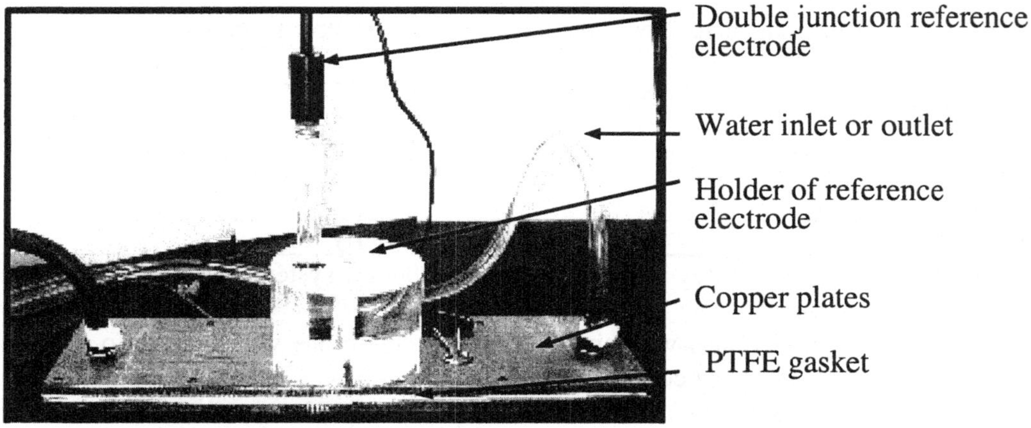

Figure ES.1 General view of the CPSF Narrow Rectangular Channel (NRC) corrosion cell.

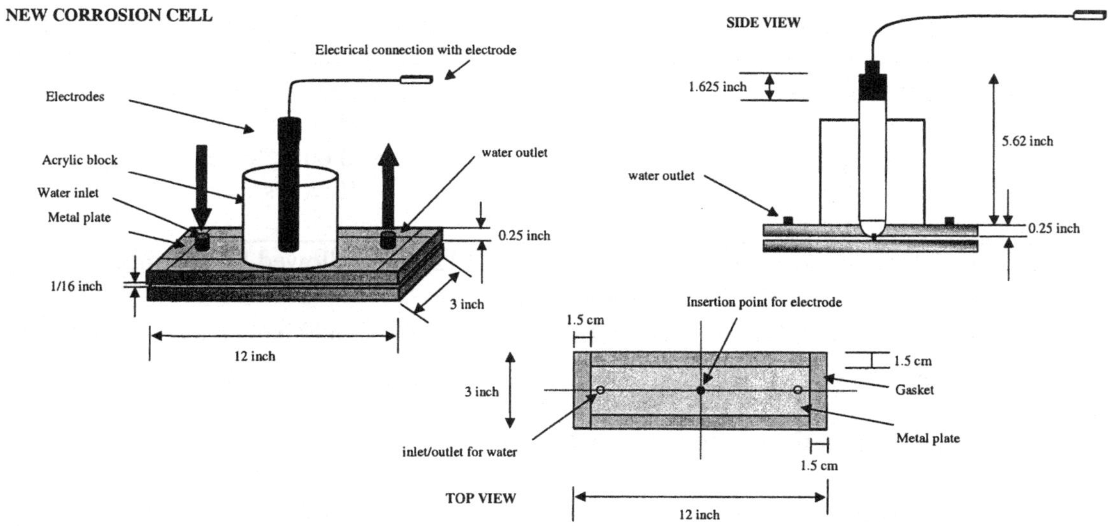

Figure ES.2 Additional dimensional details pertinent to the design of the CPSF NRC cell.

What do these new monitoring approaches tell me that current LCR sampling does not?

The existing regulatory protocol provides a tool for tracking the reduction of lead and copper to below the action level. It is not, however, the most appropriate protocol for assessing changes in customer tap lead and copper levels after corrosion control has been optimized.

Over time, the number of appropriate sampling sites is being reduced due to removal of lead service lines and/or customer unwillingness to continue with in-home monitoring, especially in homes with low lead levels. As the sample pool decreases lead level data potentially becomes skewed toward higher metal concentrations and paired data for evaluating treatment impacts is lost. Also, as water systems become optimized for corrosion control, lead and copper levels are reduced. As lead and copper levels are reduced, variability must concurrently be reduced or a greater number of samples must be collected in order to detect differences with the same degree of confidence and accuracy. Following optimization, statistical problems are intensified as variability does not decrease yet LCR compliance monitoring is performed for a reduced number of samples.

The alternative tap monitoring protocol provides:

- A reduced pool of sampling sites that will be more stable and available for continued, long term monitoring;
- A monitoring schedule that provides adequate data to be able to evaluate tap lead and copper levels in the system at a desired accuracy and confidence level;
- Monitoring at a sufficient frequency to capture changes in metals levels due to water quality changes; and
- Sufficient data to assess the variability in the lead and copper levels and adjust the monitoring protocol to address variability.

The OLCC predictive technology may, when fully developed, provide a tool that:

- Eliminates the need for widespread in-home customer tap sampling.
- Enables utilities to monitor and evaluate the corrosion potential of water several times a day.
- Reduces administrative and field activities, costs, and reduces analytical costs associated with conducting an in-home tap sampling program.
- Reduces data variability issues associated with materials, workmanship, and customer sampling.

What did field application of the alternative monitoring approaches reveal?

The alternative monitoring protocol and the CPSF OLCC predictive technology were tested in the field. Additionally, automated sampling device and pipe loop data were gathered. The following are conclusions from the projects alternative tap monitoring field activities:

- Lead level variation was generally greater than copper and the variation in copper levels was somewhat similar between all field studies.
- In general, intra-site variation was lower than inter-site variation for both metals.
- The evaluation of inter-site and intra-site data collected during the field studies provided for a more reliable determination of lead and copper levels at highly variable sites.
- In split sample activities, lead levels measured in the initial 125 mL sample were slightly elevated when compared to the 875 mL sample, and the variability in lead levels for the 125 mL samples was much higher.
- The direct comparison of alternative tap monitoring data to that of LCR compliance data for the same sites showed very similar or lower lead levels (average, median, and 90th percentile lead values) when compared to LCR monitoring data, in addition to reduced variability in the data.

The following are conclusions from the project's predictive technology field studies:

- The data from the field experiments verified the theoretical analysis and showed that the changes of E_{corr} of the corroding metal during stagnation episodes can be used to monitor metal release and corrosion processes on-line.
- pH and chlorine concentrations were observed to have affects on the values of E_{corr}.
- The relationship of the values of E_{corr} and the logarithm of the total soluble metal concentration in the CPSF cell are consistently linear and strongly correlated. The nature of this relationship is different for unchlorinated and chlorinated water.
- Because other metals showed a similar response, the CPSF method, albeit explored in detail for copper, may be applicable to monitor the corrosion of all metals of interest for drinking water industry.

The automated sampling of lead and copper was conducted in a limited fashion. The objective was to develop a prototype sampling device and collect comparative data. The data collected was used to determine if the potential exists for reducing variability using an automated device.

The preliminary comparative data collected using the automated sampler developed during this project indicated that automated sampling may potentially be of benefit to utilities based on comparison of data to manually collected samples. Further study requires the development of a refined sampler design and an extensive study conducted to determine the limitations, benefits, and reliability of such a device.

The results from this limited study show that automatically collected copper samples compare very well with manually collected samples. The lead data, though more limited (due to only one site observing measurable levels), showed that automated sampling data was less variable than manually collected samples.

The final field-tested monitoring approach was the use of pipe loops and brass blocks in the distribution system. Field tests and data were collected from two utilities that have operated their pipe loops either continuously or intermittently during the previous nine years. The data showed that the use of pipe loops and brass blocks produced statistically consistent, and comparable data to LCR monitoring data. However, pipe loops required a significant period of time to stabilize, essentially eliminating the use of pipe loops unless a utility had an operable pipe rack available.

What are the drawbacks to the new alternative in-home tap monitoring and predictive protocols developed by this project?

The alternative in-home tap monitoring and predictive protocols developed took very different approaches and as such, the limitations and drawbacks were different. Table ES.2 summarizes and compares these two protocols and the LCR compliance monitoring protocol and identifies the associated drawbacks and limitations.

Table ES.2
Comparison and summary of LCR, alternative in-home tap monitoring and CPSF NRC OLCC monitoring approaches.

Comparative Criteria	Lead and Copper Rule Protocol	Alternative Tap Monitoring Protocol	CPSF OLCC Predictive Technology
Number of sites	100	12-43 Lead; 7-24 Copper	1
Number of samples per period	100	12-43 Lead; 7-24 Copper	NA
Number of samples annually	100-200	48-172 Lead, 84-288 Copper	300-1000
Number of samples per site annually	1-2	4 Lead; 4 Copper	NA
Sample site requirements	Customer tap and stagnation period	Customer tap and stagnation period	Treatment plant and continuous flow
Additional costs beyond LCR monitoring	None	Yes	Yes
Impact of shrinking sample pool	Unable to identify required number of sample homes.	Minor impact. Number of qualifying homes reduced by 75% over LCR.	None
Customer participation requirement	Yes	Reduced	None
Staff training requirement	None	None (same as LCR)	Yes
Statistical soundness of data reflecting actual conditions	Reduced as sample pools shrink.	Increased and more indicative of corrosion control changes.	Under development
Additional equipment cost	None	$200 (chemicals)	$2,500[a]
Annual operational cost	$20,000[b,e] annually	$10,000 annually[e]	$1,200[c]
Cost per sample	$200[e]	$100[e]	$4.50[d]
Benefits /Drawbacks	• LCR specified	• Reduced variability in data • Reduced customer management • Directly comparable data • Reduced cost	• Minimal monitoring sites • Elimination of customer participation • Minimal staff requirement • Easy to operate • Low cost

a. Includes CPSF NRC, voltmeter, solenoid valve, and dedicated computer.
b. Assumes optimized system and one round (100 samples) of manual sampling.
c. Includes CPSF NRC maintenance and data management.
d. Assumes one sample per day and equipment costs annualized over five years for OLCC monitoring.
e. Includes administrative, management, collection, and analytical costs.
f. Based on reported cost estimates from Philadelphia Water Department for conducting LCR compliance monitoring and field testing of alternative tap monitoring.

What did the project conclude?

Conclusions were reached at several different stages of the project. Conclusions addressed regulator acceptance of alternative methods, protocol and technology applicability, and data validity.

The project's regulatory agency survey represented information from 33 of 50 state primacy agencies. Of the thirty-three respondents twenty-eight had addressed the LCR *Minor Revisions* sufficiently to be able to provide information pertaining to their interpretation. The following are key conclusions from primacy agency responses:

- Most state agencies would be receptive to alternative monitoring methods if they were permissible, reliable and scientifically valid.
- In the absence of an alternative method, the most likely method of verification that corrosion control has not been diminished by source or treatment changes would be was resumption of standard LCR lead and copper tap monitoring without waiting for the next compliance period.
- Other follow-up methods of corrosion control verification include maintaining standard water quality parameter monitoring and LCR lead and copper tap water monitoring during the next compliance period.
- The requirement to conduct additional monitoring could be triggered by addition of a new source water; a change in finished water pH or alkalinity; a change in oxidant or primary disinfectant in the treatment plant; an alteration in the blending of waters in distribution; or a change in chemical treatment, such as for enhanced coagulation.

Based on these responses, it could be concluded that a scientifically valid and reliable alternative monitoring method would be of value. More importantly, primacy agencies may be receptive to an alternative method.

Eleven case studies are included in this report. Some of the issues of concern identified by utilities in the cases studies included:

- The loss of customer interest in participating in the program is making it difficult to maintain sample pools of interested customers. In an increasing number of cases, this is resulting in loss of critical representative homes.
- The variability in sample data was making it increasingly difficult to optimize for corrosion control without resulting in over-treatment.
- Compliance sampling has been, and is increasingly, a drain on manpower.
- A more statistically reliable method of monitoring would be of value.

Review of historical data and identification and prioritization of variability factors in development of the alternative in-home tap monitoring protocol yielded the following conclusions:

- Several factors contribute to the potential variability measured in tap lead and copper levels including analytical variability, distributed water quality, standing time of the sample, and the age, type, and amount of lead and/or copper containing material.

- Even when these factors are controlled (to the extent possible), the variability of lead and/or copper levels measured at the tap may still be relatively large due to the complex chemical and physical factors that affect corrosion of these metals.
- The existing regulatory protocol provides compliance information, but may not be the best protocol for assessing post-optimization conditions.
- Lead levels in tap water samples can vary, both from similar sites in close proximity in the distribution system, as well as from repeat samples taken from the same site under identical conditions.
- Determining which factors to control during site selection and data collection for a particular system is essential in determining statistically valid monitoring protocols.

Findings derived from laboratory studies of corrosion monitoring using the OLCC CPSF predictive technology were that:

- The electrochemical potential (E_{corr}) of copper exhibits changes in a predictable and reproducible manner during stagnation episodes of varying duration.
- The major features of E_{corr} behavior are different in presence of oxygen only and in the presence of chlorine species.
- In the presence of oxygen only, there is a strong linear correlation between the logarithms of the total dissolved copper concentrations and E_{corr} values.
- Variations of pH, alkalinity and phosphate concentrations affect the position of E_{corr} vs. $\log(Cu_{total})$ relationships, but do not significantly affect their linearity and correlation.
- The relationship between E_{corr} and total dissolved copper concentration is valid for metal surfaces of any age.
- In the presence of oxygen, E_{corr} behaves non-monotonically during long stagnation episodes. This type of behavior can be caused by the depletion of oxygen at the surface, formation of Cu(I) species and precipitation of copper-containing solids on the metal surface.
- In the presence of chlorine, the decrease of E_{corr} after the start of stagnation episodes is accompanied by the accumulation of dissolved copper in the corrosion cell. The correlation between E_{corr} and $\log(Cu_{total})$ remains linear, but the value and sign of the slope of E_{corr} vs. $\log(Cu_{total})$ correlations are opposite of those found in the presence of oxygen only.
- In virtually all copper experiments that involved alterations of basic water chemistry (e,g., changes of pH, alkalinity, phosphate, chlorine concentration), changes of the behavior of E_{corr} could be observed immediately after the corroding surface was exposed to an altered water quality.

The major limitation of the CPSF method, at its present state of development, is that its performance has only been well explored for relatively short durations of stagnation.

Summary

This project primarily focused on the development of two corrosion control monitoring approaches: alternative tap monitoring and the novel corrosion potential stagnation/flow on-line corrosion cell. The approaches were designed to provide water utilities with tools to accurately quantify effects of corrosion control and to detect and quantify subtle changes in lead and copper levels with higher accuracy and time resolution and lower cost than the conventional Lead and Copper Rule tap monitoring protocol. Additionally the project investigated the applicability of an automated sampling device to address the issue of customer variability in data and the use of existing lead and copper pipe loops to track changes in corrosion control. The results, potential for application, and reliability of these protocols and devices was variable ranging from "utility ready and statistically reliable" to "a significant first step in the future development of a commercial device."

The alternative tap monitoring protocol showed that several factors leading to data variability can be addressed by a monitoring protocol of intensive monitoring of fewer sites. The result was reduced data variability, statistical validity, reduced staff time, and reduced cost.

The development of the on-line corrosion cell addressed many challenges associated with the initial design of a functioning corrosion cell. After several attempts the Narrow Rectangular Cell (NRC) design showed that the CPSF theory could be verified in the laboratory and in the field. In its current level of development the NRC OLCC is not widely applicable to utility use and data interpretation can be difficult for the typical operator. However, the OLCC design and findings reported by this project are a significant first step to future investigation.

Similar to the experience of the OLCC, an automated sampling device was developed to a prototype stage and tested in the field. Data collected using the devices was comparable to manually collected samples. Future investigation will require design revision and more extensive field testing.

Last, the investigation of lead and copper pipe loops as a tool for tracking corrosion control changes demonstrated that the potential for their use did exist. However, due to the time required for pipe loops to stabilize, the feasibility of their use is limited to utilities that have an operable pipe loop rack in inventory or under operation. The data provided by the pipe loop racks evaluated did not justify the time and associated financial commitment required for the development of such a tool for an already optimized utility.

CHAPTER 1
PROJECT OVERVIEW

INTRODUCTION

This Awwa Research Foundation (AwwaRF) sponsored project, was designed to explore the concerns that have been expressed by utilities that were previously optimized for corrosion control. This project strives to develop alternative methods, which are acceptable to regulatory primacy agencies, of demonstrating that corrosion controls are not adversely affected when treatment techniques and/or source waters are altered.

The intention of the alternative methodologies described in this report was to give primacy agencies and utilities new tools to show, in advance of required LCR monitoring, if changes in treatment or source waters could upset existing corrosion control applications. These methodologies are not intended to supplant Lead and Copper Rule (LCR) monitoring requirements. The objectives of the project were to:

- Develop a statistically valid alternative home sampling protocol to track the metal uptake associated with lead and copper corrosion.
- Develop an on-line accelerated testing methodology and equipment to track and predict impacts on lead and copper corrosion during treatment/source water changes.
- Perform case study investigations describing utility experience in achieving optimized corrosion control and the monitoring challenges following optimization.
- Assess regulator interpretation of the LCR *Minor Revisions* and their potential acceptance of alternative in-home tap sampling and predictive tools.

BACKGROUND

The Lead and Copper Rule *Minor Revisions* (Minor Revisions) (January 12, 2000) state that,

> *"Any water system deemed to have optimized corrosion control…shall notify the State in writing [no later than 60 days after making the change]…of any change in treatment or the addition of a new source. The State may require any such system to conduct additional monitoring or to take other action the State deems appropriate to ensure that such systems maintain minimal levels of corrosion in the distribution system."*

The intent of the *Minor Revisions* is that systems continue to operate and maintain optimal corrosion control for both lead and copper. To see that this requirement is met, primacy agencies could require additional monitoring for lead, copper and water quality parameters. The method for verification is determined at the discretion of the primacy agencies.

The flexibility provided by the *Minor Revisions* in confirming optimization was in response to issues raised about the method for monitoring lead and copper. In the *Minor Revisions*, the United States Environmental Protection Agency (USEPA) noted that the Agency did not have data to develop alternative sampling methods that would provide information with as much certainty as direct sampling at taps (April, 1996 Proposal). Comments also cited major problems with the current monitoring requirements such as lack of control over sample collection, accessibility problems, insufficient sites, and disagreement with the appropriateness of triggering Corrosion Control Treatment (CCT) requirements based on residential monitoring. In the proposed *Minor Revisions* the USEPA agreed that, while many of the cited problems were valid, no acceptable alternative to tap water monitoring had been identified. Therefore, the rationale for tap monitoring is simply that no surrogate technique had been identified that provided extrapolation to tap results with accuracy, precision, or both.

The *Minor Revisions* affect all community and non-transient/non-community (NTNC) water systems, especially those who have established an optimized corrosion control process. Any water system that has achieved approved optimized corrosion control and desires to change treatment or source water must notify the primacy agency. The primacy agency has responsibility for ensuring that the change will not negatively impact lead and copper levels at customer taps. Without alternative means to accomplish verification of continued corrosion control optimization, primacy agencies' only available option would be to require that water systems conduct additional customer home compliance sampling. This may not be timely, cost-effective, or sufficiently predictive and reliable.

PROJECT APPROACH

This project and the resulting report is divided into six distinct elements. These project elements were intended to characterize the current situation, develop potential methods for monitoring lead and copper levels, and evaluate the current state of the developed monitoring methods, as described below:

Element 1 – Primacy Agency Survey

The objective of the regulatory survey was to gather information from State primacy agencies regarding implementation of LCR requirements. The purpose of the survey was to identify barriers to alternative compliance strategies for previously optimized utilities, to identify primacy agency expectations, and to assess the willingness of primacy agencies to accept alternative compliance methods, such as those presented by the findings of this project.

Element 2 – Participating Utilities Case Studies

The objective of the project's case studies was to document utility experiences with the process of attaining corrosion control optimization and issues that have been addressed as treatment and source changes were considered and implemented. Eleven case studies are included in this report.

Element 3 – Alternative In-home Tap Monitoring Protocol

Developing statistically valid alternative in-home tap monitoring guidelines was a core objective of the project. The resulting monitoring guidelines have been developed to address the issues of sample variability, shrinking sample pools, and utility resources. The monitoring guidelines provide utilities with the tools needed to more accurately track corrosion control changes with lead and copper levels.

The first step in developing alternative protocols was the determination of key statistical factors in evaluating monitoring protocols. These factors included: summary statistics; typical variabilities; and intra-site vs. inter-site variability.

Information on these key factors was summarized from historical literature and case studies of both regulatory monitoring results and alternative distribution system monitoring completed by utilities before and after corrosion control treatment implementation.

A thorough review of the historical literature on standing lead and copper levels was completed in order to document the lead and copper level variability measured in those studies. Information from the historical literature was supplemented with project case study information. Several of the case study utilities also provided data collected using protocols that differ from regulatory requirements. The data from these alternative monitoring programs provided valuable information on site variability in lead and copper levels under shorter time periods than regulatory monitoring.

Based on the historical data, draft alternative monitoring guidelines and corresponding statistical evaluation methods were developed for field study application and eventual final recommendations. These guidelines include estimates of the number of sites and the frequency of sampling for a variety of desired accuracy and confidence levels.

Element 4 – On-line Predictive Corrosion Monitoring Protocol

This project element investigated the potential of a centralized, on-line device to monitor changes in lead and copper corrosion rates. This element of the project developed, tested and validated a novel monitoring and analytical protocol with the future potential to enable water utilities to accurately quantify effects of corrosion control and to detect and quantify minor differences in corrosion rates and metal release with accuracy, time resolution and low cost.

Implementation and verification of the On-Line Corrosion Cell (OLCC) required development of the device and generation of an adequate amount of experimental data needed to compare the performance with that of other methods. The OLCC uses the Corrosion Potential-Stagnation/Flow (CPSF) approach to evaluate corrosion rates and corrosion potential change.

The on-line predictive technology is novel. For that reason, development of the technology required both theoretical verification through modeling and detailed laboratory experiments. Development of the OLCC and the CPSF method included preliminary modeling of the relevant processes and the expansion of theoretical calculations to predict the behavior of corrosion potential, rate of metal release and the dependence vs. intrinsic corrosion and other relevant experimental settings (e.g., cell diameter, concentration of oxygen and chlorine, pH and other water quality parameters).

Element 5 – Monitoring Protocol Field Studies

Field test evaluations were conducted during the project for prototype alternative in-home tap monitoring, OLCC monitoring protocols, automated samplers and pipe loop installations. One utility tested prototype automated samplers, four utilities participated in the testing of in-home protocols, three utilities installed and tested the OLCC device, and two utilities operated pipe loops. The results of the field tests were incorporated into the final monitoring protocols of each approach. Field test reports for each of the activities are included in the appendices.

Element 6 – Proposed Monitoring Strategies

The report concludes with a synthesis of the findings from case and field studies, data evaluations, and survey responses. The result is proposed alternative lead and copper monitoring strategies for systems previously optimized for corrosion control now undertaking a change in source water or treatment practices. Included are proposed in-home tap sampling protocol and statistical evaluation methodologies, applicability of on-line predictive technology including design and operation parameters, and design and applicability protocol for automated tap sampling. This section of the report discusses the applicability and drawbacks of each approach.

A report summary is provided that discusses the need and applicability of the project's findings, additional research opportunities identified during the course of the project, and the identified next steps in assisting water systems in addressing this on-going and increasingly challenging regulatory requirement.

CHAPTER 2
SURVEY OF REGULATORY AGENCIES' LCR MINOR REVISIONS IMPLEMENTATION

INTRODUCTION

In order to achieve the goals of the project, a regulatory survey was performed. The survey gathered information from State primacy agencies regarding the implementation of LCR *Minor Revisions* requirements. The purpose of the survey was to:

- Determine the number of community and nontransient/noncommunity water systems in each state that are currently approved for optimized corrosion control.
- Assess whether primacy agencies have implemented, or plan to implement, more stringent requirements than those outlined in the Lead and Copper Rule, and the *Minor Revisions.*
- Assess primacy agency intentions requiring corrosion control optimized water systems to conduct some form of study to confirm that treatment/source water changes will not negatively impact lead and copper.

Any water system that has achieved regulatory agency-approved corrosion control optimization and desires to change treatment practices or source water, must notify regulators. Primacy agencies are responsible for ensuring that the change will not negatively impact lead and copper levels at customer taps. The *Minor Revisions* do not describe the means for demonstrating lead and copper corrosion control, leaving that determination to the primacy agency. Without alternative means to accomplish this, regulators will need to require that water systems conduct additional LCR customer home compliance sampling.

This survey was designed to assess whether primacy agencies have already determined how they will require compliance to be determined under the *Minor Revisions* and whether the primacy agency would be open to consideration of alternative monitoring approaches to verify corrosion control.

The survey was developed by the project team and field tested with the state regulating agencies in Pennsylvania and South Dakota. The survey was distributed to all fifty U.S. state regulating agencies using a contact list obtained from the Association of State Drinking Water Administrators in Washington, D.C. Agencies were provided sixty days to respond. After the response period a reminder letter was mailed and a follow-up telephone call made to each outstanding agency. A total of 33 surveys (66%) were completed and returned. The seventeen non-responsive agencies were evenly distributed across all regions of the country.

SUMMARY OF SURVEY RESULTS

The following is a question by question summary of responses.

Question #1. What is the total number of water systems within your jurisdiction that are currently optimized for corrosion control under the Lead and Copper Rule; the number of non-transient/non-community (NTNC) systems regulated under the LCR; and the approximate number of non-transient/non-community systems not required to establish optimized corrosion control treatment?

Responses to Question #1 are summarized in Table 2.1.

Table 2.1
Number of water systems regulated under the LCR corrosion control requirements

State	# of NTNC systems subject to LCR requirements	# of NTNC systems currently listed as optimized	# of NTNC systems required to be optimized	# of NTNC utilities not yet optimized
Alabama	1152	602	602	0
Alaska	637	10	318	308
Arkansas	780	56	56	0
Colorado	992	97	97	0
Connecticut	1255	176	267	91
Hawaii	132	10	10	0
Iowa	2600	1340	1400	60
Kansas	1901	972	991	19
Louisiana	1409	90	209	119
Maine	793	120	293	173
Maryland	958	209	393	184
Massachusetts	692	15	450	435
Michigan	3300	40	50	10
Minnesota	2940	921	1562	641
Mississippi	1250	30	1150	1120
Missouri	1900	27	27	0
Nebraska	790	2	90	88
Nevada	370	5	7	2
North Carolina	2948	2320	2948	628
North Dakota	346	100	114	14
Oregon	1237	202	1116	914
Pennsylvania	3472	212	1594	1382
Rhode Island	154	110	110	0
South Dakota	496	463	496	36
Tennessee	550	15	550	535
Texas	5154	188	5105	4917
Vermont	690	70	70	0
Virginia	1950	220	410	190
West Virginia	758	164	202	38
Wisconsin	2300	100	1950	1850

NTNC = Non-transient, Non-community

Question #2. Has your Agency experienced any cases of water systems changing either source water or treatment and then determining they have higher lead or copper levels following the change?

Fifty-seven percent (19) of the responding agencies reported cases of increased lead and copper levels that would require some level of follow-up monitoring under the LCR *Minor Revisions*.

Question #3. Briefly describe any additional sampling requirements implemented by your Agency that are not already described by the LCR and its *Minor Revisions*.

Thirty percent (10) of responding agencies reported that additional sampling activities were required. The following is a cumulative list; not all agencies required all of the activities.

- Daily pH and/or weekly phosphate monitoring (when phosphate addition is practiced).
- A one-year demonstration study with water quality parameter monitoring, lead and copper monitoring, and a final report for medium and small systems.
- Inclusion of ortho and total phosphate as water quality parameters when phosphate-based inhibitors are used.
- Two full rounds of follow-up sampling to re-verify corrosion control optimization when a system exceeds lead and copper action levels following a source or treatment change.
- Systems with action level exceedances following treatment or source changes are required to conduct extra sampling as determined by the state.
- Systems that exceed action levels following treatment or source changes must continue water quality parameter monitoring for two consecutive six-month periods even after meeting the action levels.
- Increased frequency of water quality parameter monitoring for process control.
- Automatic additional sampling when systems change chemical treatment.

Question #4. Briefly describe any other significant requirements that your Agency has added to the compliance requirements pertaining to the LCR and its *Minor Revisions*.

Only 13% (4) of responding states reported additional requirements. Those requirements included the following:

- Required "validation set" of samples with initial number of samples if a system exceeds action levels during reduced monitoring. Compliance is then based on the results of the validation monitoring.
- Medium size and required small systems to test for, and maintain, pH, alkalinity, and orthophosphate distribution system levels, and to maintain records of the monitoring.
- Daily pH monitoring including reporting of results.
- Making minor adjustments and resampling when corrosion control is in place, but the action level is exceeded.

Question #5. The Project Team realized that some primacy agencies may not yet be addressing the implementation of the LCR *Minor Revisions*. Questions 6-10 pertain to implementation, if your agency is unable to address these issues please stop here and return the survey.

At this point 5 of the thirty-three responding state agencies returned the survey. The remaining twenty-eight provided additional information.

Question #6. Please check the conditions below under which your Agency would likely require a corrosion control optimized water system to conduct some form of additional sampling/testing for lead and copper.

Six agencies were unable to provide specific conditions that would automatically trigger sampling, or stated that requirements are determined on a case-by-case basis. Those responding provided the following responses. The responses are ordered beginning with "most likely to require additional monitoring" and ending with "least likely to require additional testing."

- A change to, or addition of, a new source water (especially a completely new source).
- A change in finished water pH or alkalinity.
- A change in oxidant or primary disinfectant in the treatment plant.
- Alteration in the blending of waters in distribution.
- Change in optimized corrosion control, such as chemical treatment.
- Implementation of enhanced coagulation.
- A change in treatment to control DBPs and/or *Cryptosporidium*.
- A change in distribution chlorine residual type or residual level.

Question #7. Please indicate the method by which your Agency would likely require a water system to confirm that a change in treatment/source water did not adversely affect lead and copper corrosion control.

Two of the twenty-eight responding agencies were undecided. The remaining agencies provided the following responses, again beginning with "most likely method" and ending with "least likely method."

- Resume standard lead and copper tap water monitoring as would be required for compliance, without waiting for the next compliance period.
- Maintain standard water quality parameter monitoring as already specified for maintaining optimized corrosion control to confirm that no changes in water quality parameters have occurred that would require additional action.
- Wait to conduct lead and copper tap water monitoring during the next compliance period.
- Resume standard water quality parameter monitoring after the change is made to determine new water quality parameters for compliance.
- Commence compliance monitoring as if utility was never optimized.
- Monitor selected sample sites that have been historically high, as frequently as monthly for lead, copper, and water quality parameters.
- Conduct additional monitoring including pH and depending on type and timing of system change, set additional monitoring.

Of the responding state agencies none listed repeating of bench/pilot/pipe-loop studies originally conducted to determine optimized corrosion control treatment as a confirmation method.

Question #8. Is your Agency investigating, or open to suggestions for, alternative methods for use by optimized water systems to demonstrate that treatment/source water changes will not adversely affect lead and copper corrosion control?

In response, only one agency stated that they would not consider alternatives. Three agencies declined to respond. The remaining twenty-four agencies reported that they would consider using alternative methods and that approval would require review of the methodology. Four agencies provided additional comments that included:

- The Agency would consider alternatives, provided these alternatives are permissible under federal and state regulations.
- For states that have very few systems that could benefit from alternative methods the development of such methods may not be warranted.
- The Agency is studying the possibility of enforcing the current corrosion control requirements fully and dropping tap sampling completely as tap sampling has been unreliable, unscientific, and inconsistent.
- The Agency is open to suggestions of alternative methods as long as they are scientifically valid and have practical merit.

Questions #9 and #10. Would your Agency likely require any water system that plans to change its treatment/source water to notify the Agency in advance of the change rather than within 60 days after the change? If advanced notification is required, will it likely involve gaining pre-approval of the method for demonstrating that lead and copper corrosion control will not be adversely affected?

In response to the first question, three state agencies responded that advanced notification would not be required. Two responded that it may require advanced notification but were undecided. The remaining twenty-three respondents stated they would require advanced notification to approve plans that relate to source and/or treatment changes.

For those twenty-three agencies responding that advanced notification would be required, seventeen would also require that pre-approval be obtained for the alternative method to be used to verify that the source/treatment change did not adversely affect lead and copper corrosion.

SURVEY CONCLUSIONS

The project survey represented information from 33 of 50 state primacy agencies. Of the thirty-three respondents, twenty-eight had addressed the LCR *Minor Revisions* sufficiently to be able to provide information pertaining to their interpretation. Table 2.2 provides a state-by-state summary of selected survey inquiries. The following are key conclusions from primacy agency responses:

- Most state agencies would be receptive to alternative monitoring methods if they were permissible, reliable and scientifically valid.
- In the absence of an alternative method, the most likely method of verification that corrosion control has not been diminished by source or treatment changes would be resumption of standard LCR lead and copper tap monitoring without waiting for the next compliance period.

- Other follow-up methods of corrosion control verification include maintaining standard water quality parameter monitoring and LCR lead and copper tap water monitoring during the next compliance period.
- The requirement to conduct additional monitoring could be triggered by addition of a new source water; a change in finished water pH or alkalinity; a change in oxidant or primary disinfectant in the treatment plant; an alteration in the blending of waters in distribution; or a change in chemical treatment such as enhanced coagulation.

Based on these responses it can be concluded that a scientifically valid and reliable alternative monitoring method would be of value. More importantly, primacy agencies may be receptive to an alternative method. The opportunity is available for utilities to take a lead in determining what an appropriate monitoring protocol would be to verify corrosion control following treatment or source changes in advance of LCR monitoring requirements.

Table 2.2
Summary of survey inquiries by State

State	Survey Inquiries					
	Does your Agency have additional Pb or Cu sampling, or other, requirements in addition to those in the LCR and its *Minor Revisions*?	What are the conditions that would require an optimized utility to conduct additional Pb/Cu monitoring?	What is the required method of confirming that treatment/source change has not adversely affected corrosion control?	Is your Agency receptive to considering alternative methods of demonstrating optimization by previously optimized utilities?	Does your Agency require notification of treatment/ source changes within 60 days of change or in advance?	Are alternative methods of demonstrating optimization likely to require pre-approval?
Alabama	Systems practicing corrosion control that exceed AL can make minor adjustments and resample	A	1	Possibly	YES	YES
Alaska		Undecided	Undecided	YES	YES	YES
Arizona						YES
Arkansas		A,H	4	Possibly	YES	YES
Colorado		A,G	1,3,4	YES	YES	YES
Connecticut		A,B,C,D,E,F,G	3		YES	NO
Hawaii		A,G	4	NO	YES	YES
Iowa		Undecided	Undecided		YES	NO
Kansas		A	4	Possibly	YES	NO
Maryland		A	4	YES	YES	YES
Massachusetts		A,B,E,F,G	1,6	YES	YES	YES
Michigan		A,B,F,G,H	1,2,3,4	Possibly	YES	YES
Minnesota	Require "validation set" of samples with initial samples if AL is exceeded during reduced monitoring. Compliance is then based on "validation set" monitoring.	G,H	3		Utility Dependent	No Response
Missouri		A,D,G	1	YES	NO	No Response
Nebraska		Undecided	3	Possibly	NO	NO
North Carolina		A	4	YES	YES	YES
North Dakota		Undecided	3	Possibly	YES	YES
Oregon		A,B,E,G	1,4	YES	YES	YES
Pennsylvania		A,B	1	YES	YES	YES
Rhode Island		Undecided	7	Possibly	YES	YES

(continued)

Table 2.2 (cont.)
Summary of survey inquiries by State

State	Survey Inquiries					
	Does your Agency have additional Pb or Cu sampling, or other, requirements in addition to those in the LCR and its *Minor Revisions*?	What are the conditions that would require an optimized utility to conduct additional Pb/Cu monitoring?	What is the required method of confirming that treatment/source change has not adversely affected corrosion control?	Is your Agency receptive to considering alternative methods of demonstrating optimization by previously optimized utilities?	Does your Agency require notification of treatment/ source changes within 60 days of change or in advance?	Are alternative methods of demonstrating optimization likely to require pre-approval?
Tennessee		A	1,4	YES	YES	YES
Texas	Require small and medium sized utilities to monitor for pH, alkalinity, and orthophosphate levels in distribution and maintain record of results.	Undecided	3	YES	NO	Not Applicable
Utah	No Response					
Vermont		A	4	Possibly	YES	NO
Virginia		A,G	3,4	YES	YES	YES
West Virginia		A,D,E	4,5	Possibly	Utility Dependent	NO
Wisconsin		A,H	4	YES	YES	YES

1. Maintain standard water quality parameter monitoring to confirm that no changes have occurred that would require additional action.
2. Resume water quality parameter monitoring following change to determine new water quality parameters for compliance.
3. Conduct lead and copper monitoring at next scheduled compliance period.
4. Resume lead and copper monitoring as would be required for compliance without waiting until next compliance period.
5. Applies only to new source. New sources must begin completely new corrosion control compliance process.
6. Monitor selected sample sites as frequently as monthly for lead, copper, and water quality parameters.
7. Conduct additional monitoring as determined by state depending on the type and timing of change.

A. A change to, or addition of, new source water.
B. A change in oxidant or primary disinfectant.
C. A change in distribution disinfectant residual type or concentration.
D. Implementation of enhanced coagulation.
E. Alteration of blending ratios in the distribution system.
F. A change in treatment to control DBPs or *Cryptosporidium.*
G. A change in finished water pH or alkalinity.
H. Change in optimized corrosion control technique such as chemical treatment.

CHAPTER 3
UTILITY CASE STUDIES

INTRODUCTION

Parallel to the surveying of regulatory primacy agencies, the project team gathered information from participating utilities regarding their experience demonstrating optimization.

Information was gathered, and is reported, in the form of utility cases studies. The project developed case studies from the following utilities:

- Philadelphia Water Department, PA
- Hastings Utilities, NE
- Detroit Water and Sewerage Dept., MI
- City of Highland Park, IL
- Hillsborough County Water District, FL
- North Clackamas County Water Commission, OR
- Seattle Public Utilities, WA
- City of Fort Worth, TX
- Portland Water Bureau, OR
- Metropolitan Water District, NE
- Skagit County Public Utility District #1, WA

Each case study follows in its entirety and provides a utility background, lead and copper monitoring results from initial optimization monitoring, and experiences associated with demonstrating optimization following a treatment or source change.

Some issues of concern identified by utilities in the cases studies include:

- Difficulty maintaining sample pools of interested customers. In some cases this has resulted in losses of critical representative homes.
- Variability in sample data making it difficult to optimize for corrosion control without resulting in under or over-treatment and incorrect chemical addition.
- Compliance sampling has been, and is increasingly, a drain on manpower to administer the programs.
- A more statistically reliable method for monitoring would be of value.

DETROIT WATER AND SEWERAGE DEPARTMENT, DETROIT, MICHIGAN

Background

The Detroit Water and Sewerage Department (DWSD) provides potable water to approximately 4 million customers in southeast Michigan. Water is distributed to 126 communities. Surface water supplies are drawn from the Detroit River and Lake Huron and treated in five DWSD water treatment facilities that use a conventional treatment train of pre-chlorination, rapid mixing, flocculation, sedimentation, dual media filtration, fluoridation, and chlorine disinfection. The average and peak day (summer) demand requirements are 640 million gallons per day (MGD) and 1.285 billion gallons per day (BGD), respectively. Treated water alkalinity is 76 mg/L on average (1992 data), with a range of 68-110 mg/L as $CaCO_3$. In 1992, the typical finished water pH was 7.4 to 7.6.

Water quality data from the distribution system indicate that there is only a small change in pH and alkalinity from the water treatment plants to the distribution system's hydraulic ends.

The DWSD distribution system consists of more than 11,000 miles of pipe. According to 1988 utility data, distribution system piping includes lined or unlined cast iron pipe (75% of total piping, 16-inch diameter or less) and asbestos cement pipe (10% of total piping, 24 inch diameter or less). Historically, lead pipe was used extensively for service lines throughout the City of Detroit and some of the older suburbs. A 1992 survey estimated that 200,000 lead service lines were in place in the city. Galvanized piping is also common in industrial and commercial facilities, and older homes; although, the extent of its use has not been quantified. Galvanized piping can be a source of zinc, lead, and iron when corrosive conditions exist. Lead/tin solder was commonly used to join residential and commercial copper piping throughout the DWSD system before it was discontinued in 1987.

In September 1999, DWSD achieved an optimized status for corrosion control and qualified for a reduced level of monitoring. DWSD is required to maintain a minimum pH of 7.0 in the treatment plant effluents, and a minimum total phosphate dosage of 0.9 mg/L in each of the five water treatment plant effluents.

Historical Monitoring

In 1992, DWSD conducted two rounds of compliance monitoring for the Lead and Copper Rule (LCR). More than 2,500 samples were collected from 1,700 different sites that represent 97 purveyor systems. Although DWSD serves 126 wholesale customers, about 30 of these systems conduct their own lead and copper monitoring. The 1992 monitoring results (Table 3.1) show that none of the purveyor sampled systems exceeded the action level (AL) for copper. In 36 of the DWSD sampled systems, lead levels exceeded the 90th percentile AL of 0.015 mg/L. All communities that exceeded the lead regulatory AL are wholesale customers adjacent to the Detroit metropolitan area with the exception of Hamtramck within Detroit city limits. All of these service areas are mature communities containing homes constructed prior to 1950 and may contain a significant number of lead service lines.

Table 3.1
DWSD lead and copper monitoring history

	Lead (mg/L)				Copper (mg/L)	
Year	Number of purveyor systems sampled	90th percentile range (mg/L)	Action level (mg/L)	Number of systems exceeding action level	Action level (mg/L)	Number of systems exceeding action level
Round 1, 1992	97	0.003 – 0.070	0.015	36	1.3	0
Round 2, 1992	96	<0.002 – 0.324		34		0
Spring 1997*	96	<0.002 – 0.037		13		0
2000	99	0.000 – 0.077		3		0

*Corrosion control instituted in 1996

In September 1996, the addition of phosphoric acid for lead corrosion control was implemented system-wide. A second round of monitoring was conducted during the first half of 1997 in accordance with the LCR. These monitoring results indicate that by spring of 1997, significant progress was made by achieving compliance in 23 of the 36 communities that exceeded the regulatory AL in the 1992 monitoring. The 13 communities that still exceeded the AL showed a marked reduction in 90th percentile lead levels over previous monitoring rounds.

In the most recent monitoring (2000), only 3 of 127 communities in the DWSD service area had lead levels that exceeded the 90th percentile AL.

Sample Pool

During the 1992 lead and copper monitoring, nearly all sampling sites were classified as Tier 1 sites under the LCR. The majority of the sample sites were identified as either Category 1 (single family residence with a lead service line) or Category 2 (single family residence with lead-soldered copper plumbing installed after 1982). In order for the city of Detroit to obtain the required 100 samples, sample bottles were sent to 350 customers. When the required number of samples was reduced to 50 samples for 2001 monitoring, sample bottles were sent to only the 100 customers that participated in the year 2000 monitoring.

Many customers (in Detroit as well as surrounding communities) seem to have lost interest in participating in the lead and copper monitoring program. Possible reasons for this lack of interest may include the longevity of the program (8+ years) and reduced health concerns due to lower lead levels that meet regulatory requirements.

Corrosion Control Studies

During 1993-1994, DWSD completed a Lead and Copper Optimization Study in accordance with the LCR requirements. A desktop analysis was conducted to identify and compare feasible treatment alternatives to reduce lead levels at customer's taps. The recommended alternatives for pilot-scale testing included (1) increasing pH, (2) adding zinc orthophosphate, and (3) adding orthophosphate. Pilot-scale testing completed in 1993 showed that total lead uptake by water exposed to standing contact with lead pipes is much greater than from contact with lead soldered copper pipe or lead-bearing brass fixtures. The treatment alternatives studied during pilot testing were then evaluated based on treatment performance, feasibility of implementation, operational reliability and cost. Based on this analysis, orthophosphates appeared to be the optimal alternative for full-scale use as a lead corrosion control treatment. A 22-month full-scale demonstration testing was conducted to select the preferred form of orthophosphate, to select chemical feed equipment, and to develop other implementation details. Phosphoric acid was selected as the preferred orthophosphate form and yielded statistically significant reductions of lead levels at the customers' taps.

In 1999, pipe loop studies were conducted to determine an optimum pH for lead corrosion control and to evaluate sodium hydroxide addition for pH adjustment. As a result of these studies, the optimum pH range was determined to be 7.2 to 7.8. Phosphoric acid was confirmed as the preferred form of orthophosphate.

Impact on Other Regulations

The DWSD has not experienced any problems complying with other regulations while meeting Lead and Copper Rule requirements.

CITY OF FORT WORTH WATER DEPARTMENT, FORT WORTH, TEXAS

Background

The Fort Worth Water Department provides potable water to a service population of approximately 822,700 in Fort Worth, Texas and 29 nearby wholesale customers. The average and peak day demand requirements are 175 MGD and 315 MGD, respectively.

Surface water supplies include several impoundments on the West Fork of the Trinity River, and two east Texas surface reservoirs. Four conventional water treatment plants provide treated water – North Holly, South Holly, Rolling Hills, and Eagle Mountain. The treatment plants employ pre-chlorination, coagulation with ferric sulfate and polymer, sedimentation, and dual media filtration at three plants and mono media (anthracite) at the Eagle Mountain Ozonation Plant. Fluoride and lime (for pH and alkalinity adjustment) are added at all four plants. Free chlorine is used for primary disinfection and chloramines are used for secondary disinfection. An exception is the Eagle Mountain facility, where ozone and chloramines are utilized for primary and secondary disinfection.

The distribution system includes more than 2,300 miles of main constructed primarily of cast iron and lesser amounts of ductile iron, polyvinyl chloride (PVC), concrete, galvanized iron, and unlined asbestos cement. Service connections are primarily constructed of copper. Fort Worth has a lead service line replacement program in which lead services are replaced as needed when replacing old water lines and mains.

The water system maintained by Fort Worth has been optimized for corrosion control under the Lead and Copper Rule (LCR). In 1994, Fort Worth's lead and copper corrosion control plan was submitted to, and approved by, the Texas Natural Resources Conservation Commission (TNRCC). Carbonate passivation was designated as the preferred method of optimizing corrosion control. To maintain optimized conditions, finished water pH must be maintained in the range of 8.0 to 8.5 at the point of entry and within the distribution system. Finished water alkalinity at these locations must be greater than 40 mg/L as $CaCO_3$.

Historical Monitoring

Fort Worth has conducted lead and copper monitoring at customer taps as directed by the TNRCC since 1992. Samples were collected at the cold kitchen tap after a minimum of 6 hours standing time. The 90th percentile lead and copper levels were below their respective action levels during all monitoring periods, as summarized in Table 3.2.

Table 3.2
City of Forth Worth lead and copper monitoring history

Year	Number of homes	lead		copper	
		90th percentile (mg/L)	Action level (mg/L)	90th percentile (mg/L)	Action level (mg/L)
Jan-Jun 1992	100	0.012	0.015	0.710	1.3
Jul-Dec 1992	100	0.010		0.940	
Jan-Mar 1997	100	0.008		0.592	
Sept-Oct 1997	100	0.006		0.615	
Aug 1998	50	0.004		0.317	
July 1999	50	0.006		0.372	

Sample Pool

Lead and copper monitoring was initially conducted at 100 sites in the distribution system. Fifty sites contained interior plumbing of copper with 50:50 lead: tin solder constructed after 1982, and fifty sites had lead service lines.

When the reduced monitoring was initiated, 50 sites were randomly selected from the initial 100 sites, based on the customers most interested in the monitoring program. With the reduced monitoring (i.e. triennial monitoring), it is somewhat difficult to maintain the customer's interest. Unless the previous customer has moved, most customers are willing to participate when contacted. If a customer drops out of the program, an attempt is made to locate another site within the same block or area that meets the monitoring requirements (lead piping, lead pigtail, copper with lead solder, etc.). The customer service department provides customer names within two targeted areas (Como and Alta Mesa). An inspection is made of the house plumbing, water line, and connection prior to accepting the new site into the monitoring program.

Corrosion Control Studies

Fort Worth conducted a desktop evaluation of alternative corrosion control methods for its retail customer base. Carbonate passivation was selected as the preferred method of optimizing corrosion control. The formation of a carbonate-based film on interior pipe surfaces can mitigate the uptake of lead and copper into solution by being more resistant to dissolution than the pipe material itself. Minor treatment changes were implemented to control pH and alkalinity. No evaluation of corrosion control performance through demonstration testing was required for Fort Worth.

Impact on Other Regulations

Fort Worth has no issues in meeting other regulations while complying with the Lead and Copper Rule.

HASTINGS UTILITIES, HASTINGS, NEBRASKA

Background

Hastings Utilities provides drinking water on a direct service basis to 23,000 customers through 9,000 connections in south central Nebraska. Hastings is a 100% groundwater system drawing water from the Ogallala Aquifer through 25 deep wells. There is an observed correlation between water quality parameters related to corrosivity and seasonal variations in pumping rates due to irrigation. The pH of the source water varies with extreme precipitation seasons from a high of 7.8 during wet years and 7.2 in dry years, with a typical pH of 7.3. The water is classified as hard and has a natural fluoride level of approximately 0.27 mg/L. The source also contains sodium, sulfate, calcium, low levels of iron and manganese, as well as nitrates. The source water is untreated.

Hastings' distribution system is comprised mainly of cast iron and ductile iron pipe and a small amount of PVC. Service connections are typically 5/8-inch to 1½-inch copper; newer connections have been polyethylene. Approximately 40% of the utility's services were installed prior to 1942 and are connected to the mains with lead goosenecks. Home plumbing is typically copper in the service area. Sources of lead and copper in the system are typically in-home plumbing; specifically old brass faucets, lead-tin solder, and copper plumbing.

Sample Pool

Sample sites have been selected based upon service connection information and USEPA site selection protocol. The existing sampling pool consists of 201 verified addresses. Pre-testing of approximately 400 sites has been conducted consisting of a water sampling, cross-connection survey, homeowner interviews regarding water usage, and willingness to participate in sampling programs.

Hasting' original LCR sample requirement was thirty copper sample sites and thirty lead sample sites. The utility's original potential sampling pool consisted of thirty-three copper service locations, constructed between 1982 and 1986, the target period for Tier I homes. In 1942 Hastings ceased the installation of lead services due to lead shortages associated with the World War II war effort. The use of lead services did not resume following the war. Further, in 1987 Hastings enacted a lead solder ban. It is generally assumed that 50:50 lead-tin solder was used in all home plumbing systems until 1987. There is a remaining potential pool of more than 4,000 lead services; however, all of these services are pre-1942.

At this time none of the original Tier I copper sample sites remain in the sample pool. All have been lost to homeowner remodeling, water softener installations, change of ownership, and loss of participant interest. Some customers reported not being interested in participating in a study that may lead to an increase in water rates or modifications to treatment. As copper sites have become unavailable they have been replaced with lead service sites. Attrition in the lead sampling pool has been less. Lead sites are less of a concern since the potential sample pool of 4,000+ homes provide sufficient sampling sites to meet LCR requirements.

Historical Compliance Monitoring

Hastings has conducted four rounds of compliance sampling for lead and copper corrosion occurring in 1997, 2000, and twice in 2001. In both 1997 and 2000 the utility exceeded the 90th percentile AL for copper. Following a revised sampling protocol copper levels were below the AL in both 2001 sampling rounds. Hastings as never exceeded the lead AL. Table 3.3 summarizes the utilities compliance sampling history.

Table 3.3
Hastings Utility LCR compliance sampling history

		Lead			Copper		
Sampling round	Number of samples	Range (mg/L)	90th percentile sample (mg/L)	Action level (mg/L)	Range (mg/L)	90th percentile sample (mg/L)	Action level (mg/L)
1997	57	0.000-0.016	0.004		0.00-2.22	1.83	
2000	30	0.014-0.001	0.003		0.018-2.11	1.79	
2001 A (8/7-8/14)	61	0.000-0.054	0.005	0.015	0.033-3.15	1.24	1.3
2001 B (10/8-11/15)	61	0.000-0.002	0.004		0.015-2.25	0.98	

The reduction in lead and copper values for 2001 samples is attributed to Nebraska Department of Health authorized changes in sample collection procedures.

The original sampling protocol required the entire household plumbing to be stagnant for 6 to 8 hours. The samples were taken by the homeowner in the early morning to obtain the first draw sample. The point of collection was typically a kitchen or bathroom faucet. In 2001 the Nebraska Health Department authorized a revised protocol that allows for normal household use of water at all non-sampling taps. On the morning of scheduled sampling, utility staff contacted residents and tagged a faucet to be sampled so no water will be used. After 6 hours a utility staff collected a two-liter water sample. The water is with drawn slowly, filling a two-liter sample bottle in approximately three to five minutes.

Impact on Other Regulations

In 1994 Hastings Utilities conducted the *Hastings Water System Corrosion Control Study*. The outcome of that desktop study was the recommendation of two alternatives for corrosion control: pH adjustment or phosphate inhibitors. The corrosion control study noted that due to the location of the 25 water production wells, individual chemical injection facilities would be required or well production would have to be consolidated to allow for centralized treatment.

Utility evaluation of each alternative revealed that pH adjustment would have required the use of hazardous materials at wells located in residential areas and would have required extra safety precautions. A further increase in pH to a range of 7.8 to 8.0, would have increased wastewater treatment effluent ammonia levels and wastewater pH adjustment would have been necessary to control the ammonia levels.

The use of phosphate inhibitors posed the potential for biofilm growth in the unchlorinated system. Further concern was expressed relative to compliance with the Total Coliform Rule. Hastings did not have a desire to begin chlorination for cost, safety, and aesthetic reasons to control a potential increase in coliform counts.

Following review of the corrosion control study, Hastings customers and local officials expressed a concern that corrosion control treatment to reduce copper to levels mandated by the LCR may pose more overall risks than ingestion of copper at levels below the lowest observable health effect level of 5.3 mg/l.

Investigation and Compliance Efforts

Hastings staff had encountered customer complaints related to copper in the past and first draw samples had shown copper levels as high as 4.0 mg/L. These incidents were typically associated with the extreme settings and operation of home water softeners or brass fixtures. Investigation of high copper levels were typically rectified by reducing the exchange rate of the home water softeners or instructing customers to run taps until water was cool before drinking, which was normal practice for customers.

Because of these copper "incidents," Hastings became uncertain of what impact the proposed LCR would have. For that reason water testing results were brought to the attention of the Board of Public Works and City officials. It was the opinion of the Board and City officials that treatment of the source water was unacceptable unless a substantiated health risk existed.

The Board and City officials could not justify to customers the necessity of treatment for copper when copper was known to be essential for good health.

To further investigate the potential for copper corrosion in the Hastings system, approximately 500 additional samples were collected in 1992 and 1993. The results of that monitoring indicated the 90th percentile for copper would be at 1.6 mg/L, slightly above the proposed AL. Ninetieth percentile lead levels were found to be well below the 0.015 mg/L AL. Individual lead levels above the AL were investigated further. The sources of the lead were traced to homeowners collecting samples with water softeners operating and homes with homeowner completed plumbing modifications using high lead solder. In all cases where lead levels exceeded the 0.015 mg/l AL, extenuating circumstances were found to be the cause.

Due to public opposition to corrosion control treatment, Hastings staff investigated different compliance options. Those investigations lead to participation, with other Nebraska communities, in a Center for Disease Control (CDC) epidemiological study on the effects of copper in drinking water. Following the CDC study, Hastings and other Nebraska communities, requested that USEPA validate the copper AL of 1.3 mg/L. Nebraska Congressional Representatives met with USEPA officials at which time USEPA was asked to defend the copper AL of 1.3 mg/L. Local officials referenced a study by Hastings staff showing a significant decrease in second liter draws. The local officials requested that USEPA consider flushing as a form of treatment. USEPA responded that it could not accept flushing as a treatment option as it would promote wasting of water.

Another request of USEPA was the modification of the testing protocol to require a two-liter sample volume in lieu of one-liter, as the AL is based upon ingestion of two-liter of water per day. Discussions on the validity of the 1.3 mg/L copper AL lead to a copper study by the National Academy of Science. Results from that study were not conclusive enough to revise the 1.3 mg/L AL.

Additional copper studies conducted by the utility since 1993 have shown that copper levels could not consistently be reproduced in individual homes. Variations of 2.0 mg/L within homes, between sampling events were observed. Utility staff have investigated several potential causes for the variation in results including:

- piping size
- occurrence of cold water pitting
- collection techniques
- water usage prior to sample collection
- impact of extended stagnation periods
- impact of dishwashers raising pipe temperatures prior to collection
- potential impacts of microbial induced corrosion.

The utility's efforts provided valuable information, but no single conclusive cause of copper level variability was identified. It was the opinion of the utility that copper corrosion was, in part, impacted by stagnation time, microbial growth, temperature, water velocity, prior flushing, etc. Based on this, it was the opinion of the utility that it would be difficult to achieve consistent and reliable results using the USEPA sampling protocol.

Based on reported copper levels and lack of substantive justification for a 1.3 mg/L AL, the opposition from customers, elected officials, and Utility Board members was, and remains, strong against the addition of corrosion control treatment.

Currently, no corrosion-control treatment is in place. Compliance sampling for 2001 resulted in 90th percentile lead and copper levels below action levels. At this time, the utility is considered optimized for corrosion control by the Nebraska Department of Health. The next round of confirmation samples is scheduled for 2004.

Lessons Learned

Based on information gathered by Hastings staff from published literature, it is the opinion of Hastings Utilities that the implementation of corrosion control, should it become necessary in the future, would result in the over-treatment of water with respect to copper and an undue burden on customers. Compliance with the LCR, to date, has been expensive and a drain on staff resources. Utility customers have been very vocal regarding compliance with the LCR and the potential cost to rate-payers for what is perceived by them as a non-health issue. Customers that have dropped out of the program have done so for these very reasons. The loss of homes from the copper sample pool has been critical since Tier 1 sites were limited. Replacement of copper sites with lead sites creates a situation in which results can be skewed by very few homes. Due to the variability of the data it is difficult to assess treatment effectiveness and creates the potential for over treatment. The utility could benefit from a more reliable sampling protocol that addresses the decreasing sample pool issue.

CITY OF HIGHLAND PARK, ILLINOIS

Background

The City of Highland Park provides potable water to a service population of 31,000 in Highland Park and 29,000 additional people in satellite systems. The source of supply, Lake Michigan, is treated by conventional methods. The water system maintained by Highland Park has been optimized for corrosion control under the Lead and Copper Rule (LCR) and is monitored on a reduced schedule every three years. Satellite systems operated by Highland Park are also optimized for corrosion control and monitor on a 3-year schedule.

The average and peak day (summer) water demand requirements are 11 MGD and 25 MGD, respectively. Approximately fifty percent of water produced is sold to other utilities. The source of supply is treated by primary chlorination, coagulation with ferric sulfate and polymer, flocculation, sedimentation, and conventional filtration. Chloramines are applied at an average dosage rate of 1.2 mg/L total chlorine. Powdered activated carbon and potassium permanganate are added for taste and odor control during the warm weather months. Hydrofluosilicic acid is added for dental hygiene. In 1992, an orthophosphate corrosion inhibitor was added in anticipation of the LCR.

Coagulation with ferric sulfate tends to depress the pH level. Highland Park has considered using an alternate coagulation polymer to avoid this pH change, but is committed to the current treatment in order to maintain its optimized status for corrosion control. It is suspected that Highland Park may be over-feeding phosphate. However, the utility's optimization status (as well as those of the satellite utilities) would be forfeited should any feed change be undertaken. This also applies to corrosion inhibitor brands or blends not recognized as an equivalent by the Illinois EPA (there are three blends presently designated). In order to maintain the "optimized for corrosion control" status, Highland Park is required to use one of these three corrosion inhibitor chemicals at the present dosage rate.

The pH and alkalinity of treated water are typically 7.6 and 110 mg/L as $CaCO_3$, respectively. To remain optimized for corrosion control, Illinois EPA requires that the finished water have a pH greater than 7.4 and orthophosphate level greater than 0.4 ppm at the entry point to the distribution system.

Service connections are constructed of copper, lead and galvanized iron. 50:50 lead/tin solder was used in home plumbing for houses constructed from the 1960s until banned by the 1986 Safe Drinking Water Act (SDWA). No flushing or other O&M programs are implemented prior to collection lead and copper samples.

Historical Monitoring

Highland Park has conducted lead and copper monitoring at customer taps as directed by Illinois EPA since 1992. Table 3.4 summarizes the historical monitoring schedule and results.

Table 3.4
Highland Park lead and copper monitoring history

Year	Number of homes	Lead		Copper	
		90th percentile (mg/L)	Action level (mg/L)	90th percentile (mg/L)	Action level (mg/L)
Spring 1997	70	0.009	0.015	0.39	1.3
Spring 1999	31	0.005		0.34	

For samples collected during the period from 1997 to 1999, the 90th percentile lead and copper levels have all been less than the regulatory action levels. The system was declared optimized for corrosion control in 1998, and the required number of sampling sites was reduced to 30.

Monitoring is also conducted to evaluate orthophosphate levels in the distribution system. At the entry point to the distribution system, daily monitoring is conducted for orthophosphate. Ten distribution system samples are collected each quarter and analyzed for pH and orthophosphate.

Sample Pool

Highland Park requests sample participation from homeowners with "at-risk homes". These are homes that contain lead service lines or lead solder installed within 5 years of the effective date of the lead solder ban.

Participation was initiated in 1992 through print media and a newsletter. Plant staff visited interested customers and determined their risk factors and suitability for monitoring. Through this process, 75 homes were identified as acceptable candidates. Plant staff trained these customers and conducted initial monitoring at all 75 homes to maintain the customers' interest and to provide them with a service. Over time, some customers dropped out of the program and others joined. The new customers were typically residents that were concerned about lead and approached the utility about having their home tested. It was not necessary to actively solicit customers after the initial effort.

Under the reduced monitoring schedule, the sampling pool consists of a subset of the larger pool selected based on lead level history. The sampling locations that were not needed were not formally dismissed and are probably still available if needed in the future. The smaller pool is relatively stable although some attrition has occurred. Replacement homes are drawn from the original pool of 75 homes.

The sampling location that represents the 90th percentile lead level was built in 1929 and has a lead service line. It does not have "new" lead/tin solder. Utility records do not indicate whether the plumbing is galvanized or copper pipe.

Corrosion Control Studies

Neither pilot-scale nor bench-scale testing was conducted. A literature review suggested that a blended polyphosphate corrosion control inhibitor was ideal for an extended distribution system. The system was optimized by collecting samples and by adjusting the inhibitor dosage rate. An initial orthophosphate dosage rate of 0.12 ppm was used to avoid destabilizing the calcium carbonate scale.

Impact on Other Regulations

Highland Park has had no conflicts with other regulatory requirements in meeting the Lead and Copper Rule requirements.

Lessons Learned

Sample collection is a logistical challenge and a drain on utility manpower. It is difficult to persuade residents to collect samples in a timely manner. One approach that failed was dropping bottles off and asking customers to call for sample pickup. A more effective approach is to drop all bottles off on a Friday, and pick them all up on the following Monday.

In retrospect, a more thorough investigation of corrosion control reduction options may have been helpful, although expensive. The additional expense however, may have provided the utility with additional acceptable alternatives. A method of more reliably measuring the effects of a treatment change would be valuable to the utility.

HILLSBOROUGH COUNTY WATER DEPARTMENT, TAMPA, FLORIDA

Background

The Hillsborough County Water Department in Tampa, Florida provides drinking water to a population of 180,000. The source of supply is groundwater from 10 wells all situated in the same aquifer. Four wells feed the Northwest treatment plant and six wells feed the Lake Park treatment plant. Water from both treatment facilities mix in the distribution system. Treatment at both plants includes disinfection with free chlorine, fluoride addition and corrosion control. Corrosion control treatment was initiated in 1996. The water leaving the treatment plants typically has a free chlorine residual of about 2.0 mg/L. The finished water pH is 7.0 and alkalinity varies from 190 mg/L to 197 mg/L as $CaCO_3$.

Most distribution mains are constructed of PVC or concrete with some ductile iron. Customer service lines are typically PVC. No lead service lines are used in Hillsborough.

Historical Monitoring

In 1992, Hillsborough conducted two initial rounds of lead and copper monitoring. After the corrosion control program was initiated in December 1996, monitoring was conducted two times in 1997. A reduced monitoring program was initiated in 1998. Monitoring results are summarized in Table 3.5 and Figure 3.1.

Sample Pool

Homes were selected for the initial larger sampling pool in accordance with the Lead and Copper Rule criterion. The homes in the sampling pool were constructed between 1983 and 1986. The sampling pool has been maintained since the monitoring program was reduced in 1999. After the first monitoring period, 7 of the 155 participants (6 percent) decided not to continue to participate.

Table 3.5
Hillsboro County lead and copper monitoring history

Year	Number of homes	Lead		Copper	
		90th Percentile (mg/L)	Action level (mg/L)	90th Percentile (mg/L)	Action level (mg/L)
Spring 1992	115	0.008	0.015	1.91	1.3
Fall 1992	108	0.009		1.67	
Mar. 1997	115	0.003		1.03	
Sept. 1997	115	0.004		1.06	
July 1998	50	0.003		0.71	
Summer 1999	50	0.004		0.80	

A small number of the customers in the reduced monitoring program were no longer willing to participate in the monitoring program and were replaced by customers from the original sampling population. Homes from the larger pool that had the highest lead concentrations were chosen to replace those that dropped out. A few of the homes in the smaller pool were replaced with homes on the same street that were built at the same time. Customers in the larger pool that are not actively participating would most likely be willing to participate as replacements in the reduced monitoring program. A ten-dollar incentive is now being provided to participants.

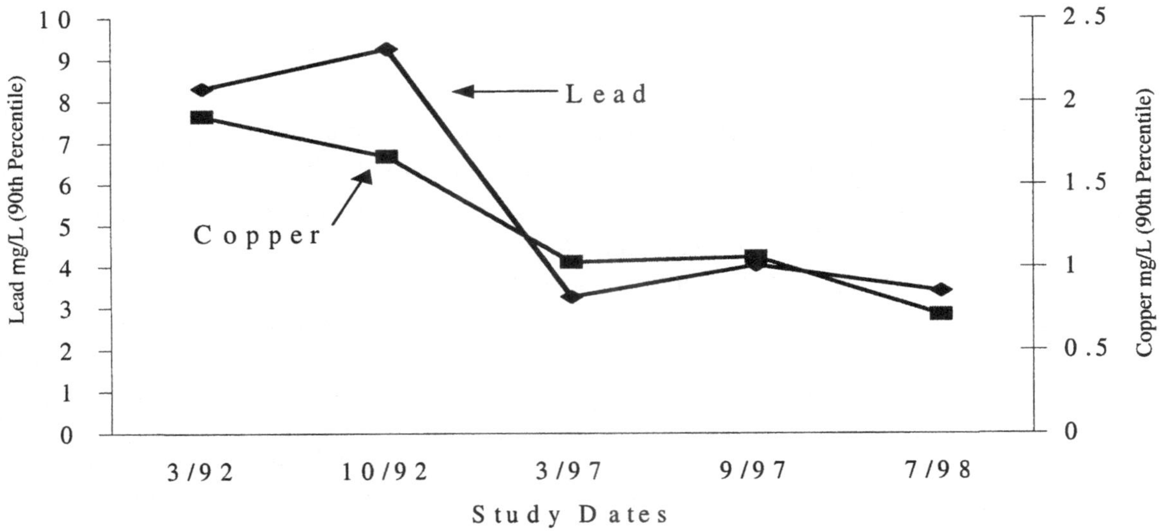

Figure 3.1 Hillsboro County lead and copper 90th percentile results, 1992 - 1998

Corrosion Control Studies

An analogous system study based on Pinellas County was the basis for Hillsborough's proposed corrosion control plan. Pinellas County, Florida has essentially the same groundwater source as Hillsborough County. Pinellas County chose to use a proprietary corrosion control inhibitor that was a blend of polyphosphate and orthophosphate. Hillsborough started with a similar blend in December 1996, and eventually determined that orthophosphate alone was equally effective. The change to orthophosphate addition occurred in June 1999. Orthophosphate is less expensive than a poly/orthophosphate blend. In retrospect, Hillsborough would probably have not started with the poly/orthophosphate blend based on chemical cost.

Impact on Other Regulations

Hillsborough has had no conflicts with other regulatory requirements while complying with the Lead and Copper Rule.

Lessons Learned

Hillsborough County Water Department has no plans to alter their current corrosion control treatment method. If corrosion control alterations did occur, it is unlikely that an exceedance of the action level would occur. The utility has observed a small but steady decline in participant interest. A method of reliably sampling with fewer homes would reduce the pressure of maintaining a large sample pool.

METROPOLITAN UTILITIES DISTRICT, OMAHA, NEBRASKA

Background

The Metropolitan Utilities District (MUD) provides potable water to a service population of 506,000. The average and peak day demand requirements are 95 MGD and 200 MGD, respectively. Corrosion control treatment has not been required as the MUD has always met the lead and copper action levels of the Lead and Copper Rule.

MUD uses surface water supplies from the Missouri River as well as groundwater supplies adjacent to the Platte River. The Missouri River supply is treated using conventional methods with an alum coagulant and lime softening. A polymer is added prior to the clarification process when the river water is muddy. Powdered activated carbon is added seasonally (typically March to June) for taste and odor control and for adsorbing atrazine. Chlorine is used for primary and secondary disinfection, and fluoride is added. The finished water has an average pH of 8.9 and alkalinity of 69 mg/L as $CaCO_3$. The Platte River groundwater supply is also treated by conventional methods using alum and lime softening. The finished water has an average pH of 9.1 and alkalinity of 131 mg/L as $CaCO_3$.

The distribution system contains predominantly cast iron and ductile iron pipe. Service connections are constructed of lead, galvanized steel and copper. Lead solder was used in home plumbing for houses constructed prior to 1986. Lead service lines were used from 1900 to 1925.

Historical Monitoring

The MUD has conducted lead and copper monitoring at customer taps since 1992 (before required by USEPA). Table 3.6 summarizes the historical monitoring schedule and results. No special source water treatment or O&M practices are implemented prior to collecting Lead and Copper Rule samples.

Table 3.6
Omaha MUD lead and copper monitoring history

Year	Number of homes	Lead		Copper	
		90th percentile (mg/L)	Action level (mg/L)	90th percentile (mg/L)	Action level (mg/L)
Feb. 1992	130	0.005	0.015	0.06	1.3
Aug. 1992	121	0.006		0.13	
Sept. 1993	66	0.005		0.05	
Sept. 1994	63	0.007		0.07	
Sept. 1995	59	0.005		0.05	
Sept. 1998	53	0.003		0.04	

Sample Pool

MUD maintains a sample pool of 144 homes for lead and copper monitoring. No turnover has been experienced in the sample pool. After two successful initial rounds of monitoring, MUD was allowed to reduce the monitoring frequency to annually for three years, then to tri-annual. The number of homes sampled was also reduced (to 25 each, lead and copper).

The 90th percentile lead level has been measured in different homes, all of which use a lead service line.

Corrosion Control Studies

The MUD has not conducted any bench-scale or pilot-scale corrosion control studies.

Impact on Other Regulations

MUD will switch to chloramines in the Fall of 2002 to control THMs. No other treatment changes are expected at this time. MUD will continue to use two-stage disinfection. Primary disinfection will remain as chlorine and will allow MUD to meet CT requirements. As no changes are expected to pH and alkalinity levels, the impact on lead and copper is expected to be minimal.

NORTH CLACKAMAS COUNTY WATER COMMISSION, OAK LODGE, OREGON

Introduction

The North Clackamas County Water Commission (NCCWC) provides potable water to a service population of 30,000 outside of the metropolitan Portland, Oregon area. The pH and alkalinity of water entering the distribution system are typically 7.6-7.9, and 20-35 mg/L as $CaCO_3$, respectively. The average and peak day demands are 3.1 MGD and 7.0 MGD, respectively.

The distribution system contains predominantly concrete-lined cast iron (60%) and ductile iron (40%) pipe. Service connections are constructed of copper and PVC. Lead solder was used in home plumbing for houses constructed between 1983 and 1985. The water system had been optimized for corrosion control under the LCR during the period 1994 to 1999. During that time, NCCWC was purchasing water that was treated with pre-chlorination, coagulation and with alum and polymer, partial sedimentation, and rapid sand filtration. The purchased water was treated with soda ash for pH and alkalinity adjustment.

Historical Monitoring

The NCCWC has conducted lead and copper monitoring at customer taps as directed by the Oregon Health Division since 1992. After the new slow sand treatment process came on-line in 1999, the NCCWC was required to verify optimization by conducting lead and copper monitoring at 60 representative homes during two consecutive six-month periods. The result of that monitoring was exceedance of the lead AL during the Spring of 2000 as shown in Table 3.7. As a result of the monitoring, the NCCWC installed a temporary caustic feed system in September 2000 to raise the water pH for corrosion control. Lead and copper levels were below their respective ALs during monitoring conducted in Fall 2000. However, the AL for lead was again exceeded in Spring 2001. The data indicates that the seasonal variations in alkalinity is the determining factor in meeting the lead action limit.

Table 3.7
NCCWC lead and copper monitoring history

Year	Number of homes	Lead		Copper	
		90th percentile (mg/L)	Action level (mg/L)	90th percentile (mg/L)	Action level (mg/L)
1992	62	0.006	0.015	0.42	1.3
1994	32	0.010		0.32	
1995	31	0.004		0.32	
1996	31	0.004		0.39	
1999	31	0.006		0.16	
Spring 2000	30	0.046		0.18	
Fall 2000	60	0.014		0.09	
Spring 2001	60	0.031		0.05	

Sample Pool

The NCCWC requests sample participation from homeowners whose houses fit Tier I or Tier III requirements, as described below:

- Tier I: Homes built between 1983 and 1985 which contain lead solder and copper home plumbing
- Tier III: Homes built after 1983 which contain copper home plumbing

Participation is initiated by mailing out approximately 80 letters to eligible homes. NCCWC then selects the required number of homes from the pool of customers willing to participate. Despite efforts to maintain a consistent sample pool, NCCWC has seen considerable turnover because of customers opting not to participate or return the letter. Only 8-10 homes have participated in all monitoring rounds since 1992. Homes which decide to discontinue involvement in sample collection are replaced by selecting another home from the list of willing participants, with Tier I homes favored over Tier III homes.

The problem of obtaining sufficient participants heightened during the fall of 2000 when the required number of homes increased from 30 to 60. The lag time in receiving response letters reached several weeks, and fewer than 60 favorable responses were received. NCCWC had to mail out additional letters to complete the sample pool, and the time required to complete all monitoring exceeded two months. To encourage participation and improve timeliness of responses, NCCWC began to offer a $10 water bill credit to those homeowners who agreed to participate.

Further, the makeup of the sample pool was altered to meet the higher required number of homes. Tier I homes have historically been targeted because lead levels were of more concern than copper. Tier III homes were requested to participate as needed to complete the 30 homes, but generally were not needed. However, with the increase to 60 homes, NCCWC requested more participation from Tier III homes to obtain enough participants. Currently, approximately 25% of requests for participation are for Tier III homes.

While the sample pool has experienced some attrition and turnover, the number of eligible homes (Tier I and III) has remained relatively constant since 1992. Several homes have expressed discontent at the need to participate every 6 months.

Corrosion Control Studies

NCCWC was required to conduct corrosion control studies in June 2000 in response to the lead AL exceedance (Spring, 2000). Based on a desktop evaluation of water quality and treatment alternatives, NCCWC chose to install a temporary caustic feed system in September 2000 to raise the pH of water from 7.4 to 8.0. Bench-scale and pilot-testing was not conducted prior to implementation.

The cost of conducting the desktop evaluation and performing engineering tasks was approximately $30,000. The capital and installation of cost of the caustic feed system was $8,000. Currently, the caustic feed system is off-line. The NCCWC is conducting a pilot study of caustic soda and soda ash. Once the piloting is complete, the NCCWC will select the appropriate treatment for corrosion control.

Impact on Conflicting Regulations

The NCCWC has not encountered any regulatory non-compliance problems as a result of installing the caustic feed system. Although, the elevated pH diminishes the disinfection efficacy of chlorine, the system is still capable of maintaining an inactivation ratio between 2 and 3 with respect to the required 1-log *Giardia lamblia* inactivation requirement after slow sand filtration.

Lessons Learned

It is proving difficult to achieve two consecutive sampling rounds below the lead action limit. NCCWC attributes this largely to the variability of lead levels. A protocol that would produce more stable results would enable NCCWC to better identify needed treatment changes to achieve optimization. Additionally, efforts to recruit and maintain sample pools has been substantial.

PHILADELPHIA WATER DEPARTMENT, PHILADELPHIA, PENNSYLVANIA

Background

The Philadelphia Water Department (PWD) provides drinking water to 1.7 million people in the City of Philadelphia. The sources of water supply are the Delaware and Schuylkill Rivers. Three conventional treatment plants (Baxter, Queen Lane, and Belmont) supply a distribution system of over 3,000 miles of water main. The finished water produced by all three of PWD's water treatment plants has consistently had lead levels less than 0.001 mg/L and copper levels less than 0.08 mg/L. Most customer plumbing consists of copper pipe.

Since 1993 PWD's strategy of maintaining a slightly above neutral pH with the addition of zinc orthophosphate has succeeded in lowering the lead and copper corrosion throughout the distribution system. Significant further reductions in the 90th percentile values are not expected.

PWD has experienced a decline in the customer base originally recruited for LCR sampling. Many customers, especially those who consistently had lead levels less than the action level, have dropped out of the program. It is thought that the inconvenience of the sampling requirements was the major reason for customer attrition. As a result, PWD may be at risk of retaining only those customers who continually test high, while losing customers who have lower results. This scenario would artificially skew the 90th percentile results to higher levels, potentially causing PWD to fail subsequent rounds of LCR sampling.

Some aberrant sampling results occur at individual homes, and may be a result of sampling error due to an unskilled sampler or due to the unknowns associated with home plumbing systems.

Historical Monitoring

Since 1992, the Philadelphia Water Department (PWD) has conducted six rounds of in home LCR sampling as summarized in Table 3.8 and Figure 3.2. In the first round of sampling in 1992, PWD was above the AL for lead.

PWD did not exceed the AL for copper in any of the six rounds of sampling. A significant drop in copper levels was observed between round one and round three (0.9 mg/L to 0.4 mg/L). Since round three, the results have been consistent (0.4 mg/L to 0.3 mg/L) indicating that the system has reached an endpoint and the plumbing systems have been passivated in relation to copper.

Table 3.8
Philadelphia Water Department lead and copper monitoring history

Year	Samples required	Homes sampled	Lead		Copper	
			90th percentile (mg/L)	Action level (mg/L)	90th percentile (mg/L)	Action level (mg/L)
Jan-Jun 1992	100	162	0.021	0.015	0.9	1.3
Jul-Dec 1992	100	143	0.015		0.8	
Jan-Jun 1997	100	118	0.014		0.4	
Jul-Dec 1997	100	108	0.011		0.4	
Jun-Sep 1998	50	79	0.010		0.3	
Jun-Sep 1999	50	59	0.009		0.3	

To evaluate lead corrosion rates at individual sampling sites, the lead data were divided into two groups. The highest value for each home in the first three rounds of sampling was plotted in a scatter graph versus the highest value for that same home in the second three rounds of sampling. Any home that did not participate in two or more sampling events in each group was eliminated. If lead levels were similar for the two groups, then the data point would fall along the line indicated in Figure 3.3. If the corrosion decreased and the second rounds' results were less than the first, then the data point would fall above the line. Figure 3.3 illustrates that corrosion decreased at most homes over the monitoring period.

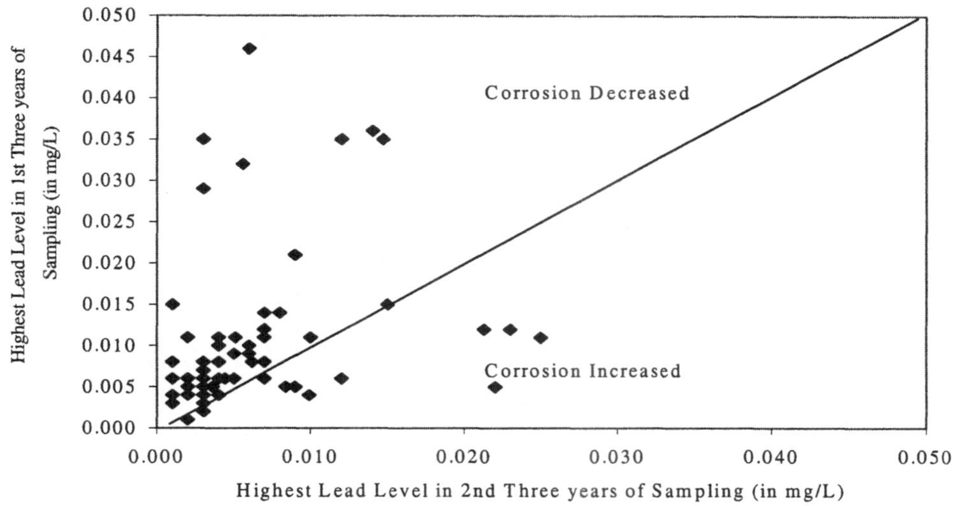

Figure 3.2 PWB lead and copper 90th percentile data summary

Sample Pool

Over the six rounds of sampling, there has been a steady decline in the number of customers willing to participate (Figure 3.4). Many customers who were initially interested in knowing the lead levels caused by their plumbing are no longer willing to go through the inconvenience of sampling especially if their results are continually below the AL.

Baxter is the largest of all three water treatment plants and its service area has had the most customer participation (38 to 42 percent of total homes sampled). Over the six rounds of LCR sampling, the Queen Lane system has lost more than two-thirds of its participants. The Belmont system served by the smallest of the City's three water treatment plants, has had the least amount of customer participation in the LCR sampling.

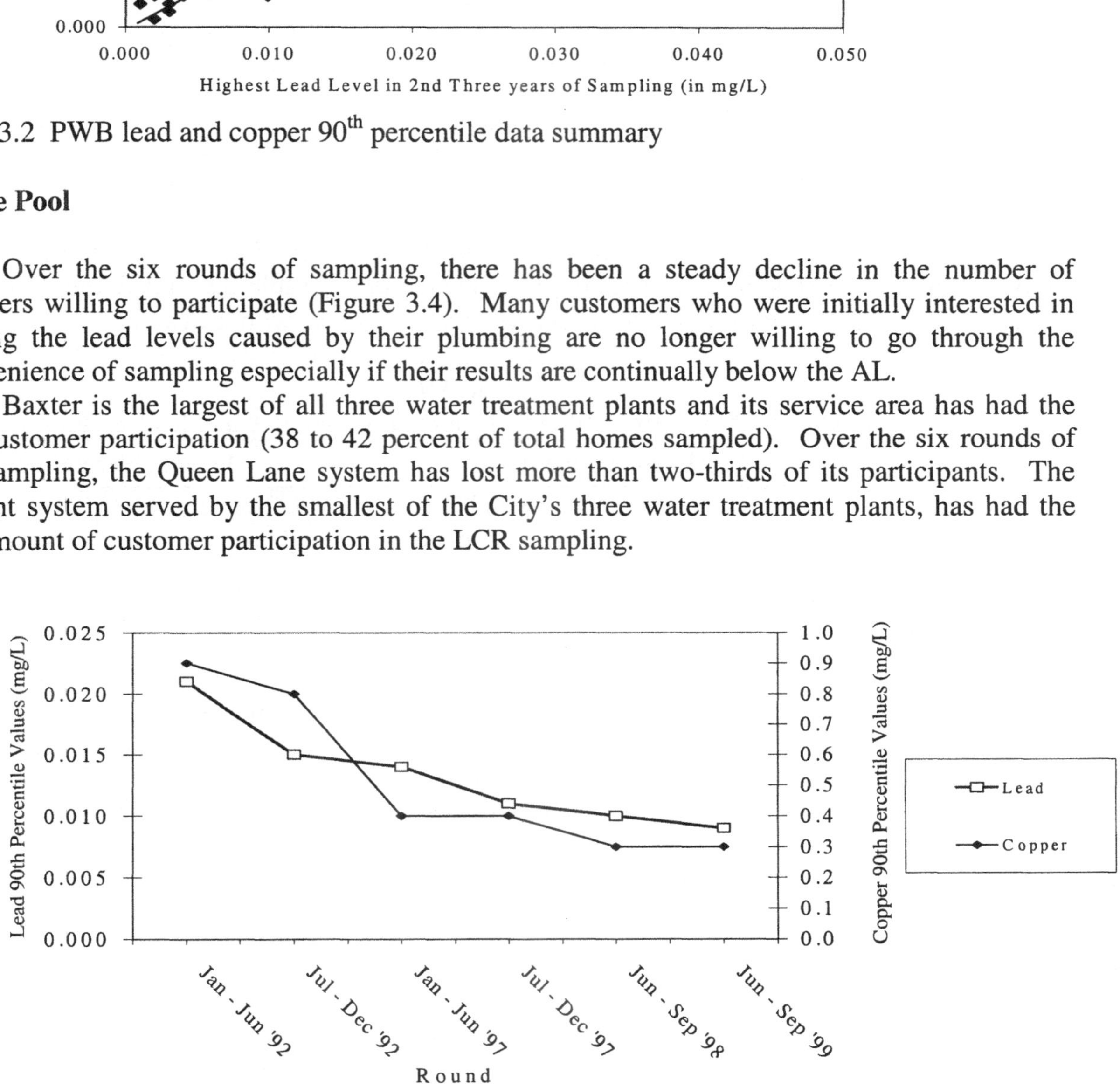

Figure 3.3 PWD evaluation of decreasing lead levels by home

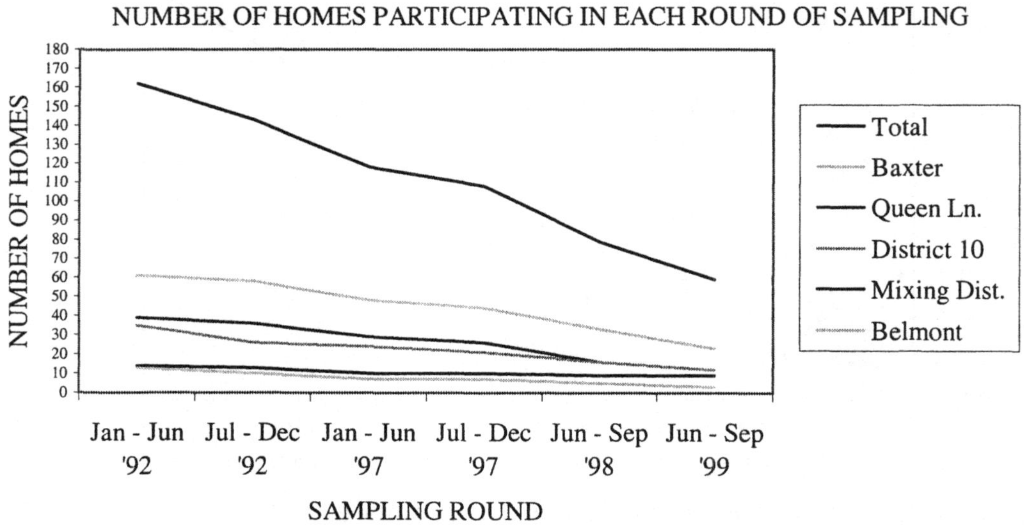

Figure 3.4 Number of homes participating in each round of PWD sampling

Some homes in the LCR sampling program have lead service lines while others have the typical copper or galvanized service line with lead solder. Table 3.9 and Figure 3.5 show an overall comparison of homes with lead service lines and homes with lead solder.

Table 3.9
PWD comparison of 90th percentile levels in lead service line and lead solder homes

	Lead service lines		Lead soldered lines	
Sampling period	90th percentile lead level (mg/L)	Number of samples	90th percentile lead level (mg/L)	Number of samples
Jan - Jun 1992	0.016	81	0.031	81
Jul - Dec 1992	0.012	72	0.019	71
Jan - Jun 1997	0.008	58	0.017	60
Jul - Dec 1997	0.009	53	0.011	55
Jun - Sept 1998	0.005	42	0.011	37
Jun - Sept 1999	0.014	30	0.009	29

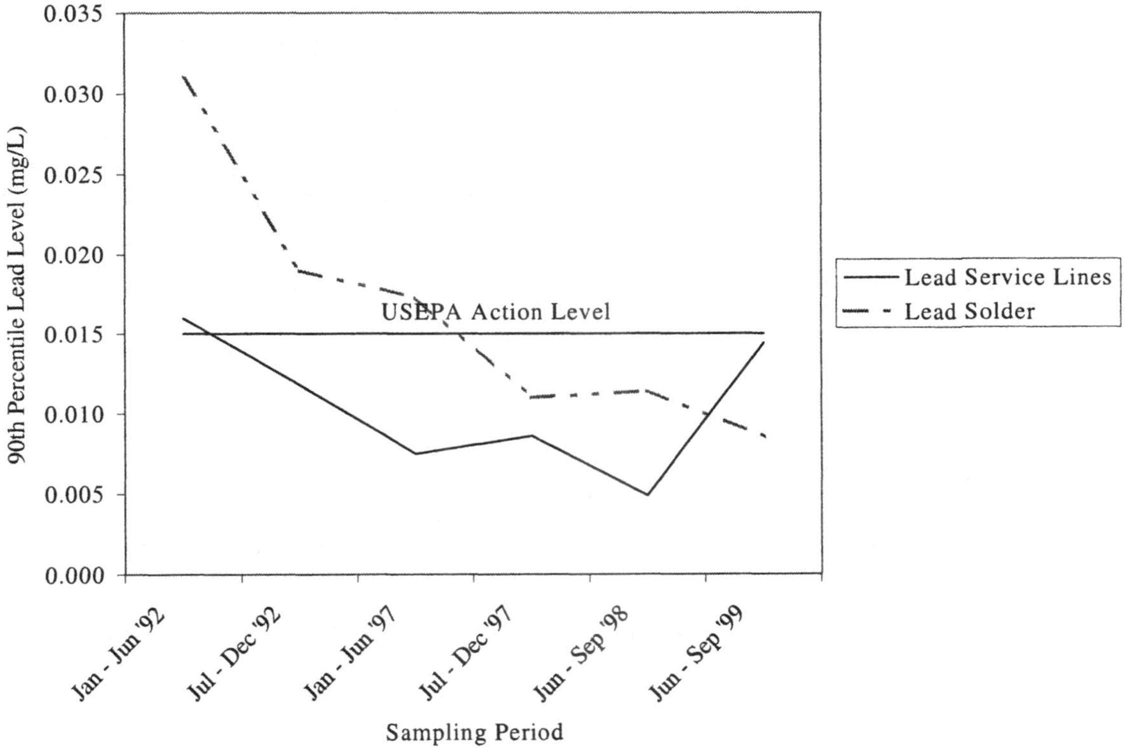

Figure 3.5 PWD 90th percentile homes with lead services lines vs. 90th percentile homes with lead solder

The increase in 90th percentile lead levels for homes with lead service lines is of concern and needs to be considered further.

Because there is uncertainty introduced to lead data by unskilled samplers, PWD decided to look at lead results with upper and lower limits of 20%. This is an arbitrary level of error used to explore the sensitivity of the 90th percentile AL to sampling variability. First, a comparison was made between the 90th percentile, the 50th percentile (median) and the average lead levels. This is shown in Table 3.10.

Table 3.10
PWD comparison of statistics for actual lead levels

Sampling period	90th percentile lead level (mg/L)	50th percentile lead level (mg/L)	Average lead level (mg/L)
Jan - Jun 1992	0.021	0.006	0.014
Jul - Dec 1992	0.015	0.006	0.009
Jan - Jun 1997	0.014	0.002	0.011
Jul - Dec 1997	0.011	0.002	0.006
Jun - Sep 1998	0.010	0.002	0.006
Jun - Sep 1999	0.009	0.003	0.009

The result for each individual sample was changed by a factor of 20% and the 90th percentile, 50th percentile, and the average were recalculated. Tables 3.11, 3.12 and 3.13 show the results for the 90th percentile, 50th percentile and average lead levels, respectively. Table 3.12 uses four decimal places to better show the slight variability in lead levels.

Table 3.11
PWD evaluation of 90th percentile lead levels with a 20% error assumption

Sampling period	90th percentile level (mg/L)	90th percentile lead level +\- 20% (mg/L)	USEPA action limit (mg/L)
Jan - Jun 1992	0.021	0.026 - 0.018	
Jul - Dec 1992	0.015	0.018 - 0.012	
Jan - Jun 1997	0.014	0.017 - 0.011	
Jul - Dec 1997	0.011	0.013 - 0.009	0.015
Jun - Sep 1998	0.010	0.012 - 0.008	
Jun - Sep 1999	0.009	0.011 - 0.007	

Table 3.12
PWD evaluation of 50th percentile lead levels with a 20% error assumption

Sampling period	50th percentile lead level (mg/L)	50th percentile lead level +\- 20% (mg/L)	USEPA action limit (mg/L)
Jan - Jun 1992	0.0060	0.0072 - 0.0048	
Jul - Dec 1992	0.0060	0.0072 - 0.0048	
Jan - Jun 1997	0.0020	0.0024 - 0.0016	
Jul - Dec 1997	0.0020	0.0024 - 0.0016	0.0150
Jun - Sep 1998	0.0020	0.0024 - 0.0016	
Jun - Sep 1999	0.0031	0.0037 - 0.0025	

Table 3.13
PWD evaluation of average lead levels with a 20% error assumption

Sampling period	Average lead level (mg/L)	Average lead level +\- 20% (mg/L)	USEPA action limit (mg/L)
Jan - Jun 1992	0.014	0.017 - 0.011	
Jul - Dec 1992	0.009	0.010 - 0.007	
Jan - Jun 1997	0.011	0.013 - 0.009	
Jul - Dec 1997	0.006	0.007 - 0.005	0.015
Jun - Sep 1998	0.006	0.008 - 0.005	
Jun - Sep 1999	0.009	0.010 - 0.007	

As can be seen in Table 3-11, using the conservative upper limit of +20%, PWD would have exceeded the USEPA AL in the first three rounds of sampling. With a lower limit of -20%,

PWD still would have exceeded the limit in the first round. If the regulation used the average of the data (Table 3.13), PWD would have passed all rounds of sampling; however, the upper limit during the first round would have exceeded the USEPA AL.

Corrosion Control Studies

Philadelphia has conducted extensive lead and copper corrosion pipe rack and jar test studies since 1989. A 1989 observational study assessed the lead and copper standing sample concentrations under the existing treatment conditions.

In 1991, PWD began a one year corrosion control pipe rack study to select corrosion inhibitors and pH ranges. This led to the addition of zinc orthophosphate at the Baxter water treatment plant (WTP) in January 1993, at 0.12 mg/L. By July 1993, all three WTPs switched from a 1:1 ZnPO4 product to a 1:3 product, dosed at 0.36 mg/L (preceded with a distribution system passivation). This systematically headed towards a unified corrosion control strategy for all three WTPs, to provide better control of 'mixed' distribution districts. In 1993, a second corrosion control pipe rack study optimized the pH and orthophosphate dosages to minimize lead and copper corrosion. PWD submitted the results of the 1991 and 1993 studies to the Pennsylvania Department of Environmental Protection (DEP) in 1994, in compliance with the LCR. Subsequently, PWD began slowly lowering the Baxter WTP's pH from 8.2 to 7.4 from 1995-1997. By 1997, all three WTPs had uniform finished pH and zinc orthophosphate dosages.

In 1994 and 1995 PWD conducted a number of jar tests and a subsequent pipe rack study investigating if non-zinc phosphate corrosion inhibitors can provide equivalent corrosion protection as zinc orthophosphate inhibitors. Results were mixed, and X-ray crystallography suggested that zinc plays an important role in passivation of mild steel coupons. Thus, Philadelphia maintained zinc addition for distribution system corrosion control.

Corrosion Control

PWD's corrosion control strategy included treatment changes at the three water treatment plants; Baxter, Queen Lane and Belmont. The treatment changes described below have lowered lead and copper corrosion in customers' plumbing.

Prior to 1993, the Baxter plant used a high pH (pH 8.0-8.2) as a means of corrosion control. Queen Lane and Belmont maintained a lower pH (7.1 and 7.0, respectively) and used 1:1 $ZnPO_4$ product at a dosage of 0.12 mg/L.

In January 1993, the Baxter plant began using a 1:3 ($ZnPO_4$) product at 0.12 mg/L of PO_4. In July 1993, Baxter switched to a 1:3 $ZnPO_4$ product at a dosage rate of 1.1 mg/L of PO_4 until the system was passivated and then lowered the dosage rate to 0.36 mg/L of PO_4 in September 1993. In 1995, Baxter began to slowly lower the pH in conjunction with the corrosion inhibitor. From 1995 until 1997, Baxter's pH was lowered from 8.2 to 7.4. In July 1996 the Baxter plant began using a 1:5 $ZnPO_4$ product at a dosage rate of 0.72 mg/L.

In July 1993, the Queen Lane and Belmont plants began using the 1:3 zinc orthophosphate product starting with a dosage rate of 1.1 mg/L until the systems were passivated, and then applied a 0.36 mg/L dosage rate. From 1993 to the present, Queen Lane and Belmont maintained a pH of 7.3 to 7.4 pH units. In July 1996, both plants began using a 1:5 $ZnPO_4$ product at a dosage rate of 0.72 mg/L.

Impact On Other Regulations

Pipe rack studies have demonstrated the efficacy of zinc orthophosphate corrosion inhibitors. Zinc added at the water treatment plant eventually enters the wastewater treatment plant, and finally the sludge processing facility. A mass balance study determined that Philadelphia must limit finished water zinc levels to 0.12 mg/L for corrosion control, since increased zinc will result in elevated sludge zinc concentrations, making the sludge unsuitable for land application.

From 1995-1998, PWD performed extensive pilot plant and jar test enhanced coagulation studies for the USEPA Disinfection/Disinfectant By-product Rule (D/DBPR). Results of this work led PWD to pursue enhanced coagulation for improved organics removal and reduced DBP formation through flocculation and sedimentation at pH 6.5. To reduce the economic impact of post-sedimentation pH elevation up to 7.4, PWD considered reducing the finished water pH to below 7.4. However this falls outside the optimum range for orthophosphate addition and is outside the state approved lead corrosion control method. Thus, in 1998, Philadelphia revisited lead and copper corrosion control by investigating elevated orthophosphate dosages at pH 7.0. This study found that an increased orthophosphate dosage, above the current 0.72 mg/L, at pH 7.0 minimized lead and copper corrosion, while further reducing mild steel corrosion. However, finished water pH must remain at or above 7.2 since PWD uses chloramines in the distribution system. Below a pH of 7.2, monochloramine will further react with ammonia to form di- and tri-chloramines, which have negative aesthetic impacts. In 1999 Pennsylvania DEP approved a drop in distribution pH; PWD moved to a 1:10 $ZnPO_4$ product with 0.12 mg/L Zn and 1.2 mg/L PO_4 to maintain lead corrosion control.

Lessons Learned

There is a utility need for a more reliable monitoring protocol to verify optimization. PWD is not faced with changing sources or treatment, but does face sample pool attrition and sample variance. It is possible for PWD to fall out of compliance with the LCR due to these factors during any given sampling. A sampling protocol providing less variable data and utilizing a small sample pool would provide PWD with a more reliable picture of corrosion control.

CITY OF PORTLAND BUREAU OF WATER WORKS, PORTLAND, OREGON

Background

The Portland Water Bureau (PWB) provides potable water to nearly 800,000 people in the Portland metropolitan area. The primary source of water is drawn from the Bull Run watershed located approximately 35 miles east of Portland. Bull Run is an unfiltered surface water source. Water is disinfected with chlorine (Cl_2) at the Headworks treatment facility, and is then injected with ammonia (as NH_4OH) and sodium hydroxide (as NaOH solution) at the Lusted Hill treatment facility located approximately 10 miles downstream of the Headworks facility.

The PWB system also has a supplemental groundwater supply located along the south shore of the Columbia River. Groundwater is primarily used as a supplemental supply during the summer months or during emergencies. When used, groundwater is disinfected with chlorine, followed by ammonia addition prior to being pumped to the distribution system. Treatment system improvements for adding sodium hydroxide injection at the groundwater pump station have been constructed.

The pH of the treated water is typically in the range of 7.3 to 7.8 pH units, and the alkalinity is typically 6 to 12 mg/L as $CaCO_3$, for the Bull Run supply. To remain optimized for corrosion control, Oregon Department of Human Services (DHS) requires that the finished water have a pH greater than 7.2 at the entry point to the distribution system and maintain a pH of 7.0 or greater in the distribution system.

Water mains are constructed primarily of cast iron, ductile iron, and steel pipe. Service connections are either copper or galvanized iron pipe.

The PWB is in compliance with all requirements set forth by the DHS Drinking Water Program. Per the agreement with DHS, the Water Bureau implements a Lead Hazard Reduction Program (LHRP) as an alternative to providing "optimal" corrosion control treatment.

Historical Monitoring

Portland has conducted lead and copper monitoring at customer taps as directed by DHS since 1992. In 1997, the PWB implemented a joint monitoring plan that includes participation of surrounding water systems using PWB supplied Bull Run water. PWB and its wholesale water customers collectively monitor at least 100 taps for lead and copper, and 25 distribution system sites for water quality parameters. The number of sample sites selected from each water system is a fraction of the total sample sites required, proportional to the system's water consumption rate. Reduced monitoring has not been implemented. Table 3.14 summarizes the historical monitoring schedule and results.

Sample Pool

The PWB joint monitoring program started with an initial pool of 131 homes in 1992. These homes were identified by scanning building permit records for water service applications from January 1, 1983 to June 30, 1985 and by inspecting home water service piping and appurtenances. The pool of targeted sampling sites is comprised exclusively of Tier 1 sites consisting of single family structures that contain copper pipes with lead solder installed after 1982. Within the City of Portland's service area, there are no known single family structures that contain lead pipes, nor are there any known lead service lines. In December 1984, the State of Oregon banned the use of lead solder and flux containing more than 0.20% lead. Pipe and pipe fittings containing more than 8.0% lead in all potable water supply systems and plumbing were banned effective July 1, 1985.

In 1997, the PWB implemented the joint monitoring plan. Homes sampled in 1997 were randomly selected from homes sampled in 1992. At least 100 samples are collected every 6 months with PWB required to collect 61 of these samples.

Table 3.14
PWB lead and copper monitoring history

Monitoring Period	Number of sites	Lead		Copper	
		90th percentile (mg/L)	Action level (mg/L)	90th percentile (mg/L)	Action level (mg/L)
Spring 1992	126	0.044	0.015	1.8	1.3
Fall 1992	125	0.053		1.3	
Spring 1997	130	0.012		0.65	
Fall 1997	129	0.013		0.68	
Spring 1998	129	0.015		0.67	
Fall 1998	127	0.019		0.67	
Spring 1999	126	0.020		0.64	
Spring 2000	125	0.015		0.66	
Fall 2000	124	0.022		0.71	
Spring 2001	144	0.017		0.61	
Fall 2001	134	0.020		0.81	
Spring 2002	133	0.013		0.42	

PWB customers who participate in the sampling program receive a $20 credit towards their water bill for each sampling period completed. Through this program PWB is able to maintain the necessary level of customer participation. Each year letters are sent to customers included on the original list of homes sampled in 1992. For example, for the second round of sampling in 2001, approximately 116 customers were sent letters of invitation to participate in sampling. Those who do not wish to participate are replaced by alternative homes from the original 1992 sampling pool.

Corrosion Control Studies

PWB initially considered four approaches for corrosion control. Based on desktop studies, a calcium carbonate approach to corrosion control was eliminated from consideration. The remaining pH adjustment, alkalinity adjustment and corrosion inhibitor alternatives were tested in a bench-scale study at the University of Washington during 1992. Twenty-eight different alternative combinations were evaluated in the two-month study. Based on the bench-scale results, four different corrosion control schemes were recommended for further pilot-scale studies.

Pilot-scale corrosion control testing using pipe loop racks was conducted from August 1993 through February 1994. The pilot plant consisted of equipment designed to model the plumbing systems and water use patterns in a home. Five test racks were used to evaluate the control (chloraminated Bull Run water from the distribution system) plus four alternatives:

(1) pH of 9.0 and alkalinity of 20 mg/L;
(2) pH of 8.0 and alkalinity of 25 mg/L;
(3) pH adjustment to 7.5 and orthophosphate inhibitor at 0.5 mg/L; and
(4) pH adjustment to 8.5 and silicate inhibitor at 10 mg/L (as SiO_2).

Based on six months of pilot study data, the target water quality conditions were determined to be a pH of 9.0 to 9.5 and an alkalinity of 20 mg/L. Phosphates were eliminated from consideration because of expected algal growth and possible bacterial growth in the open distribution reservoirs, and also due to their expected impact on wastewater discharge and treatment.

Five non-chemical measures to reduce lead and copper were also evaluated in the 1994 study, even though the LCR does not address these types of alternatives. These measures included:

(1) Flushing the tap through public education;
(2) Fixture replacement;
(3) A state-wide ban on lead in faucets;
(4) Replacement of lead-soldered copper plumbing; and
(5) A lead source identification and evaluation team

The non-chemical approach would not have provided compliance with the LCR, but supplemented the chemical treatment alternative. The recommended alternatives included tap flushing, faucet replacement, a ban on faucets containing lead, and the lead identification team. Plumbing replacement was not recommended due to its high cost and the difficulty in obtaining homeowner cooperation.

From 1994 to 1997, PWB investigated alternatives for LCR compliance. Part of that investigation included a study to determine the effects of corrosion control treatment on lead exposure through drinking water, and to determine what, if any, health benefits may result to the community from treatment. The results of this study predicted a very limited level of effectiveness of corrosion control treatment in reducing the community's blood lead levels. These findings supported the PWB's proposal to implement a Lead Hazard Reduction Program as described below.

Lead Hazard Reduction Program

With the approval of DHS, the City of Portland implemented a Lead Hazard Reduction Program (LHRP) as an alternative to providing "optimal" corrosion control treatment ("optimal" treatment is defined as increasing the pH of Bull Run water to 9.0-9.5, and increasing alkalinity to at least 20 mg/L as $CaCO_3$). The savings in capital and operating costs associated with operating at a pH of 7.3 to 7.8 versus 9.0 to 9.5 are used to fund other target interventions that are designed to achieve better public health protection from lead exposure.

PWB's commitment to the existing treatment strategy is based on the comprehensive nature of the LHRP, the significant risks from lead sources other than drinking water (e.g., lead-containing paint and dust), the demonstrated performance of the existing treatment process in effectively reducing lead and copper levels, and the uncertain incremental benefits of "optimal" treatment in achieving further reductions in blood lead levels.

The LHRP was implemented in 1997 and consisted of four components:

1. Water Treatment and Water Quality Monitoring
2. A Home Lead Hazard Reduction Program
3. Stakeholder Involvement, Public Education and Community Outreach
4. Lead in Water Testing Program

As part of the LHRP, limited corrosion control treatment provides pH adjustment up to 7.8 to meet the copper AL and reduce lead levels in water by at least 50 percent. The latest monitoring results (May 2002) show a 73% reduction in lead and copper levels as compared to results prior to implementation of pH adjustment. 90th percentile copper levels are consistently about one-half of the copper AL, while 90th percentile lead levels are slightly above or below the lead AL.

The objective of the Home LHRP is to reduce the actual or potential risks of significant lead exposure from lead-based paint and other sources in at-risk homes in the highest risk neighborhoods. This program has involved conducting in-home lead risk evaluations, providing cost-effective lead hazard reduction measures and related information to residents. The program was conducted under a cooperative agreement with, and funding to, the Multnomah County Health Department.

The public education and community outreach program is designed to be proactive in raising awareness of all potential sources of lead hazards, the potential health risks to children, and prevention measures and to conduct education and outreach in the highest risk neighborhoods.

Since June 2000, education and outreach efforts have been achieved with a cooperative agreement with the City of Portland Bureau of Housing and Community Development (BHCD). Money to fund this effort is from the support of BHCD's HUD-funded lead abatement program. When this cooperative agreement was put in place, the Home LHRP was discontinued.

The Lead in Water Education and Testing Program provides free or low cost water testing for customers in the Bull Run service area who may be at significant risk from elevated levels of lead or copper in their drinking water. Technical assistance is provided to these customers to reduce their exposure risks.

In addition to submitting Portland's LHRP to the State primacy agency, this program was also submitted to USEPA's Project XLC (eXcellence and Leadership for Communities) in March 1997. The XLC program was designed "to give regulated sources the flexibility to develop alternative strategies that will replace or modify specific regulatory requirements on the condition that they produce greater environmental benefits." Despite extensive efforts over several months, PWB and USEPA were not able to complete an agreement on this program and the XLC agreement development was discontinued in April 1999.

Impact On Other Regulations

PWB's alternative means of compliance with the LCR has not resulted in any problems in meeting other regulatory requirements.

Lessons Learned

There are many aspects of the LCR program that PWB did well and would definitely do again. Through extensive public involvement, the Water Bureau embarked on a comprehensive program to address lead hazards that pose the greatest risk to the community. The Lead Hazard Reduction Program is a risk-based, multi-media strategy of reducing risks from all sources of lead, including lead dust, lead in soil and in water. The program is comprehensive and effectively integrates outreach and education as a means to help reduce lead exposure. PWB also implemented a Lead & Copper Rule Joint Monitoring Plan in partnership with sixteen wholesale contractors. This joint monitoring provides information about treatment conditions throughout the service area.

In looking back, a slightly different path may have been taken. At the time of program development, it was assumed that as a result of limited pH adjustment, the 90th percentile lead results would generally be in the range of 20 µg/L. Instead, results have been much better than expected and have often been below the lead action level. In addition, it was assumed that Oregon DHS's approval of the program would satisfy federal regulatory scrutiny of the LHRP. Instead, the USEPA has recently indicated they may consider overruling the primacy agency's approval of the program. It now appears possible that addition of low doses of soda ash, in addition to limited pH adjustment with sodium hydroxide, may have better assured that lead results would be consistently below the action level. While such an approach would still not have theoretically satisfied the "letter" of the LCR (it would not be implementation of "optimized treatment"), it may have helped to limit USEPA's regulatory objections to the current program. Regardless of the many challenges faced, PWB continues to support the fundamental basis of the Lead Hazard Reduction Program by providing limited corrosion control treatment and addressing lead hazards that pose the greatest risk to the community.

SEATTLE PUBLIC UTILITIES, SEATTLE, WASHINGTON

Background

Seattle Public Utilities (SPU), formerly Seattle Water, provides drinking water on a direct service and wholesale basis to approximately 1.3 million consumers in the greater central Puget Sound region of western Washington. SPU operates two surface water supply sources. The Cedar River supplies 70 percent of the annual average consumption and the South Fork Tolt River provides approximately 30 percent. The Cedar River supply is currently unfiltered but receives chlorine disinfection, fluoridation, and pH adjustment with lime. The Tolt River supply is being treated by ozone, coagulation/flocculation, direct filtration, fluoridation, chlorination, and both pH and alkalinity adjustment with lime and carbon dioxide. Seattle has been treating both of its primary sources to reduce corrosivity since 1982-1983.

Over two-thirds of the transmission mains in the Seattle direct service area are cement mortar lined steel, while over 90 percent of the distribution mains are ductile and cast iron. Wholesale customer distribution materials vary significantly. Generally, the wholesale systems are comprised primarily of iron (ductile and cast iron) and asbestos cement. Most of the iron pipe in the wholesale distribution systems is cement mortar lined.

SPU operates six open distribution reservoirs within the direct service distribution system. Seattle is currently implementing a plan to cover or replace all of the open reservoirs, which should be completed by 2015. Since 1987, SPU has been converting reservoir booster chlorination systems from gaseous chlorine to sodium hypochlorite. Only two gas chlorine systems remain in use. As the reservoirs are upgraded, the remaining gas chlorination systems will be converted to sodium hypochlorite. Both the reservoir covers and the hypochlorite conversion will provide additional corrosion control benefits.

The main sources of lead in Seattle residences are the residential plumbing systems containing copper piping with lead-tin solder, and brass fixtures that contain lead. Seattle has also historically found some lead leaching from residential galvanized piping systems. Lead service lines were never installed in the direct service area or wholesale service areas. Occasionally, crews find short lead connections (< 2 ft.), which have been replaced over the last few years. The main source of copper is residential copper piping.

Historical Monitoring

Seattle and its wholesale customers jointly developed a regional monitoring program. The total sample pool was just under 400 sets of samples for each of the two six month monitoring periods in 1992. Had each district conducted sampling independently, over 1,100 samples would have been required for a population of roughly 500,000, with an additional 100 samples required for Seattle's direct service population of just under 600,000. A set is comprised of a standing and a flushed sample. The specific monitoring that was carried out included a first draw (standing) one-liter water sample and a 250 ml flushed sample from the cold water kitchen tap. The 90th percentile lead value from the regional monitoring exceeded the lead AL during both six month monitoring periods in 1992 and one round of monitoring in 1997, as shown in Table 3.15. The copper 90th percentile values were well below the copper AL in both 1992 and 1997. The 90th percentile home was likely built in either 1978 or 1979 with copper pipe and lead solder.

Seattle entered into a Bilateral Compliance Agreement with the Washington State Department of Health (DOH) in 1997. This agreement was revised in 2001. Seattle conducted a corrosion control study in January 2003. Seattle's optimization plan was implemented in May 2003; and first round monitoring was conducted in September 2003. The intent of the agreement is to focus resources on optimization of corrosion treatment facilities at the Tolt Filtration Plant. This focus on only one source was due to the differing water qualities of SPU's two sources. Initial sampling showed that samples from locations receiving strictly Cedar River water were well below the AL. Conversely, the higher lead levels were isolated to sample locations receiving Tolt river water. Monitoring in accordance with the LCR is not required until the Tolt corrosion treatment facilities have been optimized.

Table 3.15
SPU lead and copper monitoring history

Sampling period	Number of homes	Lead		Copper	
		90th percentile (mg/L)	Action level (mg/L)	90th percentile (mg/L)	Action level (mg/L)
Spring 1992	389	0.0188		0.30	
Fall 1992	354	0.0201	0.015	0.47	1.3
Spring 1997	390	0.0193		0.60	

Sample Pool

Tier 1 and Tier 2 sites, as defined by USEPA, do not exist, with the exception of for a small portion of the direct service area that is located outside of the Seattle city limits. This situation exists because a City-wide ban on lead solder took effect in 1980. The rest of King County implemented a lead solder ban in 1985. Due to the timing of the Seattle lead solder ban, the Seattle direct service area sampling pool was comprised almost completely of Tier 3 sites, while the wholesale sampling pool was comprised almost completely of Tier 1 sites.

Homes were selected by surveying customer service records from 1978 and 1979 for new connections. Postcards were sent to 3,000 customers asking for participation.

Between the first and second monitoring periods in 1992, about 10 percent of the customers dropped out of the program due to a lack of interest. The number of homes participating in the 1997 monitoring changed due to changes in the number of purveyors participating. Customers that dropped out were not replaced.

Corrosion Control Studies

From 1970 to 1983, SPU conducted a series of rigorous and detailed investigations into the internal corrosion of metallic plumbing systems within the distribution system. In response to the findings, SPU installed corrosion control treatment on the Cedar and the South Fork Tolt River supplies during 1982-1983. Initial studies conducted during 1970 to 1973 documented the aesthetic corrosion complaints, measured distribution system corrosion rates, initiated a preliminary screening of residential metal leaching, and evaluated the efficiency of pH adjustment in reducing Tolt water corrosion rates. These early studies helped define the extent of internal corrosion. During 1974 and 1975, expanded studies addressed the issues of metal leaching, and relationships between tap water burdens of toxic metals and concentrations in human tissue. From 1975 to 1978, Seattle Water and Kennedy Engineers conducted a comprehensive study on internal corrosion and developed a corrosion control strategy with two main objectives:

1. Recommendations to adjust the water chemistry of both the Cedar and Tolt supplies; and
2. Recommendations to amend the local plumbing code and to encourage the use of corrosion resistant materials.

Evaluations of alternative treatment strategies showed that the best corrosion control was obtained using the lime/soda ash combination at a pH of 9. Due to trihalomethane (THM) formation problems, chlorine efficacy, galvanized iron corrosion, and other constraints, the pH of the Tolt supply was raised no higher than 8.3. Phosphate corrosion inhibitors were not recommended due to increased potential for algal growth associated with open distribution system reservoirs. Silicate based corrosion inhibitors did not show consistent corrosion control performance or superior benefits over the pH/alkalinity adjustment strategy. Alternative disinfectants and fluoridation chemicals were also evaluated. Hypochlorite was identified for secondary disinfection. Table 3.16 shows the pH and alkalinity levels following corrosion control treatment.

Table 3.16
SPU water quality following corrosion control treatment

Source Water	pH		Alkalinity (mg/L as $CaCO_3$)	
	Pre-1982	Post -1983	Pre-1982	Post -1983
Cedar River	6.8-7.1	8.2	14-18	19
Tolt River	5.8-6.2	8.2	2	13

After corrosion treatment was initiated in 1982, full-scale treatment evaluations showed a 61 and 68 percent reduction in lead levels in residential standing samples, respectively, for the Cedar and Tolt supplies. Reductions in copper levels in residential standing samples were 50 and 74 percent, for the Cedar and Tolt supplies respectively. Likewise, a significant drop (approximately 20 percent) in customer complaints related to corrosion by-products occurred.

Additional pilot studies were conducted in 1993-1994 to evaluate the conversion of the reservoir rechlorination stations from gaseous chlorine to sodium hypochlorite. A second purpose was to evaluate an alkalinity increase at the Tolt headworks to further reduce lead solubility. The pilot study results demonstrated that both of these treatment strategies provided substantial reductions in metals leaching compared to the existing conditions.

Impact on Other Regulations

For the Tolt supply (prior to construction of the filtration plant), a higher pH could potentially impact compliance with the Surface Water Treatment Rule inactivation requirements for contact time (CT). In 1994, Seattle's finished water quality goals included a pH of 8.2. Increasing the pH to 8.6 or higher was considered as an alternative for reducing lead solubility. The CT requirements could still be met at a pH of 8.6 but would require a higher chlorine dosage (+ 0.2 mg/L) for some parts of the year. Higher chlorine residuals in the distribution system could also cause slightly increased corrosion rates to all metals and negate some of the corrosivity improvements expected by raising the pH.

With the new Tolt Treatment Facility, CT compliance impacts are not a concern. Primary disinfection for *Giardia* and *Cryptosporidium* is accomplished with ozonation, and pH adjustment occurs after this process. The interim pH and alkalinity goals are 8.2 and 13 mg/L, respectively. These may be revised following completion of the Tolt optimization study. Preliminary results from pipe loop studies indicate that the filtration plant is helping to reduce lead and copper levels due to the removal of natural organic matter (NOM). This improvement should allow SPU to meet the lead and copper ALs in future sampling rounds.

The Total Coliform Rule (TCR) was a concern for Seattle in the discussion of corrosion inhibitors due to the potential for increased biological activity caused by their use. If algal growth were stimulated within the open distribution reservoirs, continued compliance with the TCR could become difficult. To prevent this potential conflict, Seattle did not consider the use of phosphate corrosion inhibitors. Further, plans are underway to cover or replace all of the open distribution reservoirs.

SKAGIT COUNTY PUBLIC UTILITY DISTRICT #1, MT. VERNON, WASHINGTON

Background

Public Utility District #1 of Skagit County (the District) provides potable water to a service population of 50,000 in northwest Washington State, approximately 70 miles north of Seattle. The average and peak day (summer) demand requirements are 8.6 MGD and 17 MGD, respectively. The District's source of supply is Judy Reservoir, which is fed by several streams originating in the Cultus Mountain watershed.

The water system has been optimized for corrosion control under the Lead and Copper Rule (LCR) since 1998. A summary of typical source water quality parameters is provided in Table 3.17.

Table 3.17
Skagit County historical source water quality data

Parameter	Units	Range of values
Conductivity	μmhos/cm	30–60
Temperature	°C	1–21
pH	standard units	6.9–7.5
Alkalinity	mg/L as $CaCO_3$	14–16
Hardness	mg/L as $CaCO_3$	8.6–21.6
Dissolved oxygen	mg/L	9–13
Total organic carbon	mg/L	3.0–7.0
Turbidity	NTU	0.25–1.5

In the late 1980's, the District constructed a water filtration plant to comply with upcoming federal and state regulations. Initial oxidation/disinfection is provided by chlorine dioxide, primarily to control THMs. Coagulants consist of hybrid aluminum salts and a polyquaternaryamine. Carbon dioxide is injected prior to the flocculation basins in order to lower the pH and enhance coagulation. Direct filtration is enhanced with the addition of a slightly anionic filter aid. The filter media consists of one foot of silica sand and two feet of anthracite coal. Carbon dioxide is injected following filtration to lower the pH. Caustic soda is added when needed to raise the alkalinity and provide a finished water pH of 8.6 – 8.8. Chloramination is used for secondary disinfection.

The distribution system contains predominantly PVC (approximately 38% of total miles), ductile iron (approximately 32% of total miles), and asbestos cement (approximately 26% of total miles) pipe. Service connections are constructed of polyethylene. No lead service lines are used in the District. Lead solder was used in home plumbing for houses constructed prior to 1985.

Historical Monitoring

The District has conducted lead and copper monitoring at customer taps as directed by the Department of Health (DOH) since 1992. Initial monitoring resulted in a 90th percentile lead level of 0.049 mg/L. Table 3.18 summarizes the historical monitoring schedule and results.

Table 3.18
Skagit County lead and copper monitoring history

Monitoring period	Number of homes	Lead		Copper	
		90th percentile (mg/L)	Action level (mg/L)	90th percentile (mg/L)	Action level (mg/L)
Nov. 1992	65	0.049		0.37	
Spring 1998	65	0.011		0.092	
Fall 1998	62	0.012	0.015	0.109	1.3
Sept. 1999	56	0.006		0.085	
July 2000	33	0.005		0.079	

Sample Pool

A database of potential participants was developed based on the Tier requirements specified in the LCR. All sites in the initial monitoring were in Tier 1 (homes with copper pipe and lead solder constructed after 1982). Homes were not tested for the presence of lead solder, but homeowners were asked to sign a form acknowledging the use of lead solder and the age of plumbing.

When a participant's home is sold, the new homeowner is invited to participate in the program and usually agrees. If not, or if another current participant wishes to drop out, a replacement from the initial database is invited to join the sampling pool. After the first two monitoring periods, some participants did lose interest. After Tier 1 replacements were not available, DOH allowed the use of Tier 2 homes to replace participants that dropped out. When monitoring was reduced in 2000, the most cooperative participants from the larger pool were selected.

Corrosion Control Studies

Electrochemical corrosion testing was conducted to compare the corrosion control effectiveness of pH adjustment and orthophosphate addition for lead containing surfaces. The greatest reductions in corrosion rate were observed when the pH was raised to 8.5, or when the pH was raised to 8.0 and orthophosphate was added (dosage rate of 4 mg/L as PO_4). Due to a number of functional constraints, the District did not want to add phosphates to the water supply.

After conducting corrosion control testing, the decision was made to increase pH to a range of approximately 8.5 to 9.0. To maintain the desired pH range of 8.5 to 9.0 throughout the distribution system, alkalinity increases were also required. The District decided to increase alkalinity levels with the addition of carbon dioxide and caustic soda and to raise the pH to the desired range using caustic soda.

Following implementation of these treatment changes, distribution system monitoring suggested that pH and alkalinity levels were more uniform throughout the system. Alkalinity levels had approximately doubled (from 14 mg/L as $CaCO_3$ to 26 mg/L as $CaCO_3$), providing for more stable water with respect to pH and corrosion control. System monitoring data showed that it took more than a year for first draw lead levels to stabilize within the distribution system.

Lead levels at the tap have decreased substantially at most of the "high risk" homes in the District. Although the District was initially not able to meet the lead AL at pH 8.0 without alkalinity adjustment, the AL was met during subsequent monitoring due to consistently elevated pH levels (between 8.5-9.0) and increased buffering capacity. It is likely that the added alkalinity has helped to decrease lead levels by two different mechanisms; providing stabilized pH levels at the tap, and increasing carbonate levels to aid in the formation of more stable lead carbonate passivating films.

A study was undertaken by the District to determine whether elevated lead levels measured at the tap were in the soluble or particulate form. It was found that lead was primarily in the particulate form. When the pH was adjusted to 8.0 without alkalinity adjustment, elevated lead levels were mostly due to particulate lead (particulate being the difference between total and soluble lead), suggesting that stable lead carbonate films were not forming. After the pH was further increased and the alkalinity was doubled, total lead concentrations decreased and were more similar to the range of the soluble lead levels. The district determined that alkalinity adjustment was necessary to address the particulate lead fraction.

Impact On Other Regulations

The District had difficulty meeting Surface Water Treatment Rule CT requirements year-round, especially during the colder months. To address this, chlorine, chlorine dioxide and carbon dioxide were added prior to the flocculation basins to both decrease pH and increase chlorine residual throughout the treatment process. Without this process, CTs could not be met consistently without significantly increasing disinfection by-products. A method other than increasing chlorine and contact time was needed to achieve higher inactivation ratios.

DBP formation issues were a concern to the District. Total organic carbon (TOC) in the Judy Reservoir supply ranges from 3 to 7 mg/L, and a high formation potential of disinfection by-products upon chlorination exists. DBP formation issues became a higher priority to the District once the Lead and Copper Rule came into effect and the District exceeded the lead AL.

Since the District could not increase chlorine levels without increasing DBP formation, methods of decreasing pH levels throughout the treatment train were considered. Lowering the pH at the beginning of the treatment train would have the added benefit of enhancing coagulation and increasing the removal of DBP precursor materials.

Thus, it was proposed that the District inject carbon dioxide prior to the flocculation basins (if needed) in addition to the end of the treatment train where caustic soda is added. The advantages of adding carbon dioxide ahead of the flocculation basins were three-fold:

- Associated pH depression would increase coagulation efficiency to remove DBP precursors,
- Associated pH depression would increase CTs throughout the treatment process, allowing chlorine injection to be reduced, and
- Subsequent pH increases using caustic soda would provide finished water with increased alkalinity levels and thus, increased buffering capacity.

The percentage of TOC removed has increased from an average of 25% to approximately 40% since carbon dioxide injection and aluminum chlorohydrate coagulation began. The mass of TOC removed has nearly tripled from 1 mg/L to between 2.5 mg/L and 3 mg/L. Thus, even a small decrease in pH (from 6.9 to 6.5) during coagulation and flocculation appears to have significantly enhanced coagulation.

Preliminary results also indicate that Total THM (TTHM) formation within the treatment train has been reduced by approximately 33 percent. Observed decreases in TTHM formation can be attributed to both enhanced TOC removal and to the fact that fewer TTHMs are formed at lower pH levels. Prior to carbon dioxide addition, total haloacetic acid levels in the finished water ranged between 40 μg/L and 60 μg/L when TOC levels were between 3 mg/L and 5 mg/L. After carbon dioxide addition, total haloacetic acid (THAA) levels in the finished water decreased to the range of 35 μg/L to 45 μg/L even though raw water TOC levels increase to between 5 mg/L and 7 mg/L.

For surface water systems already using caustic soda or lime for corrosion control, the likelihood of compliance with several conflicting regulations can potentially be improved with relatively low expense and operator effort by injecting carbon dioxide at one or two locations within the treatment train. By comparison, if carbon dioxide were added only at the end of the treatment train (just prior to pH adjustment), only corrosion control treatment would benefit. For the District, carbon dioxide addition ahead of the flocculation basins has significantly improved water quality by enhancing coagulation, enhancing CT compliance, and improving corrosion control treatment.

Lessons Learned

The LCR monitoring program requires significant staff time to execute. The District has developed a sampling pool and a program that works well with relatively few, if any, problems. Most participants have been very cooperative.

Corrosion control is integral to the Districts treatment train. The District could not fully commit to a treatment or source change without first evaluating the ramifications of lead and copper levels and other regulatory requirements. The current LCR protocol may not be sufficient to reliably make that determination.

CHAPTER 4
ALTERNATIVE TAP MONITORING PROTOCOL FOR LEAD AND COPPER

PROTOCOL DEVELOPMENT OBJECTIVES

A primary objective of this project was to develop a statistically valid in-home monitoring protocol and data evaluation method for tracking changes in lead and copper levels at the customer tap. The alternative monitoring protocol presented in this report is intended to provide primacy agencies and utilities with a tool to more accurately track lead and copper metal release following optimization. Current regulatory monitoring requirements and the need for alternative monitoring are described below, followed by a discussion of the approach taken in developing the protocol, identification and ranking of variability factors, and a summary of historical lead and copper level data.

REGULATORY VS. ALTERNATIVE IN-HOME TAP MONITORING PROTOCOLS

The National Primary Drinking Water Regulation for Lead and Copper (Lead and Copper Rule or LCR, 1991) set the action limit (AL) for lead at 0.015 mg/L and at 1.3 mg/L for copper. The AL is a 90th percentile calculation based on samples representing overnight stagnation in customer plumbing. If lead or copper levels at the 90th percentile exceed the appropriate AL, then a treatment technique requirement is triggered that could include corrosion control treatment, source water treatment, lead service line replacement, and public education (Maximum Contaminant Level Goals and National Primary Drinking Water Regulations for Lead and Copper-Final Rule, 40 CFR, Parts 141 and 142, June 7,1991).

Corrosion control verification under the LCR involves collecting one-liter first draw samples, collected following a minimum stagnation. The LCR states this about the size of the sample:

> *"A 1-liter sample volume provides a better representation of typical drinking water consumption for an individual and a more accurate portrayal of an individual's exposure to lead and copper in drinking water. Also, a 1-liter sample represents the lead and copper contribution from not only the faucet but also from the interior plumbing of the home."*

LCR compliance sampling was designed to catch the worst-case condition of first-flush water (cold-water kitchen or bathroom faucets) in homes identified as high risk. The LCR defines high-risk locations as containing copper pipes with lead solder installed after 1982 and/or containing lead pipes or lead service lines. The distribution of high-risk locations was categorized according to a three tiered approach. Table 4.1 summarizes Tiers 1, 2 and 3 categories.

Table 4.1
Criteria for LCR Tier 1, 2 and 3 sampling sites

	Criteria
Tier 1	Single family homes that contain copper pipes with lead solder installed after 1982 and/or containing lead pipes or are served by lead service lines.
Tier 2	Buildings, including multiple family residences that contain copper pipes with lead solder installed after 1982 and/or containing lead pipes or are served by a lead service line.
Tier 3	Single family homes that contain copper pipes with lead solder installed before 1983.

In addition to the tiered categories of sample site collection, the LCR outlines criteria for utilities with lead service lines in the distribution system as:

- Utilities with lead service lines must have 50% of samples taken from sites with lead service lines and 50% of samples from sites with interior lead plumbing or lead solder installed after 1982.
- Utilities absent of lead service lines may collect 100% of samples from sites with interior lead plumbing or solder installed after 1982.

The number of required samples ranges from 100 per 6-month sampling period for large systems to 5 per 6-month sampling period for small systems. The LCR states conditions that can be met to reduce monitoring by half:

> *"EPA agrees that water systems should be allowed to reduce the number of samples and frequency of sampling if certain conditions are met, but under no circumstances should a system be allowed to entirely discontinue sampling."*

The LCR protocol requires that a large sample pool be identified, but that samples be collected at a relatively low frequency (twice/year). While the existing regulatory protocol provides a tool for tracking the reduction of lead and copper to below the action limit, it may not be the most appropriate protocol for assessing changes in customer tap lead and copper levels after corrosion control has been optimized. The regulatory protocol provides only two sets of sampling data per year which yields insufficient information regarding the variability in lead and copper levels at individual sites. Over time, the number of appropriate sampling sites is being reduced due to removal of lead service lines and/or customer unwillingness to continue with in-home monitoring, especially in homes with low lead levels. As the sample pool decreases and potentially becomes skewed to the right, paired data for evaluating treatment impacts is lost. Also, as water systems become optimized for corrosion control, lead and copper levels are reduced. As lead and copper levels are reduced, variability must concurrently be reduced or a greater number of samples must be collected in order to detect differences with the same degree of confidence and accuracy. With regulatory monitoring, the number of sample sites remains the same with each evaluation period.

Therefore, to address these monitoring issues in optimized systems, an effective alternative tap monitoring protocol should include the following:

- A pool of sampling sites that will be available for continued, long term monitoring;
- Adequate data from these sites to be able to evaluate tap lead and copper levels in the system at a desired accuracy and confidence level;
- Tap sampling of sufficient frequency to capture changes in metals levels due to water quality changes; and
- Sufficient data to assess the variability in the lead and copper levels and adjust the monitoring protocol accordingly.

DEVELOPING AN ALTERNATIVE IN-HOME TAP MONITORING PROTOCOL

The first step in developing an alternative tap monitoring protocol for lead and copper was to review and evaluate the sources of variability in tap lead and copper levels. From this information, key variability factors were identified and ranked in order of importance and impact by the Project Advisory Committee. The rankings of variability factors, along with the actual variability measured from historical monitoring studies, were used as a starting point in developing a draft alternative protocol. The rankings of variability factors provided information on site selection and sample collection recommendations.

The second step was to document the actual variability exhibited in historical lead and copper data collected under a variety of water quality conditions. This information was gathered from existing literature, case study information, regulatory monitoring, and alternative monitoring programs. The actual variability measured from historical studies provided initial estimates of the recommended number of sites and the frequency of sampling. The third step was to develop a draft alternative tap monitoring protocol and field test this protocol at participating utilities.

In step three, results from the field studies were evaluated and the results and experiences from these field studies were combined with the historical information collected to develop a final alternative tap monitoring protocol for lead and copper levels.

A discussion of the variability factors and their rankings and a summary of the historical data is presented here. The draft alternative tap monitoring collection protocol and the results of the field applications of this protocol are discussed later in this chapter.

IDENTIFICATION AND RANKING OF VARIABILITY FACTORS

Previous studies have shown that lead levels in tap water can vary significantly. This is true for both samples from similar sites in close proximity in the distribution system and served by water of similar chemical characteristics, as well as from repeat samples taken from the same site under identical conditions. The variability in tap copper levels from site to site, or between sites, has not been as rigorously discussed in the published literature. The factors that may affect the variability measured in tap lead and copper levels were identified and ranked by the Project Advisory Committee in order of significance to variability as:

- Age of material (for lead soldered copper and copper pipe)

- Water quality (pH, alkalinity/dissolved inorganic carbonate levels, temperature, chlorine residual and D.O. levels, and inhibitor levels (phosphate, silicate))
- Stagnation time
- Form of lead (soluble and particulate)
- Size and extent of lead based materials (pipe diameter, length of pipe, number of soldered joints)
- Type of materials (lead pipe, lead soldered copper pipe, brass, and copper pipe)
- Sample volume
- Sample flow
- Analytical precision

Many of these chemical and physical factors are linked. These factors can be grouped under three major categories; analytical, chemical, and physical factors (Schock 1989 and AwwaRF 1990). The following discussion summarizes the key factors associated with variation in lead and copper levels for each of these three categories. The information presented below was summarized from comprehensive published evaluations of lead and copper variability in drinking water (Schock and Levin 1988, Schock 1989, Levin et. al. 1989, AwwaRF 1990, and Schock et. al. 1995).

Chemical Factors

The most important factors influencing lead and copper mobilization are the physical and chemical characteristics of the water in contact with the lead and/or copper materials. Physical factors are difficult to separate from chemical factors, because they frequently cause chemical differences. Sources of chemical variability that contribute to the lead and copper levels at the tap include the combined effects of:

- The interactions among water constituents that can affect the oxidation rate;
- The solubility of lead and copper, and;
- The presence and properties of any corrosion passivating films on the pipe interior.

Relatively small differences in temperature, pH, alkalinity, dissolved inorganic carbonate (DIC), or other constituents can significantly affect the solubility and leaching rate of lead and copper. Solubility and dissolution rates decrease with lower temperature, diffusion inhibition, and protective coatings on the pipe.

Large changes in lead solubilty can be brought about by small pH changes. An increase from pH 6 to pH 7 can result in a change in solubility of lead by a factor of approximately 10, depending on the water composition and the nature of the passivating films that form. If pH varies by even a few tenths from one household to another, or even at the same household at different times, then lead levels may show a 3 to 5 factor difference in lead solubility (AwwaRF 1990). Generally, lead solubility is more sensitive to pH than DIC concentration. The predicted absolute minimum lead solubility point is at a pH of 9.8 and DIC concentration of approximately 4.8 mg C/L for a model where hydrocerussite is the controlling solid. This translates into a total alkalinity of approximately 28 mg $CaCO_3$/L.

Copper levels are especially dependent on the pH of the water, with lower pH levels corresponding to higher copper levels. This relationship has been documented by several studies, and high copper levels have also been observed at "dead end" locations in the distribution system where pH levels may drop considerably (Schock et. al. 1995).

Schock et. al. (1995) developed a copper solubility model and evaluated experimental data from a variety of corrosion investigations. This work showed that different controlling solids may develop depending on the age of the copper piping. Newer copper pipes (less than 20 years old) may exhibit higher copper levels than older pipes where low solubility passivating films have formed. Also, copper exhibits slow oxidation and solubilization rates. Therefore, the standing time needed to reach equilibrium conditions in the pipe may be much longer than the typical 'overnight' standing sample collected as part of the LCR protocol. Samples collected at non-equilibrium conditions can exhibit variations in copper levels depending on the levels of oxidants in the system, particularly for newer copper plumbing.

The complex chemistry of lead and copper solubility and formation of passivating films creates the potential for changes in water quality that can cause fluctuations in lead and copper levels measured at the tap. These changes can occur system-wide because of seasonal fluctuations in water temperature and chemical constituents, variations in chlorination or fluoridation practices, malfunctions of feed equipment, and alterations in water treatment (including initiation of corrosion control treatment) which changes the composition of the treated water. Therefore, maintaining consistent distributed water quality, while difficult, can be an important factor in reducing variability in standing lead and copper levels.

Changes in water quality parameters that occur within home plumbing systems can also cause differences in lead and copper levels. Water standing in pipes can undergo changes in pH, chlorine residual, dissolved oxygen level, temperature, hardness, alkalinity, and possibly other constituents resulting from various corrosion and precipitation reactions.

Physical Factors

The physical factors (in order of importance according to the ranking conducted by this report) that impact the variability seen in tap lead and copper levels include:

- Stagnation time
- Age of materials
- Amount and type of lead and/or copper material exposed to the water
- Water use and volume of water collected
- Form of lead and copper
- Quality of workmanship

Stagnation Time

When collecting samples for lead and/or copper it is important to standardize the stagnation time. Small differences in the amount of time water is left to stagnate in lead or copper containing pipe may cause considerable variability in the lead or copper concentration. The sensitivity of the lead concentration to time is greatest in the time range of zero to approximately 3 hours for ½-inch to ¾-inch pipe; these time ranges approximate the standing times represented by "random" daytime samples. Copper levels can continue to rise beyond a 16 hour standing time, particularly if the copper pipe has not aged sufficiently (Schock et. al. 1995). The current regulatory monitoring protocol for both lead and copper levels requires a minimum 6 hour standing time before samples can be collected. Therefore, it is unlikely that copper levels are at equilibrium under typical LCR "overnight" stagnation conditions, unless passivating films are well developed.

Age of Materials

The age of materials will impact the corrosion of lead and/or copper materials. Studies have shown that newer lead-based solders will leach lead at relatively high rates, with a general decline after the solder is five years old due to the surface being depleted of lead and/or surface films building up (Neff et. al. 1987). Lead leaching from brass faucets will also decline over time, but lead pipe can leach lead throughout its service life. Copper dissolution rates depend on the aging of the passivation films that form on the surface of the copper piping. For copper, the highest corrosion rates and solubilities are associated with the newest piping, i.e. less than 20 years old (Schock et. al. 1995).

Amount and Type of Material

Many plumbing systems contain a mix of piping materials, such as lead service lines with copper plumbing in the house, copper sections joined by lead/tin solder, and a variety of faucet materials. The number of lead soldered joints in the pipe and the ratio of the area of the exposed soldered joint to that of the exposed copper pipe will impact the lead levels measured at the tap. The diameter of a lead pipe is also an important factor in raising, or lowering, the lead content of standing samples. Lead levels would be expected to rise more quickly in stagnant samples from pipes of small diameter when compared with pipes of larger diameter (Kuch and Wagner 1983, Schock and Wagner 1985).

Water Use and Collection Volume

The number of people residing in the household and the water use/water flow patterns may introduce a source of variability in lead and copper levels measured at the tap, as will the velocity of flow during the sampling event. When the faucet is turned on during sampling, a condition of "plug" flow is created for the water closest to the soldered joints or lead pipe. As the water is drawn through the pipe, the shape and length of the plug of water is altered by the friction of the water against the pipe wall. Conditions of turbulent or laminar flow impact the mixing, dilution, and shape of the plug. In pipes with several joints, the plugs will tend to blend somewhat because of the mixing. Some household systems have changes in pipe diameter in the flow path to the faucet, creating an additional opportunity for altering the nature of flow and the amount of mixing. Long passages of pipe before the sampling outlet will result in considerable mixing of the plug and lower lead levels in the sample than would have been observed in water in contact with the lead source. Also, higher sampling velocities may scour the inside of the pipe causing release of particulate metals.

The volume of water drawn for a sample is particularly important as a potential source of lead and copper variability. Significant amounts of lead and copper can originate from contact with the brass faucet material. The internal volumes of the fixtures vary with design but are usually about 90 mL to 120 mL for kitchen faucets. Bathroom faucets are likely to be much smaller (AwwaRF 1990). The contribution of the water contained within the fixture to the concentration of the metal found for the whole sample will vary with the total sample volume. Smaller sample sizes (i.e. 100 mL or 250 mL) collected from standing water may have higher lead concentrations than larger size samples, such as 1-liter samples, because they target the lead contribution from the faucet and not from other lead sources in the household.

Form of Lead and Copper

A very important variable in the measurement of lead and copper in potable water is its physical state (i.e. in dissolved-soluble form or particulate-insoluble form). Several sources are possible for particulate lead in water samples. Lead particles can originate from the precipitation of lead solids, such as passivation films or corrosion products, or from removal of existing deposits by scouring or sloughing off during water flow. Relatively large pieces of solder on the interior or a pipe can be dislodged and travel down the pipe to a bend, elbow, restriction, or the screen in the faucet. Lead could be absorbed on foreign particles, such as iron oxides or calcium carbonate particles. Hulsmann (1990) found that physical disturbances in mains, such as cleaning, scraping, relining, and maintenance work can result in increases in particulate lead.

"Dissolved" lead has traditionally been differentiated from "particulate" lead in field and laboratory studies by filtration of the sample through a membrane filter of 0.4 or 0.45 microns pore size. This procedure is complicated by the tendency of dissolved lead to absorb onto, and be retained by, various filter materials (Schock and Gardels 1983). The occurrence of particulate lead is not consistently related to dissolved lead levels.

Because the uptake of particulate lead into the water is often caused by physical means, its contribution is difficult to predict with any certainty. A large amount of the variability in household lead levels can be attributed to this problem. Vibration of the pipe when water faucets are turned on, or scouring of film or corrosion product from the pipe during water flow can introduce particulate or sorbed lead into the sample. A sampling program that includes a 0.45 micron filtration may dampen some of the lead level variations caused by the particulates but would not accurately reflect usage or exposure conditions of the customer.

The release of particulate copper, particularly the rare problem of 'blue water', appears to be much more complicated, and current research has not yet identified specific causative factors. Edwards et. al. (2000) hypothesized that a combination of microbiological and chemical factors are involved and that an initiation time is required before high concentrations are developed. When blue water problems occur, the water can typically contain from 3 – 10 mg/L Cu, of which practically all can be removed using a 0.45 μm pore size filter.

Quality of Workmanship

The quality of the workmanship of pipefitting can also have an affect on lead and copper levels. If substantial amounts of high lead solder or flux are left on the interior of copper pipe, lead levels in the water can be increased. Conversely, poorly made joints where the lead solder does not evenly and completely fill the gap between pipe sections can result in lower lead levels than well-made joints. Poor workmanship in pipefitting has also been shown to contribute to higher copper corrosion rates (AwwaRF 1990).

Analytical Factors

The analysis of water samples for lead and copper concentrations is subject to variability caused by several factors. These include random response errors from the analytical instrument, procedural errors with instrument calibration, dilutions, and sample manipulations. The presence or absence of potential interferences in the samples also impacts the analytical variability. These error sources can be isolated by laboratory managers or researchers for each laboratory performing the analyses and can be controlled to relatively low levels using quality control procedures. Quality control procedures can also be used to yield estimates of laboratory specific precision and accuracy for the various analytical methods employed. In general, the relative standard deviation of duplicate analyses for lead and copper will increase as the sample concentration decreases (i.e. the lower the level being measured, the greater the relative analytical variation in the values). When analyzing lead and copper levels, laboratories can select the most appropriate analytical technique for the expected levels, evaluate the laboratory precision for these methods and conduct appropriate laboratory quality control procedures to reduce the analytical variability.

Ranking Variability Factors

The Project Advisory Committee was tasked with ranking the described factors as high, medium, or low. A "high" ranking indicates that a particular factor has a relatively larger impact on the variability measured in lead or copper levels, whereas a "medium" or "low" ranking indicates that factor has less of an impact. The age of the material received the highest rankings, followed by distribution system water quality variations, stagnation time, and form of lead and/or copper. Water usage and laboratory quality control received the lowest rankings. Table 4.2 presents a summary of the ranking results.

HISTORICAL DATA REVIEW

A variety of regulatory data sets, along with special studies and pilot plant data were evaluated to determine the actual variability observed in historical monitoring data. All evaluated data sets employed the regulatory collection protocol, (i.e. 1-liter standing samples after a minimum 6 hour standing time). The reviewed data sets were divided into data groups representing "inter-site studies" and "intra-site studies". "Inter-site studies" are data collection programs in which one sample is taken from several different sites during a monitoring period. These studies provide information on the variability in lead or copper levels among several different sampling locations. "Intra-site studies" are data collection programs in which several samples are taken from one site over a specific time period. "Intra-site studies" provide information on the measured variability of lead and copper levels from an individual site over time.

The data for each inter-site monitoring program consisted of one lead and/or copper sample collected from several discreet sites over a specific monitoring period. The majority of inter-site studies were from LCR compliance monitoring data, therefore the monitoring period could range from several weeks to six months. The monitoring programs categorized as "special studies" were monitoring programs that followed the LCR protocol with respect to site selection and data collection, however they contain data that was collected at a greater frequency than that

Table 4.2
Factors impacting variability in lead and copper levels measured at the tap

Category	Factor	How factor can be controlled in monitoring protocol	Rankings (# of votes for each ranking)			
			Low	Medium	High	Final rank
Analytical	Laboratory quality control and analytical precision	• Specifying appropriate analytical methods • Proper QA/QC measures in the laboratory • Reporting analytical precision for lead and copper methods used	3	1	--	8
Chemical	Distribution system or within site water quality variations, i.e. pH, alkalinity (DIC), phosphate, etc…	• Select sites where distributed water quality is consistent, i.e. not in blending zones or dead-ends	-	2	2	2
Physical	Stagnation time	• Standardize stagnation time for all samples	-	2	2	2
	Age of material	• Select monitoring sites where the lead and/or copper sources are the same or of similar age	-	1	3	1
	Amount and type of lead and/or copper material	• Select monitoring sites with similar plumbing configurations, i.e. pipe I.D.'s, number of soldered joints, length of copper pipe and/or lead service line, similar faucet types	-	3	1	6
	Water usage	• Select sites that may exhibit similar overall water usage, i.e. same number of residents, same age of residents, etc.	2	2	--	7
	Form of lead and/or copper (soluble vs. particulate)	• Identify sampling procedures to minimize particulate lead and copper • Select sites less prone to particulate metals levels	1	-	3	2
	Quality of workmanship	• Select sites with good workmanship	1	2	1	5

required for LCR compliance purposes. The monitoring programs that were categorized as "pilot" contained data generated from pipe loop studies that were designed and operated similar to LCR monitoring protocol sites, or brass block evaluations of standing lead and copper levels. Historical data consisted of monitoring studies that were evaluated and listed in the Final LCR Rule (1991) that incorporated monitoring protocols similar to the LCR requirements.

The range of values for the summary statistics for lead level and copper levels from both inter-site and intra-site evaluations are presented in Tables 4.3 through 4.6. The coefficient of variation (CV) is a measure of the variability, and is defined as the standard deviation divided by the particular data set. For lead and copper levels, the CV values calculated indicate that inter-site variation was greater than intra-site variation, and that lead level variation was generally greater than copper variation.

SUMMARY OF VARIABILITY FACTORS

Several factors contribute to the potential variability measured in tap lead and copper levels. This is critical because, statistically, the greater the variability, the larger the sample population size needs to be for reliable results. Analytical variability can be controlled somewhat through selection of appropriate analytical methods and proper laboratory procedures. Variations in distributed water quality can cause variations in tap lead and copper levels, so selecting sites where the variation can be minimized would help to reduce the variability of standing metals levels. Differences in the standing time of the sample, metal age, metal type, and amount of lead and/or copper containing material at the sampling site can also result in variations in lead and copper levels. Even when these factors are controlled (to the extent possible), the variability of lead and/or copper levels measured at the tap may still be relatively large due to the complex chemical and physical factors that affect corrosion of these metals.

While the existing regulatory protocol provides compliance information, it may not be the best protocol for assessing post-optimization conditions, (i.e. changes in lead and copper levels after a water quality change in a previously optimized system). Previous studies have shown that lead levels in tap water samples can vary significantly, between both similar sites that are in close proximity in the distribution system and are served by water of similar chemical characteristics, as well as between repeat samples taken from the same site under identical conditions. This variability is primarily due to chemical and physical factors that govern the solubility of lead and copper in water, and may be altered with the implementation of treatment or source water changes. Understanding the inherent variability in metals levels and determining which factors to control during site selection and data collection for a particular system is essential in determining statistically valid monitoring protocols for assessing changes in lead and copper levels in systems where treatment or source water changes have occurred.

FIELD STUDY OBJECTIVES

The field studies summarized in this chapter were designed to apply alternative monitoring techniques in a utility setting. The results of these studies were integral in determining the real-world applications and the development of final recommendations for this report. The complete field studies are included in Appendix C.

Table 4.3
Range of values for summary statistics for inter-site lead programs

Type of study (# of studies evaluated)	Number of samples	Mean (µg/L)	Median (µg/L)	90th percentile, (µg/L)	Standard deviation	CV min - max (std. dev/mean)	CV 25th - 75th percentile values
LCR (60) compliance	4 – 1610	1.0 – 35.0	0.0 – 16.0	3.0 – 80.0	1.2 – 57.0	0.6 – 3.9	1.0 – 1.6
Pilot studies (16)	5 - 378	6.0 – 211.0	5.0 – 149.0	10.0 – 416.0	1.7 – 284.0	0.2 – 1.9	0.4 - 0.7
Special studies (13)	9 – 112	2.0 – 32.0	2.0 – 27.0	3.0 – 86.0	0.7 – 33.9	0.2 – 3.2	0.5 – 1.7
ALL DATA	4 - 1610	1.0 – 211.0	0.0 – 149.0	3.0 – 416.0	1.2 – 284.0	0.2 – 3.9	0.7 – 1.5

Table 4.4
Range of values for summary statistics for intra-site lead programs

Type of study (# of studies evaluated)	Number of samples	Mean (µg/L)	Median (µg/L)	90th percentile, (µg/L)	Standard deviation	CV min-max (std. dev/mean)	CV 25th - 75th percentile values
Pilot studies (13)	23 – 45	6 – 138	5 – 122	8 – 279	3 – 104	0.5 - 0.8	0.6 - 0.8
Special studies (9)	6 – 10	1 – 13	1 – 4	1 – 93	0 – 30	0 – 2.2	0.3 - 0.6
Historical studies (19)	4 – 5	8 - 91	5 - 73	16 – 19	6 – 73	0.4 – 1.0	0.5 - 0.8
ALL DATA	4 – 45	1 – 138	1 – 122	1 – 279	0 – 104	0 – 2.2	0.5 - 0.8

Table 4.5
Range of values for summary statistics for inter-site copper studies

Type of study (# of studies evaluated)	Number of samples	Mean (mg/L)	Median (mg/L)	90th percentile, (mg/L)	Standard deviation	CV min-max (std dev/mean)	CV 25th and 75th percentile values
LCR compliance (57)	4 - 422	0.03 – 1.36	0 – 1.38	0.07 – 2.19	0.03 - 0.78	0.4 – 2.2	0.6 – 1.1
Pilot studies (14)	5 - 412	0.01 – 0.39	0.01 – 0.34	0.02 - 0.73	0.01 - 0.24	0.1 – 2.8	0.3 - 0.6
Special studies (13)	9 – 112	0.05 - 0.86	0.05 - 0.56	0.08 – 2.51	0 - 0.88	0.1 – 1.0	0.6
ALL DATA	4 – 422	0.03 – 1.36	0 – 1.38	0.02 - 2.51	0 - 0.88	0.1 – 2.2	0.5 – 1.0

Table 4.6
Range of values for summary statistics for intra-site copper programs

Type of study (# of studies evaluated)	Number of samples	Mean (mg/L)	Median (mg/L)	90th percentile (mg/L)	Standard deviation	CV min-max (std. dev/mean)	CV 25th - 75th percentile values
Pilot studies (13)	23 – 56	0.01 - 0.25	0.01- 0.25	0.02 - 0.38	0.1 - 0.16	0.4 - 0.7	0.4 - 0.6
Special studies (9)	6 – 10	0.02 - 0.14	0.02 - 0.14	0.03 - 0.19	0.01 - 0.04	0.1 - 0.5	0.2 - 0.3
ALL DATA	6 – 56	0.01 - 0.25	0.01 - 0.25	0.02 - 0.38	0.01 - 0.16	0.1 - 0.7	0.25 - 0.5

Alternative In-Home Tap Monitoring

The field study alternative in-home tap monitoring protocol, based on the ranking of variability factors and a review of the variability seen in historical lead and copper data sets, was implemented at four field study utilities: Philadelphia, PA; Hastings, NE; Oak Lodge Water District, WA; and Detroit, MI. The results of this monitoring were used to develop the final alternative in-home tap monitoring protocol recommendations.

Automated Sampling

In order to reduce the potential for "human error" during collection of tap water samples by customers, a prototype automated sampler was designed. Five of these automated samplers were field tested by the Philadelphia Water Department separate from the alternative tap monitoring protocol. The results of the field studies provided comparative data between automated and manual sampling in customer homes and a determination of the feasibility of using automated sampler in customer homes. This field study is presented in its entirety.

ALTERNATIVE TAP MONITORING FIELD STUDIES

Introduction

The draft alternative tap protocol developed in this chapter was tested at four utility locations. The following summarizes the draft protocol that was recommended to each of the participating utilities, describes the actual field study monitoring that occurred at each location, and presents the overall results and findings from this effort.

Data Collection Protocol

A tap monitoring protocol was designed to generate data with which to evaluate variability in tap lead and copper levels and provide input to a final alternative monitoring strategy. The protocol addressed what type of sites should be selected, the number of sites which should be sampled, how many samples to collect at each site, the length of the evaluation period for data collection and the corresponding frequency of data collection. These recommendations were based on a ranking of factors affecting variability in tap lead and copper levels and how these factors might be controlled in a data collection protocol. The number of sites and frequency of sampling recommended in this protocol were based on the variability measured from a wide variety of lead and copper monitoring program data gathered from utility and historical sources. The draft protocol recommended to utility participants of the alternative tap monitoring field studies and the programs implemented are summarized in Tables 4.7a and 4.7b.

Site Selection Criteria

The Final Lead and Copper Rule specified sample site locations that are biased towards sites with higher risk of elevated lead levels. These sites do not necessarily correspond to sites that may experience elevated copper levels. Copper levels may be higher in newer construction and from sites at dead end locations, where pH may have decreased. The age of the copper

Table 4.7a
Comparison of recommended protocol with actual field study alternative tap monitoring programs for Hastings and Oak Lodge Water District (NCCWC)

	Recommended draft alternative monitoring protocol		Actual field study monitoring programs			
	Lead	Copper*	Hastings		OLWD	
Site Selection Criteria	• Single family residences with lead solder installed prior to 1982 or lead service lines	• Single family residences with copper piping • Three age categories: >20 years ≥10 but ≤ 20 years < 10 years	Single family residences with lead solder 8 sites >20 years (from 23 – 58 years old) 2 sites < 20 years (5 and 16 years old)		Single family residences with lead solder 9 sites from 18 – 20 years old 1 site 28 years old	
	Lead	Copper*	Lead	Copper	Lead	Copper
Minimum No. of Sites	25	10	10	10	10	10
Minimum No of Samples/site	10	4	6	6	10	10
Evaluation period	4 – 10 weeks		6 weeks		5 weeks	
Data Collection and Analysis Criteria	• 1000 mL Samples (split sampling with initial 125 mL followed by 875 mL encouraged) • Analyze total lead and copper levels • Minimum 6 hour stagnation times. All samples should have approximately similar stagnation times		• 1000 mL samples • Total lead and copper analyzed • Minimum 6 hour stagnation		• 1000 mL samples • Total lead and copper analyzed • Minimum 6 hour stagnation	

* Sample numbers would apply to each age category

Table 4.7b
Comparison of recommended protocol with actual field study alternative tap monitoring programs for Philadelphia and Detroit

	Recommended draft alternative monitoring protocol		Actual field study monitoring program			
	Lead	Copper*	Philadelphia		Detroit	
Site Selection Criteria	• Single family residences with lead solder installed prior to 1982 or lead service lines	• Single family residences with copper piping • Three piping age categories: >20 years ≥10 but <20 years < 10 years	Single family residences with lead solder 14 sites >20 years 5 sites ≥10 years but <20 years 4 sites < 10 years 2 sites with mixed age piping		Single family residences with lead solder and LSL – 16 sites Single family residences with lead solder and copper service – 4 sites 3 sites >20 years 3sites ≥10 years but < 20 years 3 sites < 10 years 11 sites with no age information	
	Lead	Copper*	Lead	Copper	Lead	Copper
No. of Sites	25	10	25	25	20	20
No of Samples/site	10	4	10	10	10	10
Evaluation period	4 – 10 weeks		10 weeks		7 weeks	
Data Collection and Analysis Criteria	• 1000 mL Samples (split sampling with initial 125 mL followed by 875 mL encouraged) • Analyze total lead and copper levels • Minimum 6 hour stagnation times. All samples should have approximately similar stagnation times		• Majority of samples were1000 mL. • Consecutive 125 mL followed by 1000 mL samples collected during two rounds of sampling • Total lead and copper analyzed • Stagnation times from 5.5 to 9 hours		• Consecutive 125 mL and 875 mL samples • Total lead and copper analyzed • Stagnation times from 5.5 to 9.5 hours	

*Sample numbers would apply to each age category.

piping may also have a dramatic affect on the copper levels measured. Therefore, it was recommended that the alternative tap monitoring protocol incorporate sites that target both lead levels and copper levels.

Utilities were asked to select sites for lead level analyses that were single family residential locations at homes with lead solder installed prior to 1982 and/or homes with lead service lines. Sites selected for copper level analyses were to be single family residential locations categorized into three groups based on the approximate age of the interior copper piping:

- Greater than or 20 years old,
- Less than or equal to 20 years old but greater than or equal to 10 years old, and
- Less than 10 years old

Utilities were instructed to select copper level sites in one or all three of these categories, however the recommendations for the number of sites which should be sampled and the number of samples from each site would be applied to each category.

The sites selected for this alternative monitoring program could be subsets of the sites used for regulatory monitoring, however to reduce the variability in lead and copper levels, all attempts should be made to select sites that:

- Were not located in dead-end/low flow areas,
- Had the same approximate age of interior plumbing,
- Had similar faucets (both age and type) where the water sample is collected, and
- Were located in areas that exhibit relatively consistent distributed water quality.

Number of Sites, and Frequency of Sampling, and Recommended Sampling Period

The variation exhibited in lead and copper levels (from the historical studies discussed previously in this chapter) was used as a starting point in developing the number of sites and frequency of sampling to recommend for the draft protocol. The number of samples was based on the highest accuracy levels (i.e. tightest band around the mean) that would still correspond to somewhat realistic sample numbers given the range of coefficients of variation (CVs) calculated from actual lead and copper monitoring data. Minimum numbers of sample, were recommended to the utility participants and they were encouraged to sample from even more sites in order to provide greater statistical accuracy in the results.

The minimum number of sites recommended for lead levels was twenty-five, with each of these sites sampled ten times over the evaluation period. For copper levels, the minimum number of sites recommended was ten (per age category), sampled four times over the evaluation period. For example, if a utility selected both copper sites that were less than ten years old and copper sites that were greater than twenty years old, then efforts should be made to obtain ten sites in each category. Individual sites could be used for both lead level and copper level analyses.

For each of the sites chosen for data collection, samples were to be collected several times during a specified evaluation period. Evaluation periods of 6 months to 1 year may be applicable in utility situations where distributed water quality is relatively stable and there is little seasonal variation. However, shorter evaluation periods would provide more timely data with which to make comparisons and determine if lead and/or copper levels have changed because of water quality changes in the system. The recommended evaluation (sampling) period for this field study data collection effort was 10 weeks.

Sample Collection Methods and Analytical Requirements

Utilities were encouraged to collect consecutive 125 mL and 875 mL, first-draw tap water samples from the cold-water kitchen or bathroom tap and analyze these samples for total lead and copper. Collecting consecutive samples would allow the faucet contribution to lead and copper levels to be determined. To reduce the variability in the data, utilities were asked to have samples collected after the water had been motionless in plumbing pipes for at least six hours and to ensure that the stagnation times from all monitoring sites be approximately the same. Samples were to be collected by either system personnel or residents who were instructed on the sampling procedures required. Analyses for lead and copper was to be conducted by laboratories that were certified by USEPA or the State.

Actual Field Study Data Collection Programs

The City of Hastings conducted six rounds of sampling in the spring of 2002 at 10 sites in the distribution system. These sites all contained ¾-inch to 1-inch diameter copper services that were from five to fifty-eight years old. Eight of the sites had service lines that were greater than 20 years old and 2 of the sites had services that were less than 20 years old. One-liter standing samples were collected at the kitchen tap from the ten sites approximately weekly and analyzed for total lead and copper levels.

The Oak Lodge Water District (OLWD) conducted ten rounds of sampling at ten single-family residences in their system as part of this project. The sampling took place over a five-week period during the winter of 2001, and all of the sites contained copper services. Nine of the sites were Tier 1 with lead soldered copper premise piping installed between 1982 to 1984. One site was Tier 3 and the plumbing was installed in 1974. One-liter samples were collected at the kitchen tap from these ten sites on a biweekly basis.

The Philadelphia Water Department (PWD) conducted ten rounds of sampling over a ten week period from twenty-five single family residential sites in the winter of 2001. All 25 sites contained copper piping and the majority of locations contained piping over 20 years old. For the first and last sample rounds, a split 125-mL followed by a one-liter sample was collected. For the remainder of the sample rounds, a one-liter first draw sample was collected. The stagnation time was recorded for all samples. Stagnation times were greater than 5.5 hours at all sites, and less than or equal to 9 hours at most locations. There were some longer stagnation times recorded however, some as high as 46 hours. Therefore, only those samples where the standing times were less than or equal to 9 hours were included in the analyses of the data.

The alternative tap monitoring field study completed by Detroit Water and Sewer Department consisted of consecutive 125 mL followed by 875 mL samples collected at 20 residential sites over a seven week period in the spring of 2002. The approximate age of the interior plumbing was not available for over half of the sampling sites, and the remaining sites were split evenly between sites less than 10 years old, between 10 and 20 years old, and greater than twenty years old (three in each category). Average stagnation time for the samples was 7 hours, with all samples collected after the water had been standing from between 5.5 and 9.5 hours. Detailed individual field study descriptions are included in Appendix C.

Summary of Results

Results from the field studies provided information on variation in lead and copper levels from controlled monitoring studies, the impact of piping age on copper levels, faucet contributions to lead and copper levels, and comparisons to lead and copper results from regulatory monitoring.

Variation in Lead and Copper Levels

Tables 4.8 and 4.9 present the summary statistics for all of the lead and copper data collected from the four field studies. The range of CVs calculated for the data from each field study are also presented in Tables 4.8 and 4.9 to provide an indication of the overall variability measured at each utility. Higher CV values indicate more variability in the data (i.e. higher standard deviation/mean).

Table 4.8
Summary statistics for lead levels from alternative tap monitoring field studies

Utility	N	Mean (μg/L)	Median (μg/L)	90th Percentile (μg/L)	Standard Deviation	CV (Std. Dev/Mean)
Hastings	60	21.8	21.2	27.0	6.1	0.3
OLWD	100	17.3	10.0	38.5	19.0	1.1
Philadelphia	186	1.8	1.0	3.0	3.4	1.9
Detroit	143	5.1	3.0	8.5	4.8	0.9

Table 4.9
Summary statistics for copper levels from alternative tap monitoring field studies

Utility	N	Mean (mg/L)	Median (mg/L)	90th Percentile (mg/L)	Standard Deviation	CV (Std. Dev/Mean)
Hastings	60	1.10	1.06	1.89	0.65	0.6
OLWD	100	0.09	0.09	0.17	0.05	0.6
Philadelphia	186	0.16	0.15	0.32	0.10	0.6
Detroit	143	0.03	0.02	0.07	0.03	0.8

The CVs in Tables 4.8 and 4.9 indicate that lead level variation was generally greater than copper variation, and that variation in copper levels was somewhat similar between all field studies. Lead level variation as measured by CV was typically higher than copper level variation, with the exception of lead level results from Hastings.

Tables 4.10 through 4.13 display the summary statistics for inter-site and intra-site evaluations of the data collected at each field study location. Inter-site studies involve analysis of metals levels collected at different sites during individual sample rounds. Ranges of data in

Tables 4.10 and 4.12 show the range of inter-site data from all sample rounds. Intra-site studies involve analysis of metals levels collected at one sample site over several rounds of monitoring. Ranges of data in Tables 4.11 and 4.13 show the range of intra-site data from all sample sites. In general, intra-site variation was lower than inter-site variation for both metals. Lead level variation was greater than copper level variation at both sites. The two utilities that exhibited elevated lead levels also displayed relatively consistent variation between sites (typical CVs of approximately 1.0 for OLWD and 0.25 for Hastings).

Table 4.10
Summary statistics for lead levels for all inter-site alternative tap monitoring field studies

Utility	N	Mean µg/L	Median µg/L	90th percentile µg/L	Standard deviation	CV Min-Max	CV 25th – 75th percentile
Hastings	10	16-25	15-24	21-38	3.3-9.2	0.2-0.4	0.2 - 0.3
OLWD	10	9-33.4	7-27.5	26-91	9.5-33.7	0.8-1.2	0.9-1.0
Philadelphia	3-20	1.3-3.5	1 – 1.5	2-9	0.6-8.6	0.5-2.1	0.5 – 1.4
Detroit	3-19	3-7.1	3-3.75	3-17.7	0.3-9.7	0.3-1.5	0.4-0.8

Table 4.11
Summary statistics for lead levels for all intra-site alternative tap monitoring field studies

Utility	N	Mean µg/L	Median µg/L	90th percentile µg/L	Standard deviation	CV Min-Max	CV 25th – 75th percentile
Hastings	6	16-30.5	19-29.6	23.5-49	2.8-11	0.1-0.4	0.2 - 0.3
OLWD	10	0.9-54.8	1-48.5	2-86.5	0.7-26.2	0.2-1.7	0.3 - 0.8
Philadelphia	2-8	1-17.3	1-15	1-32	0-10.7	0-1.0	0 - 0.6
Detroit	3-11	3-13.1	3-14.3	3-22.3	0.04-7.8	0-1.3	0 - 0.7

Table 4.12
Summary statistics for copper levels for all inter-site alternative tap monitoring field studies

Utility	N	Mean mg/L	Median mg/L	90th percentile mg/L	Standard deviation	CV Min-Max	CV 25th – 75th percentile
Hastings	10	1.04-1.2	0.9-1.1	1.9-2.3	0.6-0.8	0.59-0.61	0.5 -0 .6
OLWD	10	0.07-0.10	0.07-0.11	0.13-0.19	0.05-0.07	0.5-0.7	0.6 - 0.7
Philadelphia	3-20	0.11-0.19	0.09-0.16	0.2-0.46	0.09-0.13	0.5-0.8	0.7 - 0.8
Detroit	3-19	0.02-0.04	0.02-0.03	0.02-0.13	0-0.05	0.5-1.2	0.6 - 0.8

Table 4.13
Summary statistics for copper levels for all intra-site alternative tap monitoring field studies

Utility	N	Mean, mg/L	Median, mg/L	90 percentile, mg/L	Standard deviation	CV Min-Max (Std. dev/mean)	CV 25th – 75th percentile
Hastings	6	0.3-2.4	0.3-2.4	0.3-3.0	0.05-0.6	0.1-0.9	0.1 - 0.2
OLWD	10	0.03-0.18	0.04-0.19	0.05-0.2	0.01-0.04	0-0.5	0.1 - 0.3
Philadelphia	2-8	0.03-0.42	0.03-0.42	0.06-0.48	0.01-0.13	0.1-0.6	0.1 -0 .4
Detroit	3-11	0.02-0.11	0.02-0.11	0.02-0.18	0-0.03	0.01-0.5	0 -0 .3

The evaluation of inter-site and intra-site data collected during the field studies provided for determination of lead and copper levels for highly variable sites. In one case, it was discovered that the interior piping in the home had been replaced just five months earlier, and presumably the faucet was replaced as well. This site exhibited average lead levels greater than 30 μg/L and average copper levels greater than 2.2 mg/L, both of which were significantly higher than the lead and copper results from the other sites selected for this alternative monitoring program. In other instances, no explanation for the higher metals levels at an individual site could be determined.

Figures 4.1 through 4.4 display the mean lead and copper values measured from both the historical data and the four field studies versus the CV. When compared using the method detailed in Appendix A, the variation measured in the field studies was slightly lower than was seen in the historical data sets discussed previously in this chapter. This could be due to the use of fewer sample sites with more controlled site selection and sample collection procedures in the field studies which may have minimized variation, or the fact that there were more historical data sets evaluated and therefore a greater potential for variation across utilities.

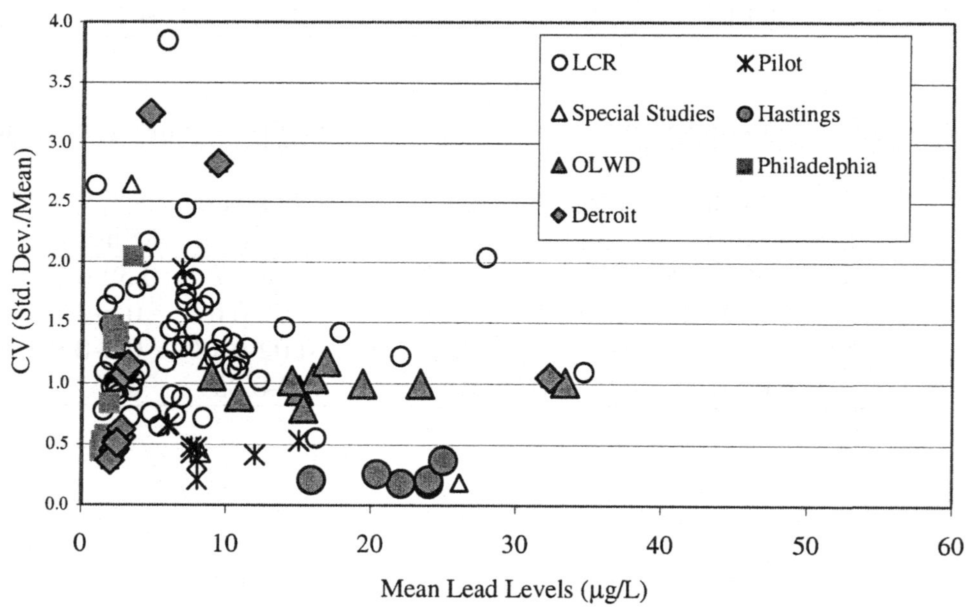

Figure 4.1 Coefficient of variation vs. mean lead levels for inter-site lead studies

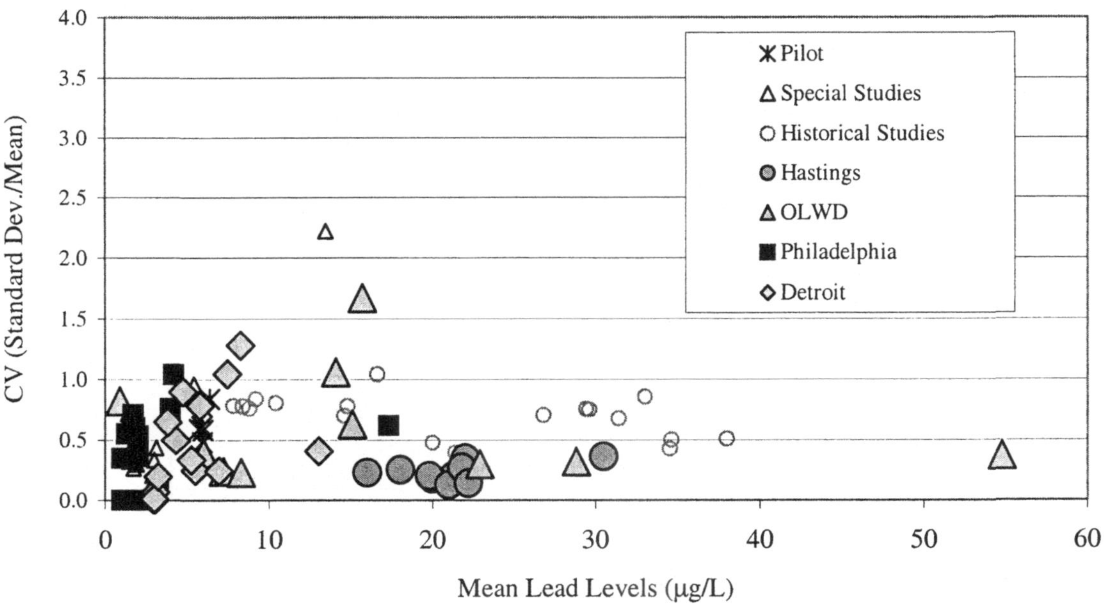

Figure 4.2 Coefficient of variation vs. mean lead levels for intra-site lead studies

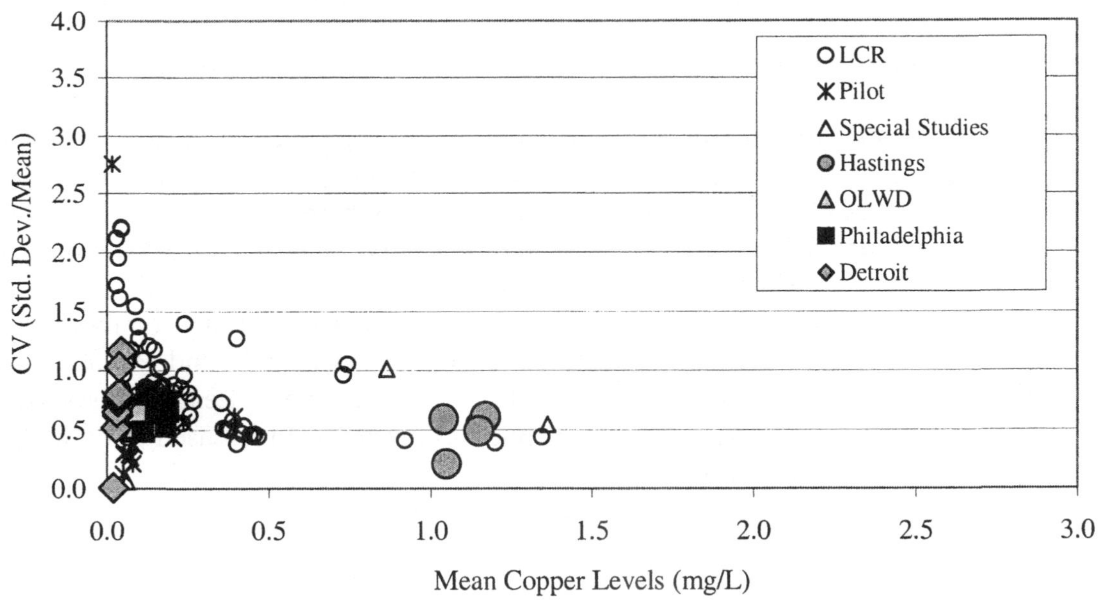

Figure 4.3 Coefficient of variation vs. mean copper levels for inter-site copper studies

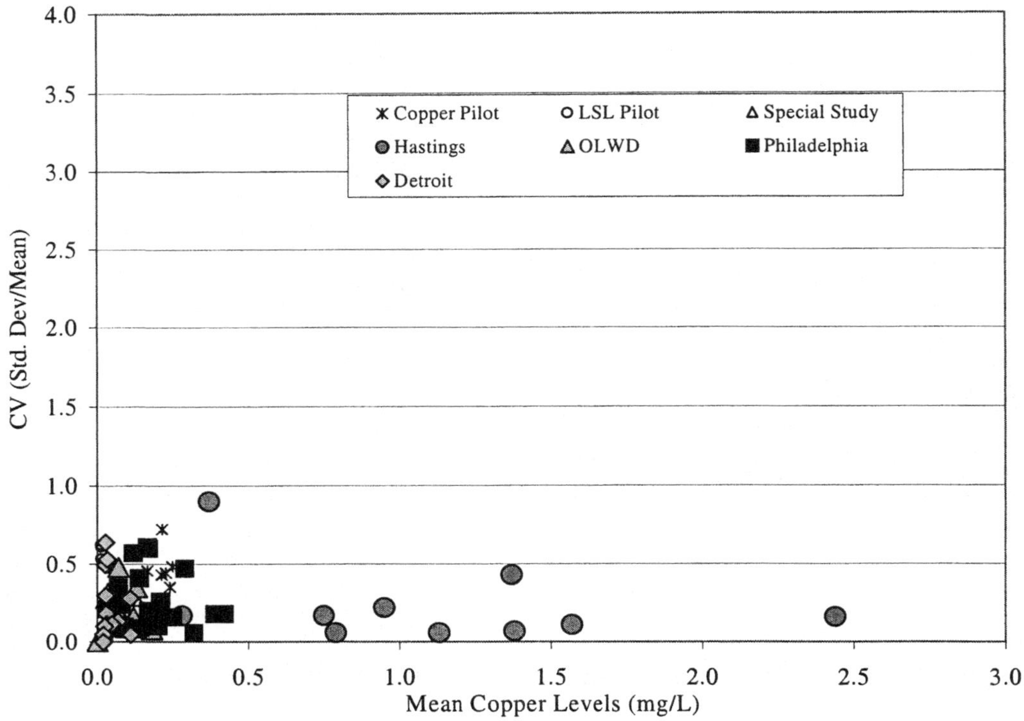

Figure 4.4 Coefficient of variation vs. mean copper levels for intra-site copper studies

Impact of Piping and Faucet Age on Lead and Copper Level

A small number of sites in the field study monitoring programs were selected that contained newer piping, however the numbers were not adequate to develop summary statistics and evaluations of variability in lead and copper levels for the categories of piping age recommended for the copper level analyses. Therefore, only analysis of age of piping versus copper levels was completed where there was sufficient data.

Three utilities (Hastings, Philadelphia, and Detroit) recorded the age of piping at their sample sites. Hastings identified the age of the copper service lines and Philadelphia and Detroit recorded the age of the interior piping at their sampling sites. Lead levels showed no correlation with piping age, but copper levels declined as piping age increased at both Hastings and Philadelphia (Figures 4.5 and 4.6).

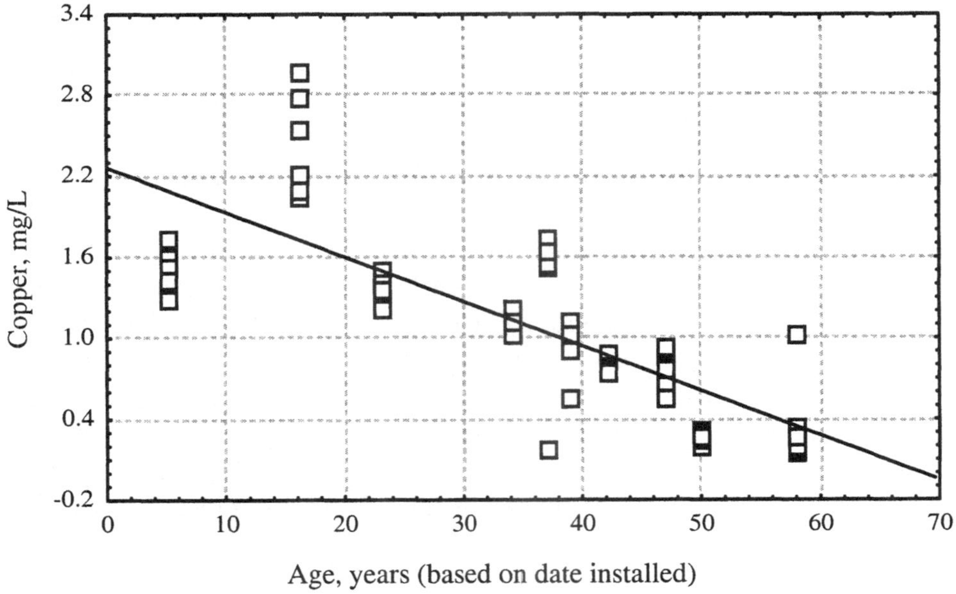

Figure 4.5 Copper levels vs. age of service piping – Hastings, NE

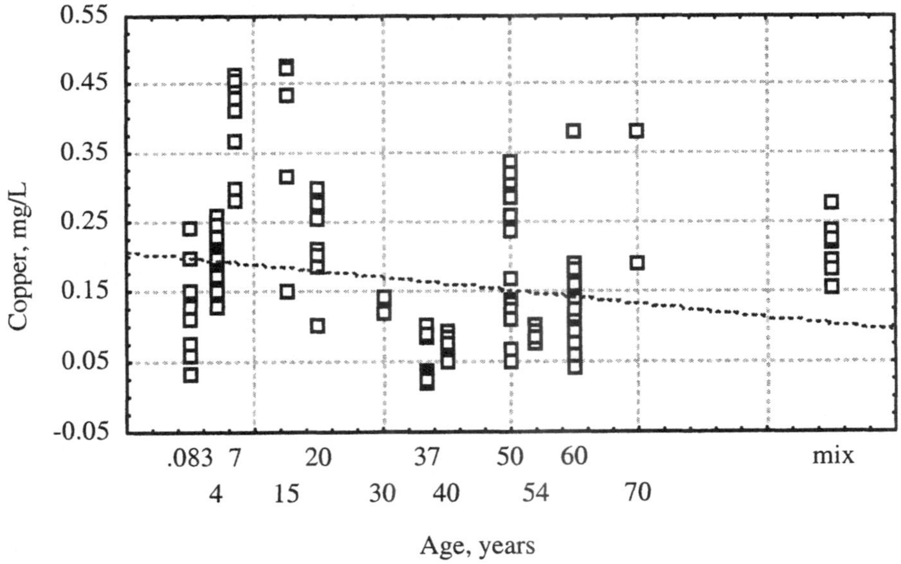

Figure 4.6 Copper levels vs. age of interior piping – Philadelphia, PA

There was insufficient data from Detroit to evaluate the impact of interior piping age on lead or copper levels (i.e. the majority of sites sampled for the alternative monitoring program did not have interior piping age available). However, Detroit recorded the age of the faucet at most sites. An evaluation of faucet age versus lead and copper levels for both 125 ml and 875 mL samples indicated that both lead and copper levels declined with faucet age.

Faucet Contributions to Lead and Copper Levels

Philadelphia and Detroit conducted sequential split sampling as part of their alternative tap monitoring programs. In Philadelphia a split 125-mL followed by a one-liter sample was collected during the first and last sample rounds, to evaluate the faucet contribution to standing lead and copper levels. In Detroit, consecutive 125 mL followed by 875 mL samples were collected at every site.

Table 4.14 is a summary of lead and copper level results from the split sampling conducted in Philadelphia and Detroit. Results in Philadelphia indicated that lead and copper levels for the initial 125 mL and second 1000 mL samples were similar. Lead levels in the second 1000 mL sample were more variable than lead levels measured in the initial 125 mL sample, but copper levels exhibited similar levels and variability for both sample volumes.

In Detroit, lead levels measured in the initial 125 mL sample were slightly elevated when compared to the 875 mL sample, and the variability in lead levels for the 125 mL samples was much higher. Copper levels were also higher in the 125 mL samples, however in general, all copper levels were extremely low (< 0.1 mg/L).

Table 4.14
Summary of lead and copper levels from sequential sampling

Utility	Sample	N	Mean	Median	90th percentile	Standard Deviation	CV (Std. Dev. /Mean)
				Lead Levels (μg/L)			
Philadelphia	125 mL first draw	67	3.5	1.0	5.0	8.5	2.4
Philadelphia	1000 mL second draw	66	4.3	1.0	5.0	15.3	3.6
Detroit	125 mL first draw	143	6.7	3.0	10.0	18.6	2.8
Detroit	875 mL second draw	143	4.9	3.0	9.0	4.2	0.9
				Copper Levels (mg/L)			
Philadelphia	125 mL first draw	67	0.17	0.14	0.30	0.14	0.9
Philadelphia	1000 mL second draw	66	0.15	0.12	0.29	0.11	0.7
Detroit	125 mL first draw	143	0.03	0.02	0.06	0.02	0.7
Detroit	875 mL second draw	143	0.03	0.02	0.07	0.03	0.8

Comparisons to Regulatory Monitoring Data

Comparisons were made between the field study results and LCR monitoring completed for each utility. Tables 4.15 and 4.16 display the summary statistics for lead and copper levels measured during these monitoring programs.

Table 4.15
Comparisons of summary statistics for lead levels from alternative tap monitoring field Studies and LCR data

Sampling program	N	Mean (μg/L)	Median (μg/L)	Min - Max	90th percentile (μg/L)	Standard deviation	CV(Std. Dev. /Mean)
Hastings							
1993 LCR	188	1.7	1.0	0 – 15	3	2.1	1.2
1997 LCR	57	1.9	1.0	0 – 16	5	2.9	1.5
2000 LCR	29	2.1	1.4	1 – 13.5	3	2.3	1.1
2002 Field Study	60	21.8	21.2	12.7 – 49.0	27	6.1	0.3
OLWD							
1992 LCR	62	3.4	2	2 – 17	6	2.5	0.7
1994 LCR	32	3.6	2	1 – 16	10	3.7	1.0
1995 LCR	31	2.5	1	1 – 17	4	3.4	1.4
1996 LCR	31	2.1	1	1 – 11	4	2.0	0.9
1999 LCR	31	3.2	2	0 – 8	6	2.4	0.8
2000 LCR Spring	61	13.8	5	0 – 100	46	20.5	1.5
2000 LCR Fall	62	6.9	3	0 – 134	14	17.3	2.5
2001 LCR	60	12.5	7.5	0 – 59	33	14.4	1.2
2002 Field Study	100	17.3	10	0 - 92	38.5	19	1.1
Philadelphia							
1992 LCR Spring	162	14.2	6	1 – 205	22	28.2	2.0
1992 LCR Fall	143	8.6	6	1 – 70	15	9.6	1.1
1997 LCR Spring	118	11.0	2	0 – 270	15	37.5	3.4
1997 LCR Fall	108	5.5	2	0 – 144	11	15.3	2.8
1998 LCR Summer	78	6.4	2	1 – 95	11	14.7	2.3
1999 LCR Fall	59	8.6	3	3 – 131	10	19.5	2.3
2002 LCR	62	7.9	2	1 – 109	14	18.5	2.3
2002 Field Study	186	1.8	1	0 - 32	3	3.4	1.9
Detroit							
1992 LCR Spring	89	10.8	7	1 – 85	21	12.01	1.1
1992 LCR Fall	104	8.8	4	1 – 83	116	14.97	1.7
1997 LCR Spring	118	5.1	2	0 – 112	12	12.22	2.4
1997 LCR Fall	122	5.8	2	2 – 246	8	22.34	3.9
1998 LCR Fall	102	4.5	2	2 – 81	8	8.27	1.8
1999 LCR Spring	104	3.5	2	2 – 18	7	3.29	0.9
2000 LCR Summer	119	3.4	2	0 – 40	7	4.29	1.3
2002 Field Study	143	5.1	3	3 - 38	8.5	4.8	0.9

Table 4.16
Comparisons of summary statistics for copper levels from alternative tap monitoring field studies and LCR data

Sampling program	N	Mean (mg/L)	Median (mg/L)	Min – Max	90th percentile (mg/L)	Standard deviation	CV (Std. Dev. /Mean)
Hastings							
1993 LCR	188	0.67	0.45	0 - 3.93	1.63	0.65	1.0
1997 LCR	57	0.73	0.47	0 - 2.22	1.82	0.70	1.0
2000 LCR	29	0.71	0.46	0.02 - 2.11	2.06	0.71	1.0
2002 Field Study	60	1.10	1.06	0.17 - 2.97	1.89	0.65	0.6
OLWD							
1992 LCR	62	0.23	0.20	0.01 - 0.55	0.42	0.13	0.6
1994 LCR	32	0.19	0.16	0.04 -0.56	0.32	0.14	0.7
1995 LCR	32	0.18	0.16	0.02 -0.55	0.32	0.12	0.7
1996 LCR	30	0.20	0.16	0.02 - 0.72	0.41	0.17	0.9
1999 LCR	31	0.08	0.07	0.02 - 0.24	0.16	0.06	0.8
2000 LCR Spring	61	0.09	0.08	0 - 0.42	0.18	0.07	0.8
2000 LCR Fall	62	0.04	0.04	0 - 0.14	0.09	0.03	0.8
2001 LCR	60	0.03	0.03	0 – 0.10	0.05	0.18	0.7
2002 Field Study	100	0.09	0.09	0 - 0.20	0.17	0.05	0.6
Philadelphia							
1992 LCR Spring	162	0.35	0.18	0.02 – 1.79	0.94	0.39	1.1
1992 LCR Fall	143	0.30	0.18	0.02 – 1.25	0.75	0.28	1.0
1997 LCR Spring	118	0.18	0.13	0.03 – 0.88	0.39	0.16	0.9
1997 LCR Fall	108	0.18	0.13	0.01 – 0.86	0.40	0.16	0.9
1998 LCR Fall	78	0.15	0.112	0.02 - 0.61	0.35	0.12	0.8
1999 LCR Fall	59	0.15	0.11	0.03 - 0.55	0.37	0.12	0.8
2002 LCR	43	0.11	0.09	0.01 - 0.32	0.25	0.08	0.7
2002 Field Study	186	0.16	0.15	0.02 - 0.48	0.32	0.10	0.6
Detroit							
1992 LCR Spring	89	0.10	0.04	0 -0.69	0.32	0.13	1.4
1992 LCR Fall	104	0.06	0.03	0 - 0.31	0.16	0.06	1.0
1997 LCR Fall	79	0.04	0.01	0 - 0.80	0.10	0.10	2.2
1997 LCR Fall	122	0.04	0.02	0 - 0.51	0.07	0.07	2.0
1998 LCR Fall	102	0.04	0.02	0 - 0.47	0.13	0.06	1.6
1999 LCR Fall	104	0.03	0.01	0 - 0.28	0.07	0.05	1.7
2000 LCR Summer	100	0.04	0.01	0 - 0.63	0.08	0.10	2.2
2002 Field Study	143	0.03	0.02	0.02 - 0.17	0.07	0.03	0.8

Lead levels measured from alternative monitoring at Hastings were much higher and the variation much lower than the most recent LCR data collected at that utility. Lead levels and the variation in lead levels from the alternative monitoring at OLWD were very similar to lead levels from regulatory monitoring completed by that utility in 2000 and 2001. In Philadelphia, the alternative monitoring lead levels were much lower than regulatory results from 2002, and the variation in lead levels from alternative monitoring was also lower. Of the four field study locations, OLWD and Detroit selected sites for their alternative monitoring that were a subset of the sites used for regulatory monitoring. The results from these field studies indicated very similar lead levels (average, median, and 90th percentile lead values) when compared to LCR monitoring data. The variation in lead levels was much lower in the Detroit field study results than the Detroit regulatory monitoring, however.

Copper levels measured during alternative monitoring in Hastings were similar to regulatory monitoring results from sites that contained lead soldered copper and the variation in copper levels was also the same. Copper levels from alternative monitoring were different from regulatory monitoring results from sites with lead service lines, however. Results at OLWD indicated that copper levels from alternative monitoring were similar to regulatory monitoring results with similar variability. Copper levels measured in Philadelphia during the alternative monitoring program were similar to the most recent regulatory monitoring completed, and the variability was lower than regulatory sample results. In Detroit, the alternative monitoring copper levels were similar and displayed less variability than regulatory monitoring data.

AUTOMATED SAMPLER FIELD STUDY

Introduction and Approach

The collection of samples by customers, regardless of utility provided instruction, introduces a level of uncertainty that becomes increasingly important as sample pools shrink and the impact of "suspect" data can have a greater impact on monitoring results.

To address this issue, a prototype automated sampling unit was designed and constructed to provide sampling of customer taps with minimal participation from the customer. Figures 4.7 and 4.8 show the sampler and timer components.

Each unit was designed with a chrome-plated aerator diverter to allow tap use during the daytime without sampling unit disconnection. The units were designed with a compact footprint of 13" x 22". Each unit was designed with plexi-glass enclosures to protect sensitive electrical components.

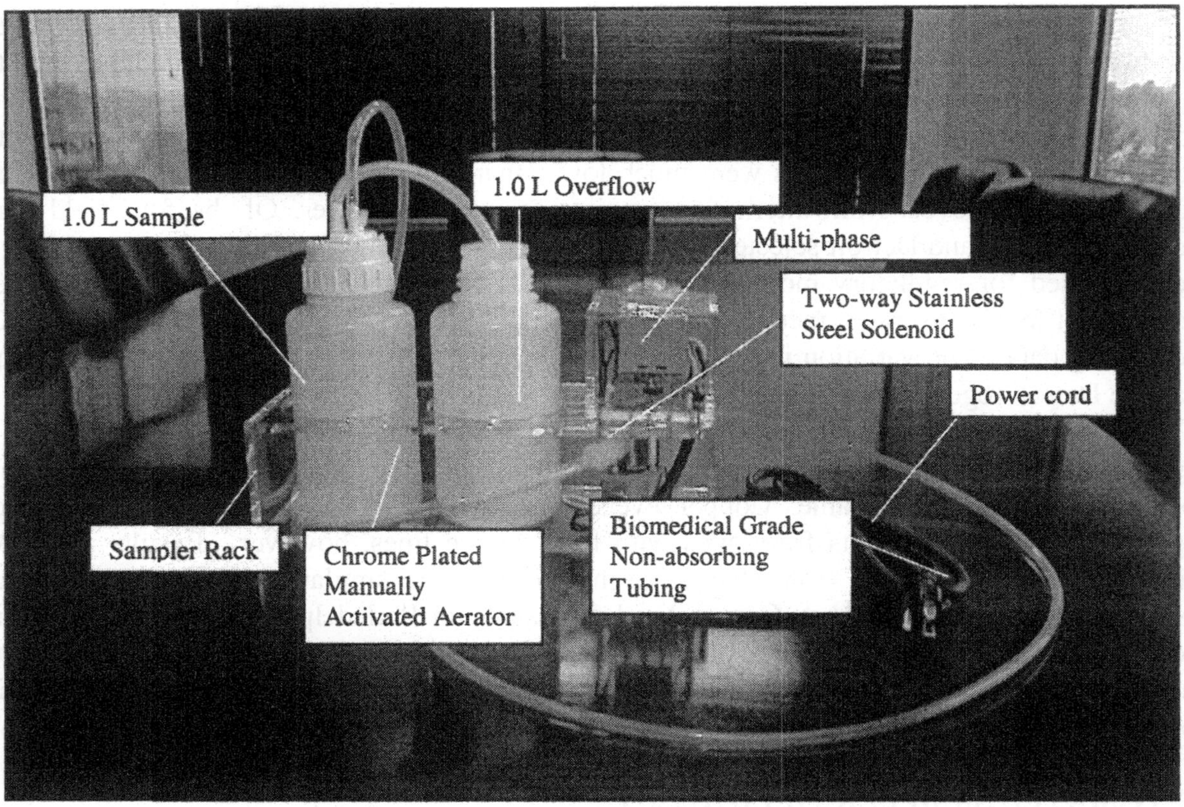

Figure 4.7 Automated sampling apparatus

Figure 4.8 Automated sampler multi-phase timer face

Five automated units were constructed and field-tested within the Philadelphia Water Department. Since the units were a prototype level of sophistication they were installed in "controlled" customer homes. For the purposes of this field activity, "controlled" was defined as homes of utility employees or non-utility employee customer homes that had demonstrated a level of competence in past LCR compliance sampling and cooperation with utility activities. Each unit was installed and calibrated by utility staff. Calibration was required to establish the appropriate length of time the sampler would draw a sample, relative to the tap flow rate, in order to provide a 1.0 L sample without overflow. Home-owners were instructed in the steps to activate the sampler following final tap use and the handling of samples following collection.

The operative elements of the sampling units were a two-way solenoid valve operated by a multi-phase timer that opened the valve at a pre-selected time and then closed the valve after a predetermined period sufficient to fill a one-liter sample bottle. Following installation and calibration the following steps were followed by customers to properly energize the unit after last use of the tap:

1. Turn on faucet. To the extent possible match tap flow to that used during calibration.
2. Pull the tap aerator/diverter pin out to direct flow to the sampling unit and release knob. Water flow should be sufficient to maintain the diverter pin in the open position without drips from the faucet. If, upon release of the diverter pin, the diverter closes and flow returns to the faucet, tap flow should be increased and the step repeated.
3. Plug in apparatus to power source.

Home-owners were instructed to collect eight samples in alternating automated/ traditional fashion over a ten day period. This approach was designed to provide sample comparison of an automated sample followed by a traditional manual sample the next day. Table 4.17 summarizes the sample tap locations and sampling conditions.

Automated Sampler Field Study Results

Each automated sampler unit required separate calibration in order to provide the proper sample volume. All collected samples were analyzed for lead and copper levels. All five locations were measurable for copper, but only one location (AWLC-108) produced measurable lead levels. The remaining four locations exhibited lead levels of less than 0.001 mg/L. Tables 4.18 and 4.19 show summaries of data collection by site.

Table 4.17
Summary of automated sampling location and sampling conditions

Location identification	Faucet location	Frequency of tap use	Sample period*	Collection protocol notes
AWLC- 107	1st Floor bathroom: Least used bathroom	Daily	8 days	Automated samples collected on first four days followed by four consecutive days of manual samples.
AWLC – 108	Basement bathroom sink: Used a few times annually	Monthly	11 days	Samples collected in alternating fashion beginning with automated.
AWLC- 121	Kitchen Sink	Daily	46 days	Samples collected in alternating fashion beginning with manual. Length of time between a manual and the comparative automated sample was six days.
AWLC –201	Kitchen Sink	Daily	6 days	Samples collected in alternating fashion beginning with manual. First four samples collected on consecutive days. Last four samples collected in two days with 10 and 14 hours stagnation times.
AWLC- 206	Kitchen Sink	Daily	52 days	Samples collected in alternating fashion beginning with manual. Comparative (manual-automated) sample sets #1 and #3 were collected 48 hours apart. Comparative sample sets #2 and #4 were collected with a 20 and 16 day period between samples, respectively.

*All sampling conducted between 9/21/02 and 12/8/02

Table 4.18
Automated sampling copper summary data

Copper sample site	Mean copper concentration (mg/L)		Range (mg/L)	Std. deviation
	Manual	Automated		
#107	0.100		0.09-0.13	0.02
		0.081	0.07-0.09	0.01
#108	0.170		0.15-0.19	0.02
		0.129	0.08-0.15	0.03
#121	0.049		0.04-0.10	0.01
		0.106	0.09-0.14	0.01
#201	0.523		0.34-0.64	0.16
		0.561	0.49-0.60	0.07
#206	0.157		0.14-0.17	0.01
		0.128	0.06-0.18	0.06

Table 4.19
Automated sampling lead summary data

Lead sample site	Mean lead concentration (μg/L)		Range (μg/L)	Std. deviation
	Manual	Automated		
#107	4.0		1.0-1.4	0.7
		1.0	None	0
#108	62.0		20.0-126.0	50.0
		29.0	21.0-46.0	12.0
#121	1.0		1.0-2.0	1.0
		1.0	None	0
#201	1.0		None	0
		1.0	None	0
#206	1.0		None	0
		1.0	None	0

Figure 4.9 shows a comparison of lead levels from manual and automated sample collection at location AWLC-108. Manual sampling provided a 1.0 L initial draw sample. The automated sample provided a sample volume between 800-1000 mL initial draw sample.

All samples showed a lead level in excess of the 15.0 μg/L action limit. The data showed a greater variability for manually collected samples. The variability is also notable when considering that all eight samples were collected in a manual-automated manner over an 11-day period.

Comparative copper samples were collected at all five field locations. Figures 4.10 – 4.14 show a comparison of side-by-side data.

All of the reported copper levels were well below the 1.3 mg/L action limit. The variability in the data both between data points of the same collection method and between the two data methods was significantly less than that exhibited by the lead data.

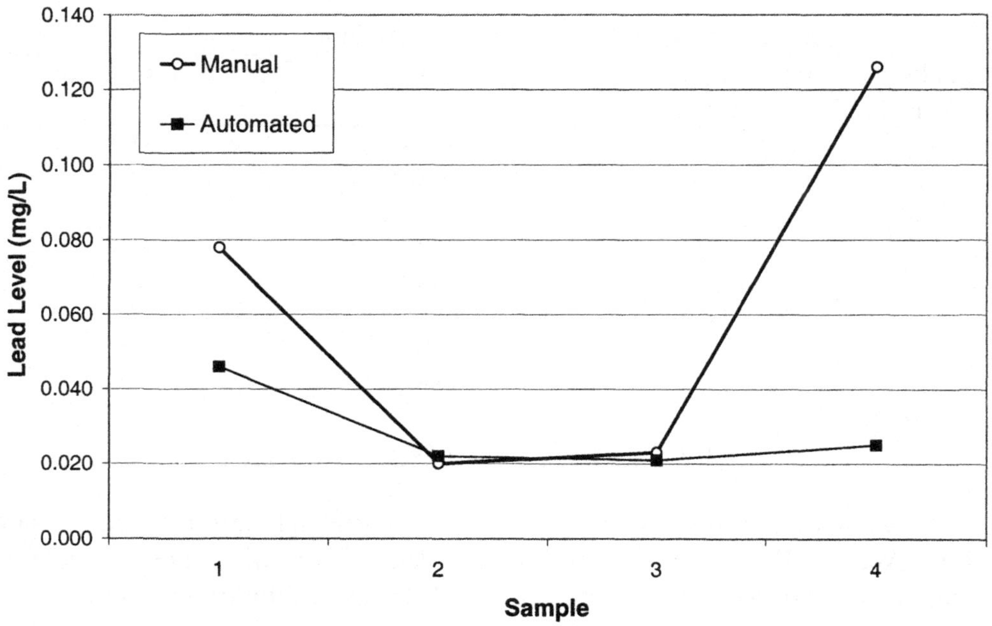

Figure 4.9 Site #108 manual vs. automated lead sampling

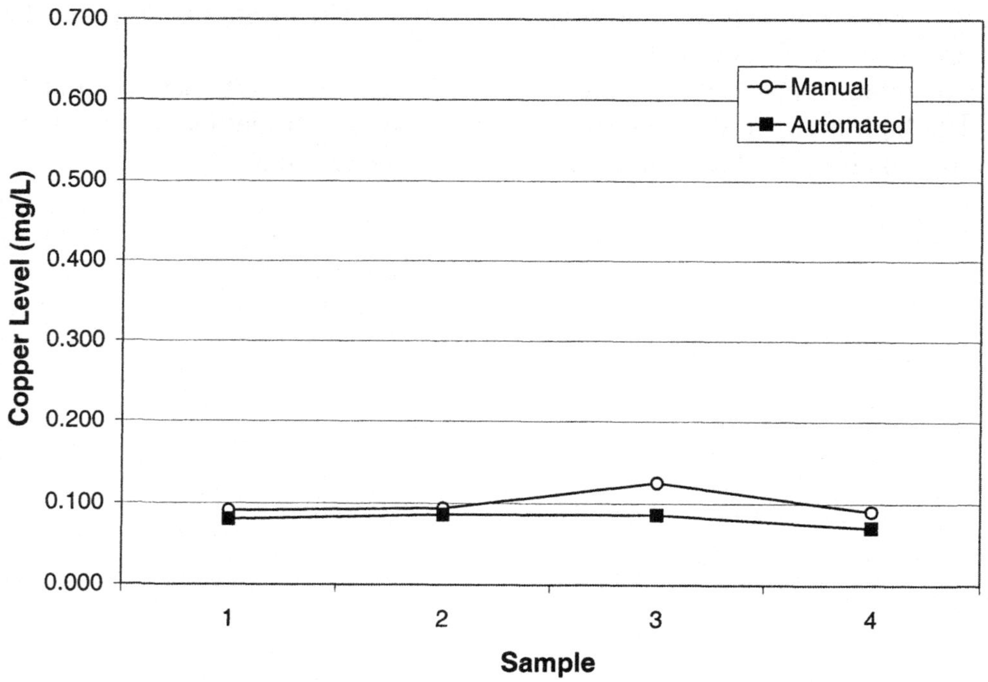

Figure 4.10 Site #107 manual vs. automated copper sampling results

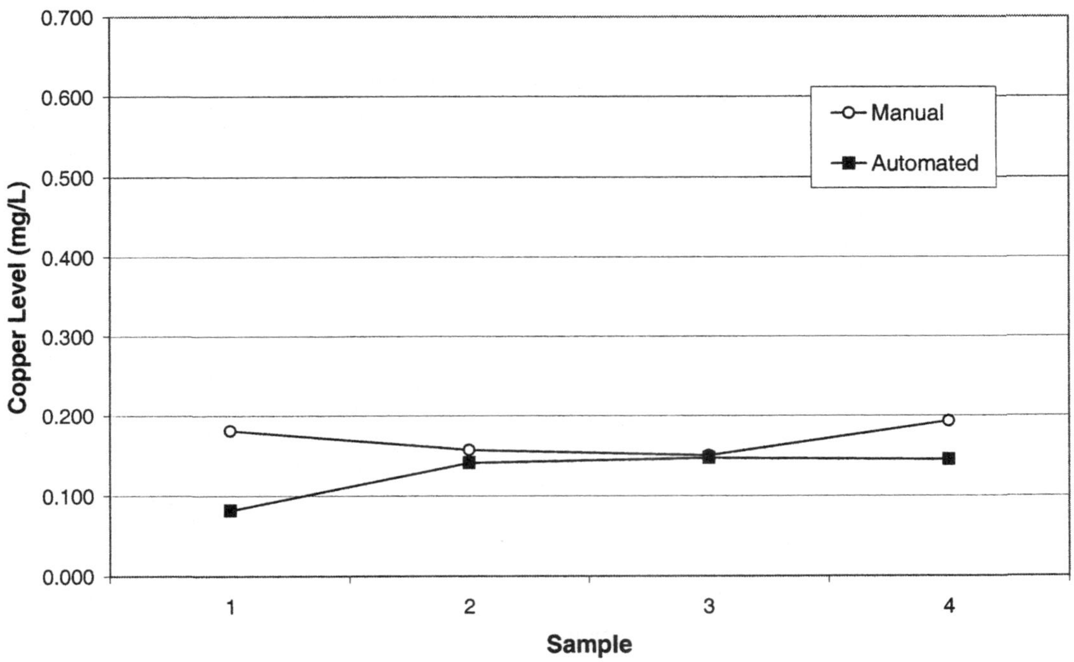

Figure 4.11 Site #108 manual vs. automated copper sampling results

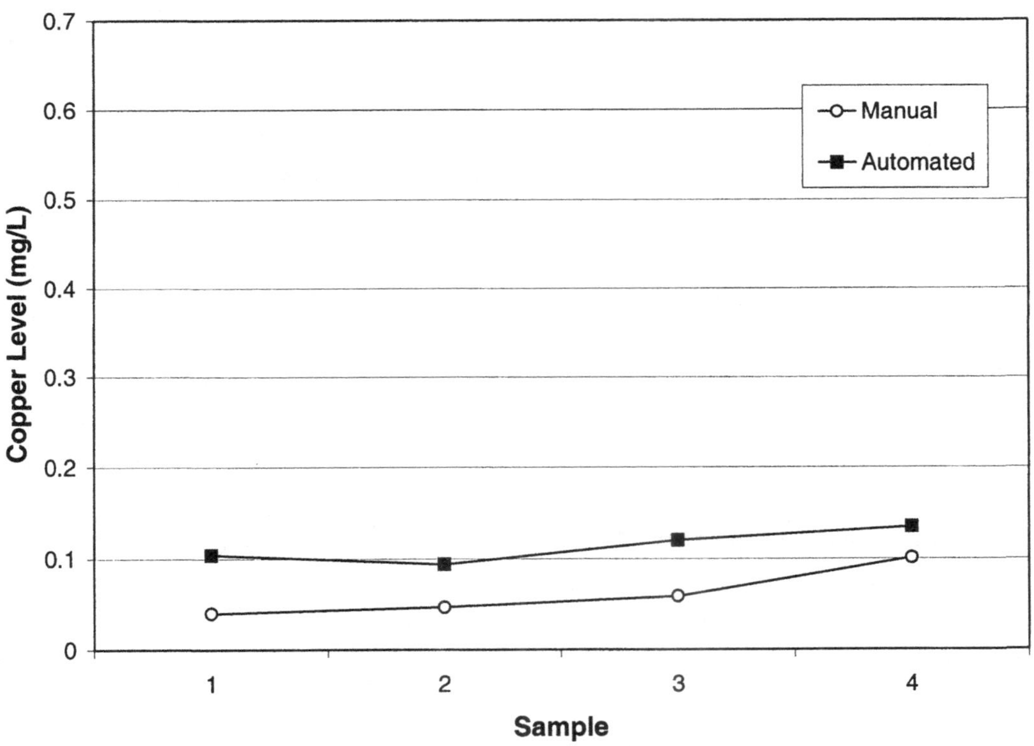

Figure 4.12 Site #121 manual vs. automated copper sampling results

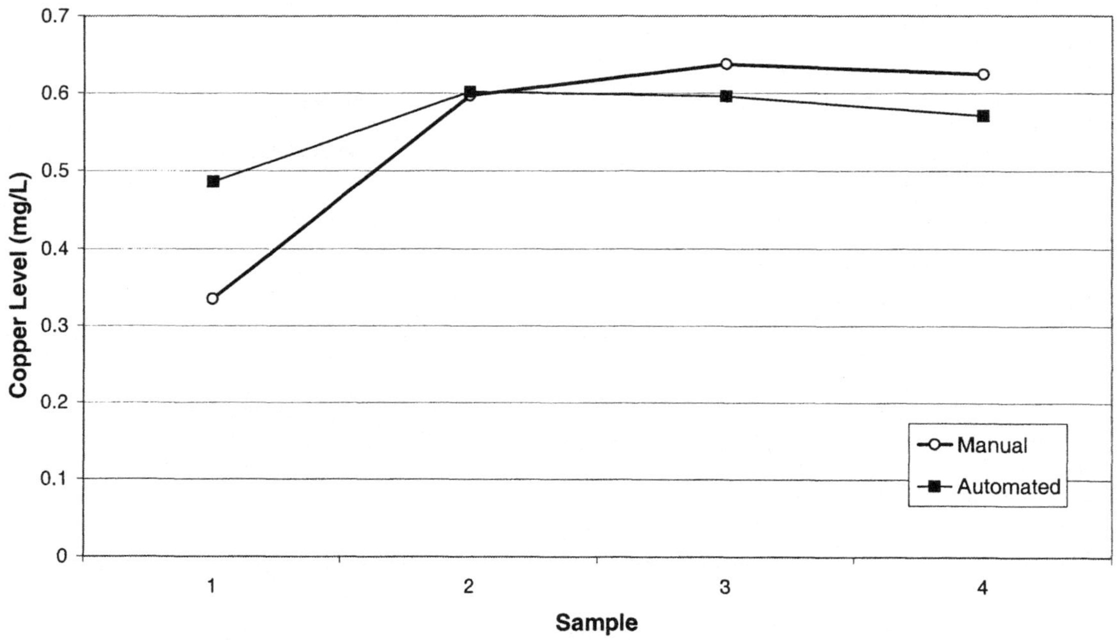

Figure 4.13 Site #201 manual vs. automated copper sampling results

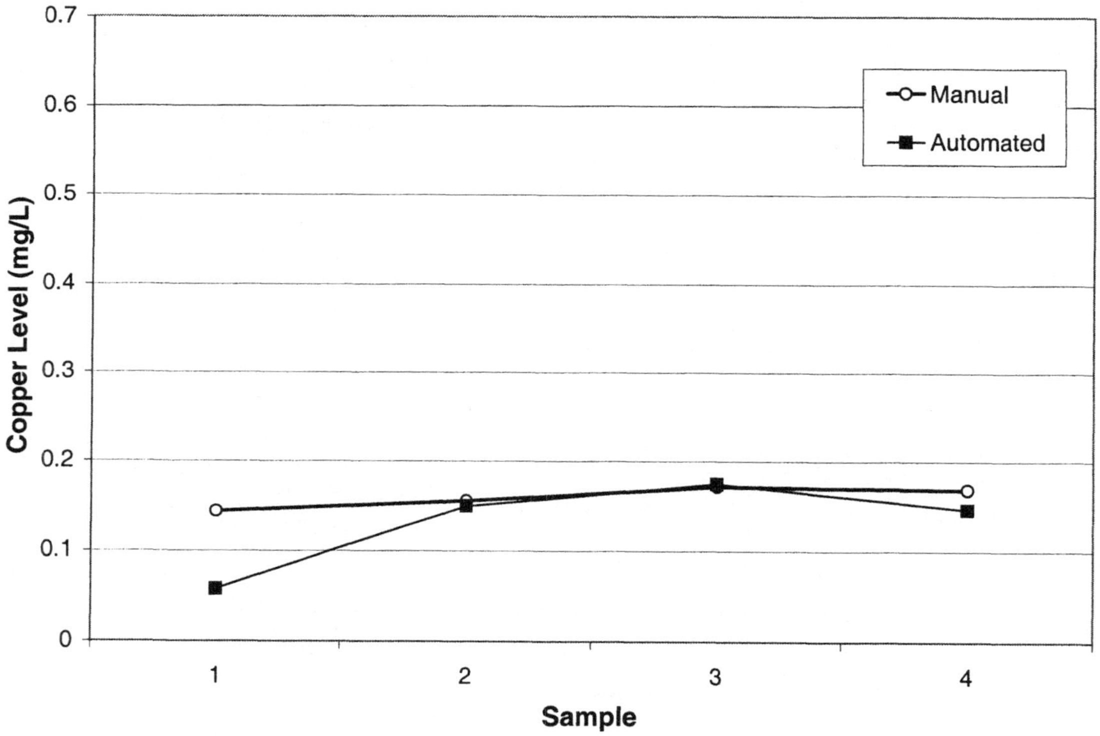

Figure 4.14 Site #206 manual vs. automated copper sampling results

Conclusions of Automated Sampler Field Studies

The automated sampling of lead and copper sites was conducted in a limited fashion. The objective was to develop a prototype sampling device and collect comparative data. The data collected was used to determine if the potential exists for reducing variability using an automated device.

The prototype sampler was designed with "off-the-shelf" equipment from scientific and equipment supply houses. This approach provided a reliable and functional design. Additional study in this area would require the refinement of the sampler design. Some of the refinements noted by field study utility staff included:

- Tubing in the sampler needs to be both flexible and non-absorptive of lead and copper. The prototype device was constructed with biomedical grade tubing that did not absorb metals, but was a stiff material. A tubing material that was both non-absorbing and flexible could not be found in supply houses.
- Measurement of sample volume based on actual volume rather than time would eliminate the potential for over or under collection.
- A more user-friendly diverter design would be beneficial. The pin diverters were prone to closing with changes in water pressure. It was initially difficult to determine a flow rate that held the diverter in an open position while not causing dripping from the tap.
- Any sampler design will need to account for the variability in tap size and design.
- Any sampler design would need to be sturdy enough to withstand repeated field use.
- Field teams reported that the prototype was both too delicate and bulky.
- The field teams reported that the circuit breakers in remodeled homes were sensitive to the samplers. This resulted in the circuit breakers being tripped when the solenoid was activated.

The comparative data collected using the automated sampler indicated that automated sampling may potentially be of benefit to utilities. The next step would be the development of a refined sampler design and an extensive study conducted to determine the limitations, benefits, and reliability of such a device.

The results from this limited study show that automatically collected copper samples compare very well with manually collected samples. The lead data, though more limited, showed that automated sampling data was less variable than manually collected samples. Whether the difference in variability was the result of sampling techniques or lead concentration variability in the plumbing was beyond the scope of this project.

CHAPTER 5
DEVELOPMENT OF PREDICTIVE TECHNOLOGY FOR LEAD AND COPPER CORROSION MONITORING

INTRODUCTION

Another primary objective of the project was to develop, test, and validate a novel analytical protocol that would enable water utilities to accurately quantify effects of corrosion control and to detect and quantify minor differences in corrosion rates and metal release with high accuracy, in a short time frame, and at reduced cost. In order to achieve these objectives, the following project elements were completed: a general review of current electrochemical corrosion models, development of an On-Line Corrosion Cell (OLCC), laboratory experimentation, and field application using the final corrosion cell design.

The first portion of this chapter is a general review of current electrochemical corrosion models. It includes an examination of the current models of metal release kinetics, current theories of the electrochemical corrosion process, and current methods for monitoring electrochemical corrosion.

The second portion of this chapter focuses on the development of an OLCC. Once the final OLCC was developed, experiments were performed in the laboratory in order to determine the behaviors of significant corrosion-related parameters. The objectives of these experiments were to determine patterns of behavior for the corrosion potential (E_{corr}), the concentrations of metal released, and the correlation between them. Water quality parameters, phosphate concentrations, and chlorine concentrations were all varied during the experiments to determine the effects on the corrosion-related parameters.

The third portion of this chapter is a summary of field studies that were designed to apply alternative monitoring techniques in a utility setting. The results of these studies were integral in determining the real-world applications and the development of final recommendations for this report. The objective of the field studies was to test the viability of the technology in identifying features of the behavior of corrosion potential (E_{corr}) in drinking water containing residual chlorine, identify trends of metal release, and explore correlations between total metal concentration and E_{corr} values. Field testing was implemented at: HDR Laboratories, Bellevue, WA; San Francisco, CA; and Pinellas County, FL. The complete field studies are presented in Appendix D.

CORROSION THEORIES AND PREDICTIVE TECHNIQUES

Metal Release Kinetics and Modeling

The fundamental difference between electrochemical corrosion and metal release, or uptake, from solid phases is that electrochemical corrosion is controlled by electron exchange between the metal and oxidant (oxygen or chlorine species) and can be mathematically modeled as such. Metal release is governed by the metal's solubilities and equilibria with relevant soluble species rather than by electronic fluxes through a corroding surface.

The concentrations of soluble lead and copper species, when in equilibrium with their solids, can be determined based on theoretical solubility models (Edwards et. al. 1996b, Hidmi and Edwards 1999, Marani et. al. 1995, Schock and Gardels 1983, Schock 1989, Schock 1990, Taylor and Lopata 1984). The kinetics of dissolution of solid phases, or the uptake of soluble species, is a much more complex process, especially for copper. Three main types of metal release, as a function of time, are as follows:

- *Type 1: Monotonic exponential (asymptotic) increase of metal concentration.* This is the typical functional dependence of metal concentration vs. time and virtually the only pattern observed for lead release (Clement et. al. 2000, Kuch and Wagner 1983, Lytle and Schock 2000, van der Leer et al. 2002).
- *Type 2: Non-monotonic behavior of metal concentrations.* This is characterized by an initially rapid increase of soluble metal, followed by a stabilization of soluble metal concentrations, and then a gradual decrease and, in some cases, virtual disappearance. This has been observed for copper only (Merkel et. al. 2002, Merkel et. al. 2001, Lytle and Schock 2000, Werner 1992). The described phenomenon has been primarily related to the occurrence of Cu^{+} disproportionation to Cu (0) and Cu^{2+} or a reverse reaction of Cu^{2+} to form Cu^{+}.
- *Type 3: Non-monotonic behavior with the effects of stagnation time.* This third type of metal release kinetics occurs during stagnation (Merkel 2003, Merkel et. al. 2002, Merkel et. al. 2001). In this case, the non-monotonic behavior of copper concentration described above is followed by an additional phase, during which the concentration of copper slowly increases. This phase has been associated with a slow dissolution of Cu^{2+} solids (e.g., malachite) following the disappearance of Cu^{+} formed in the preceding phase of copper corrosion (stagnation).

A representation of these types of behavior is shown in Figure 5.1.

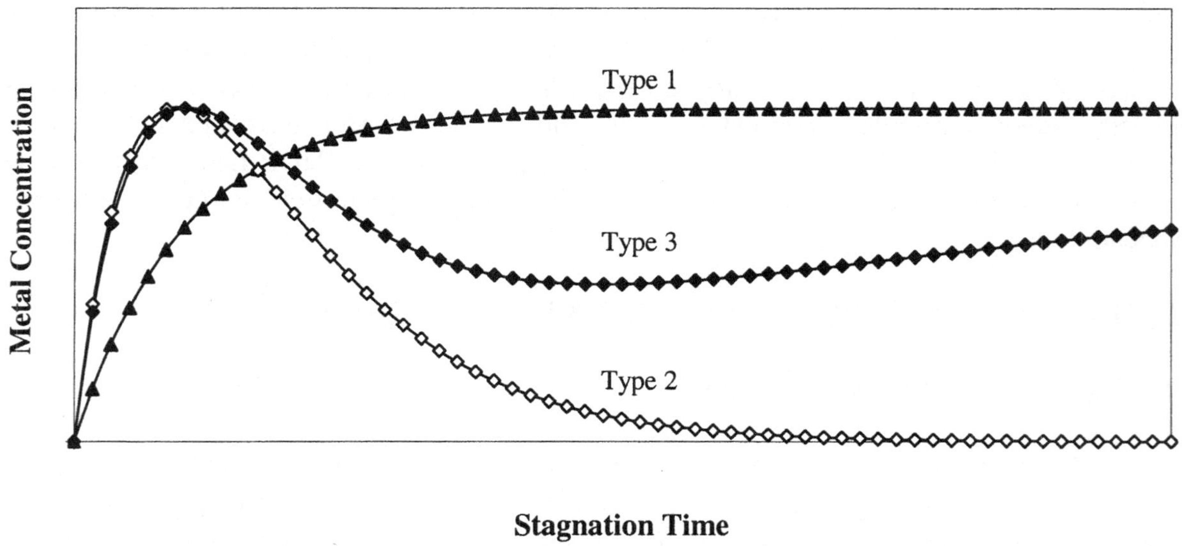

Figure 5.1 Representation of three main types of metal release

The behavior of metal concentrations during stagnation of drinking water in household plumbing is a complex function of the chemical and physical factors described in Chapter 4. The formal kinetics of metal accumulation during stagnation has been investigated in some detail (Kuch and Wagner 1983, Lytle and Schock 2000, Merkel et. al. 2002, Van Der Leer et. al. 2002, Werner 1992), but more remains to be done to provide a comprehensive theory of these processes. The need for further investigation is especially pronounced in regards to non-monotonic behavior as affected by stagnation time and the formation of colloidal particles (Britton and Richards 1981, de Mora et. al. 1987, Hulsmann 1990, Swistock, Sharpe and Robillard 1993). In fact, no general theory of colloidal mobilization of lead- or copper-containing particles appears to exist and even the best-fit solubility models explain only about 18% of lead and 35% of copper variation in the 90th percentile (Edwards et. al. 1999). It has been observed that providing direct estimates of lead and copper concentrations at the tap remains a difficult task (Lytle and Schock 2000).

Lead and Copper Corrosion Pathways

Despite limitations, the described equilibria and kinetic models of metal release have had a notable measure of success. This is especially true in predicting trends (rather than absolute metal concentrations) associated with changes of pH, alkalinity, and the presence of phosphate inhibitors. For a complete analysis, it is also necessary to examine the electrochemical processes constituting the primary cause of oxidation of lead and copper containing materials.

At a corroding metal surface, the electrons from the oxidized metal are consumed by the oxidants. At equilibrium, the number of electrons released by the metal equals the number of electrons consumed; therefore the half-reactions of oxygen reduction and metal oxidation are coupled. The electrochemical potential at which this coupling occurs is called the corrosion potential (E_{corr}). The graphic position of E_{corr} is defined by the properties of the metal, water quality parameters and concentrations of the oxidants (dissolved oxygen and chlorine species). A more detailed discussion concerning the values of E_{corr} in drinking water, its relationship to concentrations of oxidants, and its behavior during episodes of stagnation and flow is presented in Appendix B. An extensive discussion of the theory of electrochemical corrosion of metals goes beyond the scope of this report and can be found in a variety of sources (e.g., Internal Corrosion 1996).

Oxygen is the major oxidant in potable water and lead and copper release is almost always associated with oxygen consumption. Species of chlorine are thermodynamically more potent oxidants than dissolved oxygen (the standard electrochemical potentials of the O_2/H_2O and $HClO/Cl^-$ couples are 1.23 V and 1.48 V, respectively), but the molar concentration of chlorine species is considerably less than that of dissolved oxygen and this diminishes their contribution to electrochemical corrosion. The effects of chlorine on lead release do not seem to be clearly discernible (Lee, Becker and Collins 1989, Reiber 1991), while copper release tends to increase in the presence of either free or combined chlorine (Boulay and Edwards 2001, Hong and MacAuley 1998, Zhang et. al. 2002).

Electrochemical corrosion fundamentally involves electron transfer from the metal to the oxidant and it can be quantified using a measure of electron flux, or corrosion current (i_{corr}). When the exchange of electrons between the metal and oxidant is distributed uniformly, general corrosion takes place. The outflow of electrons from the metal surface can become localized (notably, in the presence of sulfides) and pitting corrosion can occur. The latter is frequently associated with high rates of copper release and rapid pipe wall penetration (Edwards, Ferguson and Reiber 1994, Jacobs et. al. 1998).

Another type of corrosion takes place when two dissimilar metals corrode simultaneously while in direct electrical contact. This is the case for lead-tin solder in contact with copper tubing and for the lead phase in leaded brasses and bronzes (lead is immiscible with, and forms a separate phase dispersed in, copper-based alloys, American Society for Metals 1979). This type of corrosion is known as galvanic corrosion and results in a higher rate of oxidation of the less noble metal (lead in contact with copper or brass) than is observed for the metal without galvanic coupling (Gregory 1990, Reiber 1991, Walker and Oliphant 1982). The standard electrochemical potential (E_0) of the Cu/Cu^{2+} couple (E_0=+0.34 V) is more positive than that of the lead Pb/Pb^{2+} couple (E_0=-0.13 V) (CRC Handbook of Chemistry and Physics 1990). This is offset by the fact that the piping surface area of copper is generally much higher than that of lead solder. Therefore, the potential of the solder surface is approximately equal to E_{corr} of copper. Similarly, the lead phase in leaded brasses and bronzes is galvanically coupled with the copper-based matrix. In these conditions, the surface lead in lead-tin solder and leaded alloys undergoes rapid oxidation, as evidenced by high rates of lead leaching at the beginning of exposure (Paige and Covino 1992, Schock and Neff 1988).

The properties of corroding surfaces change continuously until a steady state condition is reached. This occurs because electrochemical corrosion is associated with multiple processes including localized or general oxidation of the metal substrate, release of soluble metal, deposition of solid phases and reduction of oxygen to hydroxyl at the surface (leading to an increase of local pH). The steady state of a corroding surface can be defined as the state in which, at any moment in time, the amount of oxidized metal produced by electrochemical corrosion is the same as the amount lost due to dissolution. For surfaces with low-solubility phases, the amount of surface scales neither decreases nor increases within a considerable period of time (months to years) and the rate of overall metal release is expected to be very low in these cases. The issue of how fast a corroding surface reaches a steady state is central to predicting the corrosion in drinking water distribution networks.

Typically, the rate of corrosion (i_{corr}) and metal release both decrease with time following the initial period of the metal's exposure to drinking water; meaning, the surface undergoes passivation at that time. The duration of this phase depends on flow conditions, nature of the corroding material, prevailing type of corrosion (general, pitting, or galvanic coupling), and water quality parameters. As a result of these complex effects, the duration of the phase tends to vary widely (Snoeyink and Wagner 1996). It has been observed that stabilization of corroding lead and copper surfaces may take many years. In contrast, galvanically coupled lead in solder and leaded copper alloys undergo passivation within several weeks and possibly even after a few days (Reiber 1991, Schock and Neff 1988).

It has been hypothesized that during the passivation of a metal experiencing general corrosion, corrosion products accumulate on the surface, gradually cover the metal and make it less accessible for oxidants. The precipitation of calcium carbonate has been frequently seen as an important factor contributing to passivation. Correspondingly, parameters useful in evaluating the saturation capacity of water during calcite phase (Langelier and Ryznar indexes) have traditionally been considered good indicators of water corrosivity. However, this does not mean that the Langelier and Ryznar indicies are "corrosion" indicies, in fact, they are not. They are solubility indicies.

Recent data indicate that the accumulation of corrosion scales such as malachite on the surface does not, by itself, impede oxygen transfer to the surface (Merkel et al. 2002), nor is the formation of calcite a necessary step in the passivation of copper or lead-containing materials (Boffardi and Sherbondy 1991, Edwards et. al. 1996, Lee, Becker and Collins 1989). Evidence shows, for copper, passivation is caused by changes of the properties of cuprite, a semiconductor oxide always present on the surface of copper. The crystalline structure, ionic and electron conductivity of the surface layer of cuprite, rather than its thickness, appear to control the rate of copper corrosion (Baitz and Follner 2002, Merkel 2003).

Monitoring Corrosion in Drinking Water

Corrosion rates can be quantified using a number of electrochemical methods (Reiber, Ryder and Wagner 1996). All of these methods apply a disturbance of the electrochemical potential of the metal and require specialized instrumentation, such as potentiostats, to acquire and process the data. The most common techniques employ measurements of electric current values while electrochemical potential is linearly scanned from a more negative E_{corr} (cathodic polarization) to the more positive E_{corr} (anodic polarization). This procedure is referred to as a potentio-dynamic scan (PDS) and is used to generate data necessary to ascertain the rate of corrosion of a freely corroding surface at E_{corr}. PDS can also be performed over a range as much as +/- several hundred mV around E_{corr}.

Polarization scans can be performed in a narrow range of potential called linear polarization (LPS) close to the corrosion potential of approximately ±15 mV from E_{corr}. The main parameter provided by LPS method is so called polarization resistance (R_p), which is generally inversely proportional to i_{corr}. The equipment for LPS scans tends to be less sophisticated than that necessary for other electrochemical methods, but the values of polarization resistance do not allow precise quantifying of i_{corr} values.

PDS experiments yield more reliable i_{corr} values, but affect the state of the corroding surface because anodic and cathodic polarization at potentials far from E_{corr} accelerates oxidative and reductive processes. Also, the interpretation of PDS data is not free of artifacts and can yield erroneous i_{corr} values if, for example, the electrochemical process is affected by mass transfer limitations, or more than one electrochemical process takes place at the surface.

Alternative methods that have been employed both in laboratory and field conditions to determine i_{corr} values include electrochemical noise technique and electrochemical impedance spectroscopy (EIS). These novel approaches have been used in studies of corrosion in drinking water less extensively than PDS and LPS. However, they can provide more detailed information about the nature of corrosion processes. One use of the electrochemical noise technique is as a tool for detecting and monitoring pitting corrosion (Leban et. al. 2001, Schmitt et. al. 2001). EIS provides more diverse information and can be used to quantify the rate of growth of solid phases, resistivity of semiconductor surface layers (especially important in the case of cuprite), and effects of diffusion control limitation (Frateur et. al. 1999, Webster et. al. 2000, Zhang et. al. 2002). The interpretation of the data provided by EIS and electrochemical noise technique is model-specific and the use of improper or incomplete theoretical models for a corroding surface can lead to artifacts and misinterpretation of the experimental data. Overall, the performance of both techniques has not been explored sufficiently to be widely recommended and practically implemented. There is little doubt, however, that in the future the use of these and other novel electrochemical methods will expand.

Summary

This discussion outlined the principles of operation and performance of electrochemical methods in broadest detail. In all cases, the most important parameter relevant to monitoring metal release is the value of i_{corr}. If that parameter is determined correctly and the corroding surface has reached a condition in which the amount of oxidized metal equals to that lost by dissolution and/or colloidal mobilization (steady state), i_{corr} values can be used to directly predict metal release.

The use of i_{corr} values to quantify metal release is limited to cases when it can be assumed that no accumulation of surface scales is taking place. This requirement can be met at two widely divergent conditions: either at very low pH values, at which lead and copper solids are unstable and only soluble species are formed; or for surfaces that have undergone passivation, accumulated surface solids, and have reached a steady state. For all situations in between these two extreme cases, i_{corr} values can be employed as indicators of trends in metal release rather than its absolute measure.

At present, no uniformly applicable methodology exists to determine whether a corroding surface has reached a steady state and/or whether this steady state will or will not be affected by variations of water quality or hydraulic conditions.

Continuous measurements of both metal release and corrosion rates, preferably carried out in situ, are desirable in probing the surface's stability and, eventually, assisting a utility in complying with relevant regulations. The performance of either electrochemical techniques or metal release observations, when they are carried out separately and independently from each other, may not be adequate for the goal of in situ quantification of corrosion in drinking water. Of additional concern is the need to employ adequately trained personnel and have relatively sophisticated equipment to collect samples and/or acquire and process electrochemical data. In the latter case, the interpretability and precision of electrochemical data when determining i_{corr} values is model dependent, and the values of i_{corr} directly quantify the rate of metal release.

The following section of this Chapter discusses alternative approaches to monitoring metal release in situ and presents experimental data that show the benefits and drawbacks.

DEVELOPMENT OF CORROSION POTENTIAL STAGNATION/FLOW CORROSION CELL

The primary objective of the predictive technology portion of the project was to develop and test a corrosion cell suitable for on-line monitoring of corrosion and metal release processes in drinking water using the corrosion potential stagnation/flow (CPSF) method. Applying the CPSF concept requires that the behavior of E_{corr} during stagnation episodes be ascertained in great detail, and high precision. Simultaneous with the E_{corr} measurements, metal concentrations in water samples withdrawn from the corrosion cell need to be determined and correlated with the corresponding E_{corr} data.

The understanding of the fundamentals of the CPSF method and the relationship of equilibria and kinetics of heterogeneous and homogeneous processes in the corrosion cell became clearer as the project progressed. Correspondingly, the design of the CPSF cell underwent a transformation. Four different designs of CPSF cells were prepared and tested in the laboratory. Some of the major attributes of these cells considered in subsequent designs included the reproducibility of results, portability, and minimum electrode error.

The data presented in this report have been predominantly generated with the final design of the CPSF corrosion cell. However, because much effort was invested into its development and some indicative data were generated using other types of the CPSF cells, each cell design is described in this chapter.

Initial CPSF Cell Design

Following the exploratory phase of E_{corr} measurements, a first prototype of a corrosion cell was designed specifically for CPSF measurements. The cell employed a system of 12 in. long copper pipes with diameters ranging from 1/8 in. to 1 in. Standard single-junction reference electrodes were positioned outside of the pipes. The reference electrodes were used to acquire E_{corr} data in experiments with all versions of the CPSF cell, except the final version. At the start of each experiment, the internal solution in the electrode was changed. The electrodes were connected to the corroding pipes using a piece of 1/16 in. neoprene rubber tubing (Figure 5.2).

Figure 5.2. General view of the CPSF cell version I

These electrodes consisted of a mercury/mercury(I) chloride element surrounded by saturated KCl solution contained in an outer tube of the electrode. The electrochemical contact between the electrode's internal solution and the outside electrolyte (drinking water quality) is made through a ceramic junction located in the tip of the electrode. The KCl solution found in this ceramic junction and the outside electrolyte form a conductive ionic bridge between the reference electrode and the corroding surface. Due to the design of the electrode, there was always some leakage of chloride from the electrode to the outside solution.

The idea behind the design of the first version of the CPSF cell was that the rate of E_{corr} changes would be notably different with electrodes with different diameters and that a complete depletion of DO inside the corroding pipes with small diameters could be achieved within a reasonably short stagnation time.

It was soon determined that oxygen easily permeated into the cell through plastic and rubber tubing employed to connect the cell to the water source. Therefore, contrary to the initial CPSF theory, significant depletion of DO (manifested by a significant decrease of E_{corr} values after some stagnation time) was not seen. However, E_{corr} changes at the beginning of stagnation were similar to those predicted by the theory. The second drawback of the cell was the external position of the reference electrodes. The CPSF theory stipulates that the tip of the reference electrode be as close to the corroding surface as possible, especially for drinking water qualities that have low conductivity. The critical importance of this requirement was not entirely recognized at this phase of the CPSF development. As a result, E_{corr} data obtained using the first CPSF cell version were indicative of the later results, but did not have adequate precision.

Second Version of CPSF Cell

Once it was realized that short copper pipes with varying diameters did not yield E_{corr} data expected for stagnation episodes of varying duration, the design of the cell was changed. In the second version of the CPSF cell, a 36 in. long piece of ½ in. copper tubing was used. Five single-junction reference electrodes were positioned along its length. Two electrodes were placed at approximately ½ in. from the edges of the pipe, one electrode in its middle, and two remaining electrodes between the central and end reference electrodes. The initial concept of the design was to examine whether all electrodes would give identical trends because it was theorized that E_{corr} of the corroding pipe should be identical at any place on its surface.

Immediately following the start of operations with the second CPSF cell version, it became apparent that E_{corr} vs. time data for the electrodes positioned at the edges of the pipe and in its middle were consistently different. The locations at the edge of the pipe did not experience substantial change in E_{corr} during stagnation while those in the middle of the pipe mostly behaved as predicted by the CPSF theory (monotonically increasing with time to a maximum). Repetitions of stagnation experiments with the second CPSF cell version confirmed the observation.

It was determined that E_{corr} values measured along the same corroding surface are not necessarily the same. E_{corr} values depend on the local concentration of DO (which varies between the pipe edges and the middle of the pipe) and soluble copper ions, as well as the properties of the diffusion layer, which also vary between the edges and middle of the pipe.

Based on these observations, it was concluded that three new criteria must be met with the CPSF design. First, the design of the CPSF must ensure that no edge effects influence the behavior of E_{corr}. Second, during stagnation, any influx of oxygen to the cell from the atmosphere must be prevented. Third, the tip of the electrode should be positioned as close to the corroding surface as possible.

Third Version of CPSF Cell

Following the conclusions from the second version of the CPSF cell, a third version of the CPSF cell was developed and termed the CPSF Super Long Coil Copper Cell. In this design, a 25 feet long piece of ¼ in. copper tubing was used. A standard single-junction reference electrode was inserted in the mid-section of the pipe. Its tip was positioned so it would be as close to the internal radius of the pipe as possible. However, due to the significant curvature of the internal surface this was difficult to control. The third version of the cell is shown in Figure 5.3.

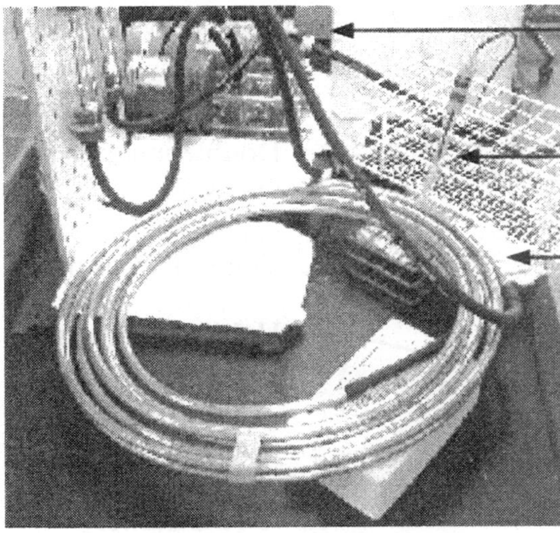

Figure 5.3 General view of the CPSF Super Long Coil Copper Cell

Due to the length of the CPSF super long coil copper cell, it more accurately resembled a water distribution pipe in typical household plumbing. Initial data obtained with it were reproducible and the relevant trends were in accord with the general CPSF theory. Following that, two more CPSF super long coil copper cells for laboratory use were prepared. One super long coil copper cell was used for almost a year.

The CPSF super long coil copper cell design still had drawbacks. The first was related to the mechanical stability of the reference electrode. Because the reference electrode was inserted into a thin wall of the pipe, the pipe itself did not provide adequate stability and additional support was required. It became evident that the design of the CPSF super long coil copper cell was not feasible for field studies.

When the design and tests of the CPSF super long coil copper cell were underway, it was also determined that leakage of chlorine from the internal compartment of the single-junction reference electrode caused pronounced changes to the visual appearance of the pipe (the morphology of the surface directly adjacent to the electrode tip). This necessitated the use of double-junction reference electrodes, whose design and geometrical size made it practically impossible to insert them in the thin wall of ¼ in. copper pipe. This difficulty resulted in the ceasing of experiments with the CPSF super long coil copper cells.

Final Version of the CPSF Cell: Narrow Rectangular Channel (NRC) Cell

To make the system more suitable for field deployment and simplify the theoretical analysis of E_{corr} data, another design of the CPSF cell was implemented, the narrow rectangular channel (NRC) cell (Figure 5.4). Contrary to previous versions, the cell was made of two rectangular 12 inch × 3 inch copper plates separated with a 1/16 inch thick Teflon gasket glued around at the edge of the metal plate. The thickness of the metal plates was ¼ inch. The design creates a narrow rectangular channel through which the water can flow. The internal volume of the CPSF cell exposed to corrosion during stagnation was approximately 20 mL. Total metal surface area exposed to corrosion is about 240 cm^2. The reference electrode of the cell was supported by an acrylic block. The electrode was positioned so its tip just touched the inner surface of the upper copper plate. Overall, the cell's design allowed easy assembling of the cell at the time of its commissioning or decommissioning.

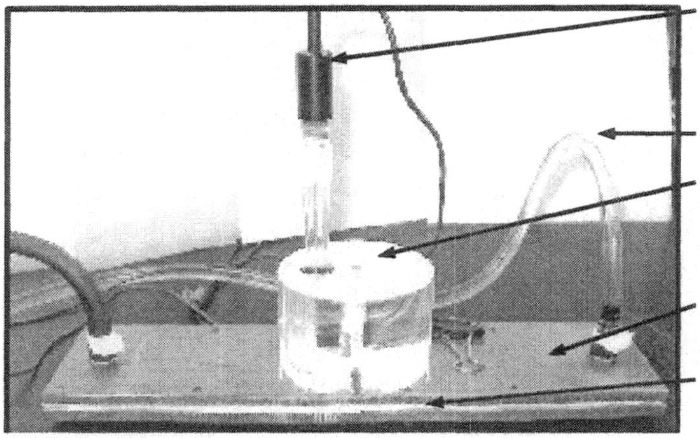

Figure 5.4 General view of the CPSF Narrow Rectangular Channel (NRC) corrosion cell

The geometric parameters of the cell are shown in Table 5.1. Additional details of the cell design are shown in Figure 5.5. Four CPSF NRC cells were prepared for laboratory and field tests. The new design improved the durability and portability of the cells. It was found that the reproducibility of E_{corr} measurements with the NRC cells was considerably better than with any other CPSF cell designs.

Table 5.1
CPSF Narrow Rectangular Channel (NRC) cell specifications

Parameter	Value
Length of metal plate in contact with water	27.31 cm (10.75 in.)
Width of metal plate in contact with water	4.45 cm (1.75 in.)
Height of the cell (water)	0.16 cm (0.06 in.)
Volume of water in the cell	20 mL (1.18 in^3)
Area of the metal plate in contact with drinking water	245 cm^2 (38 in^2)
Diameter of acrylic block to support the reference electrode	7.62 cm (3 in.)
Diameter of hole for the tip of reference electrode	0.15 cm (0.059 in.)

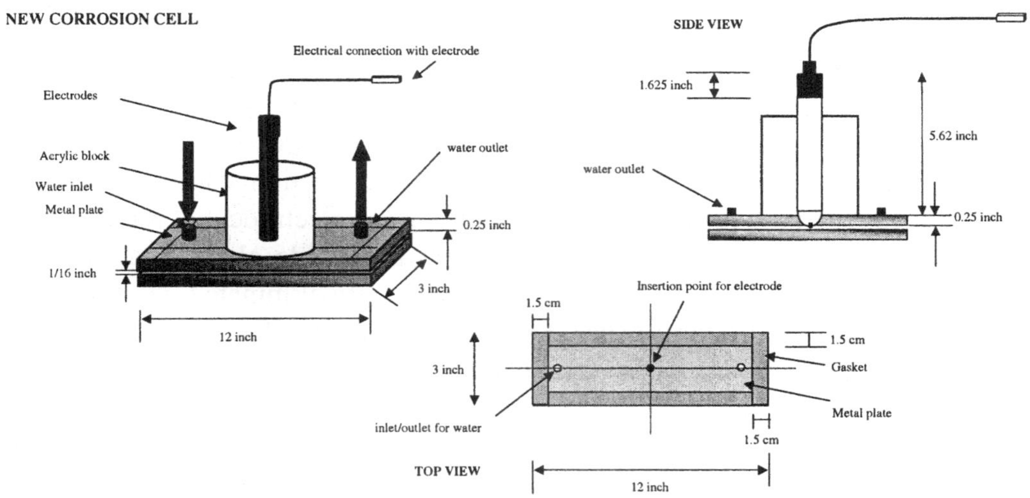

Figure 5.5 Dimensional details pertinent to the design of the CPSF NRC cell

Initially, single-junction reference electrodes were employed to measure E_{corr} in the NRC cells. However, two additional problems were identified when single-junction reference electrodes were used. First, it was recognized that due to the differences of the ionic strengths of the internal solution (saturated KCl) of the reference electrode and the drinking water quality, "junction potential" on the border between the ceramic junction and drinking water developed (Koryta 1975). Its magnitude is not always possible to evaluate theoretically, and it may vary from 1 mV to greater than 15 mV. The existence of the junction potential introduces a systematic error in E_{corr} measurements and, as such, can be accounted for. An additional complication associated with the existence of the junction potential was that, due to a slow leakage of KCl into drinking water during stagnation and associated changes of the ionic strength, the value of the junction potential can also change with time, compromising the consistency of E_{corr} measurements.

A second, more serious problem, was found somewhat later in the project. When the first CPSF NRC cell was disassembled after approximately two weeks of use, it was determined that chloride leakage from the electrode affected the visual appearance of the surface adjacent to the electrode tip. That part of the surface had a much darker color and its overall appearance was notably different from that of the rest of the electrode (Figure 5.6).

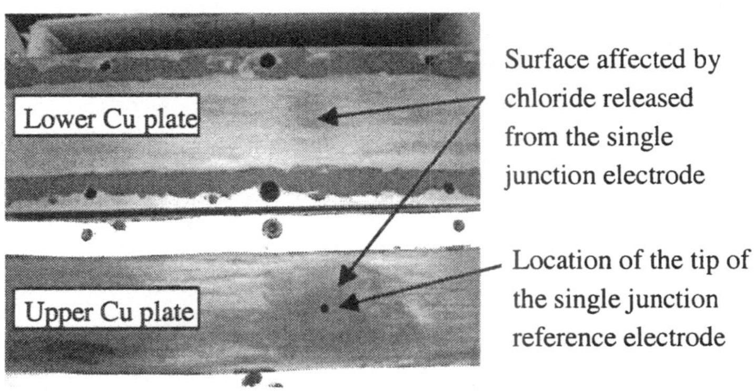

Figure 5.6 Effects of chloride leaching from a single junction reference electrode on the visual appearance of corroding copper surfaces in the CPSF NRC cell

It was determined that the chloride leakage, which is not necessarily of much importance in experiments with constantly flowing water, had to be prevented in CPSF experiments involving long stagnation and accumulation of high doses of chloride in the part of water close to the reference electrode. To eliminate this problem, a double-junction standard calomel electrode was used. The concept is the same as that of a single-junction electrode, but the KCl solution that surrounded its electrochemical unit is separated from an outside electrolyte by an additional reservoir with a second junction. The latter component largely eliminated chloride leakage from the internal reservoir to the drinking water. Double-junction standard calomel electrodes were used with all four CPSF NRC cells throughout the study in both laboratory and field experiments.

The dates of commissioning and types of water qualities used in the experiments with the four CPSF NRC cells are shown in Table 5.2. Note that all four CPSF NRC cells were initially exposed for at least a week to the synthetic water quality during testing in the laboratory.

Table 5.2
Summary of commissioning and use of CPSF NRC corrosion cells

Cell Designation	Date of commissioning	Water qualities used (in sequence of experiments)*
C-1	March 2002	Synthetic water, Pinellas County water, HDR water
C-2	April 2002	Synthetic water (pH variation and chlorine dose)
C-3	June 2002	Synthetic water (variations of alkalinity and phosphate)
C-4	June 2002	Synthetic water, HDR water & San Francisco water

*Detailed water qualities are described in Table 5.3

Experimental Setup and Mode of Operations

The experimental setup for CPSF measurements is shown in Figure 5.7. Water was pumped through the CPSF corrosion cell at a constant flow rate of 1.0 L/min using a peristaltic pump. In most of the experiments carried out in laboratory conditions, the water was recirculated using a 20-liter container. Water in the reservoir was exchanged once a week (in some cases more frequently). During each water exchange, the container and all connecting plastic tubing were flushed with organic-free water.

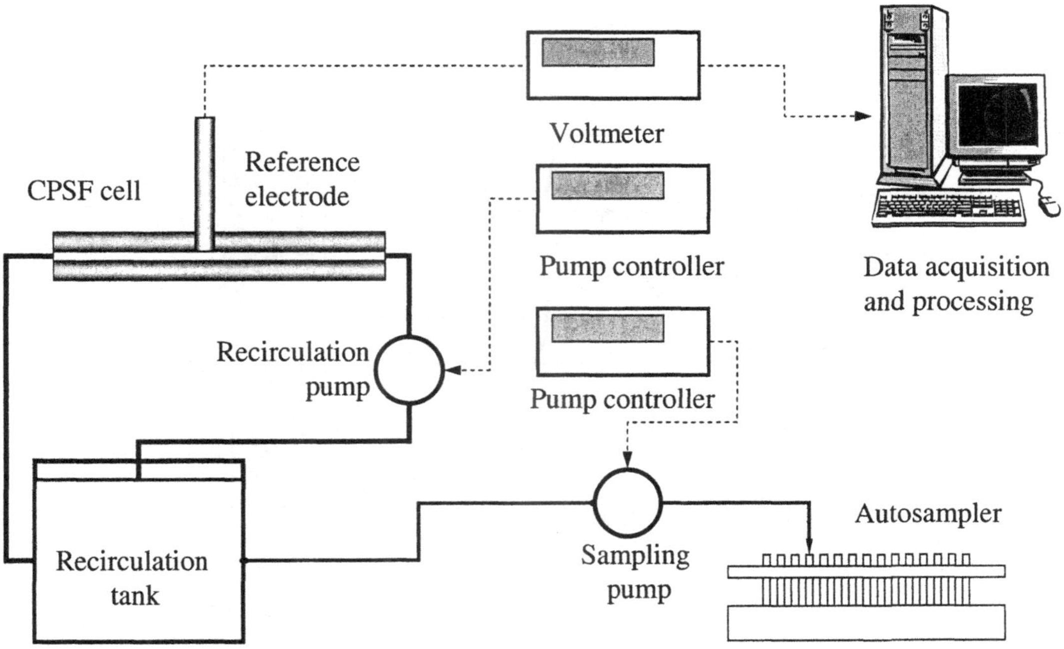

Figure 5.7 Experimental setup for E_{corr} measurements in laboratory conditions

Water samples from the reservoir were taken with regular frequency to monitor the release of copper during stagnation and flow episodes. E_{corr} data were acquired using a multi-channel, high-precision data acquisition multimeter connected to a computer. Software was utilized to control the acquisition of E_{corr} data, which were exported to, and manipulated with, commercial spreadsheet software. In some experiments, an ion selective electrode was also used to collect online data on the accumulation of copper in the recirculation reservoir.

Stagnation-Flow Sequences

A controlled sequence of stagnation and flow episodes was one of the key features of CPSF experiments. Two basic sequences were employed. The main sequence consisted of one 8 hour-long stagnation period between multiple 30 minute stagnation/flow periods. This flow pattern was designed to mimic typical household water usage.

Figure 5.8 also shows the sampling times, which were set to occur just before and after selected stagnation periods.

Alternative stagnation-flow sequences with varying durations of stagnation periods were also utilized. In one such sequence employed in the field experiments (Figure 5.9), stagnation times of 5, 10, 15, 30, 45, 60, 90, 120 and 180 minutes with 30 minutes of flow between them were used. This sequence was used in experiments carried out to examine effects of stagnation time on E_{corr} and metal release. In the measurements with varying stagnation times, the water was not recirculated. Samples of known volume (150 mL or 1 L) were withdrawn from the corrosion cell at the end of each stagnation period and analyzed for total metal.

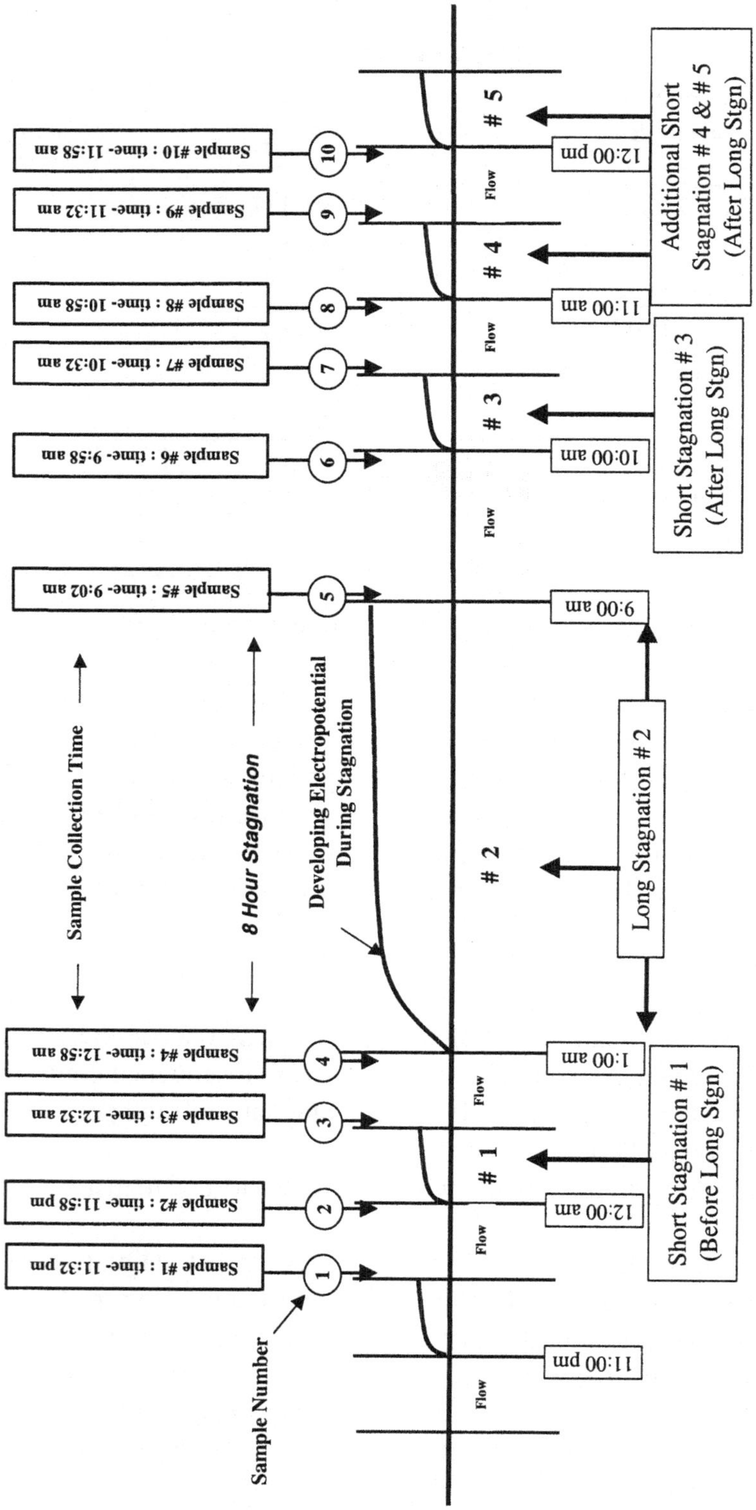

Figure 5.8 Stagnation/flow sequence that included one 8-hour stagnation and multiple 30-minute stagnation episodes. Designed to mimic household water usage.

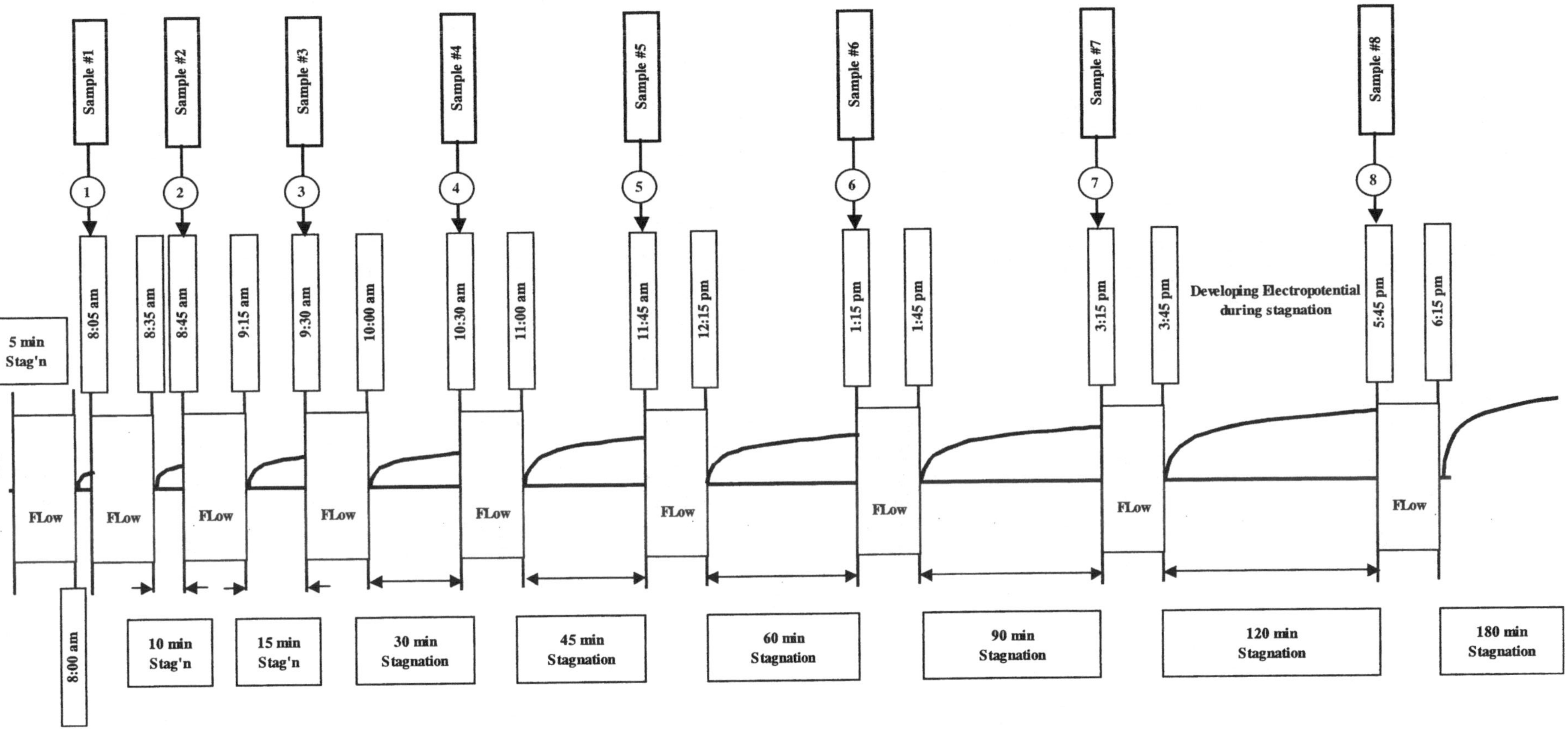

Figure 5.9 Stagnation/flow sequence with duration of stagnation periods varying between 5 to 180 minutes.

In all cases, E_{corr} data were collected continuously. In the case of main stagnation/flow sequence (Figure 5.8), E_{corr} data were processed that corresponded to the long stagnation period, two short stagnation periods prior to it, and three after it. In the case of experiments with alternative stagnation flow sequences, all relevant E_{corr} vs. time profiles were acquired.

Sample Collection

Water samples were collected at the start and end of selected stagnation periods and processed to determine metal concentrations. The timing of sample collection was controlled with an autosampler and occurred as shown in Figure 5.8 and Figure 5.9. All sampling lines were flushed before collecting a 5 mL sample from the outlet of the CPSF cell. The first 5 mL of water withdrawn from the reservoir was used to flush connecting tubing, and the second 5 mL was dispensed into 5 mL plastic test tubes.

For continuous flow systems, water samples were collected manually at the end of selected stagnation periods. To ensure collection of the whole water sample exposed to corrosion during the stagnation, 1.0 L sample was taken during the field tests. In the laboratory CPSF experiments that employed varying stagnation times, the volume of sample withdrawn from the system was 150 mL. The actual volume of water in contact with the metal surface at a given time during corrosion was about 20 mL for the NRC cell.

Water Quality Variations

A portion of the experiments evaluating different water qualities was carried out using synthetic water. The composition of the synthetic water and the corresponding range of water quality parameters (pH, alkalinity, concentrations of free chlorine and phosphate) examined in the laboratory are shown in the first column of Table 5.3. Unchlorinated water shipped from Pinellas County, Florida was also used in experiments preceding CPSF field tests at that site, but is not represented in the table.

Table 5.3
Typical water quality parameters for laboratory experiments

Parameter	Synthetic Water	Pinellas County Water (chlorinated)	HDR Water (chlorinated)	San Francisco Water (chlorinated)
pH	6.5 – 8.5	7.7	6.5-8.5	8.6
Alkalinity (mg/L as $CaCO_3$)	15 - 100	200	16	20
Hardness (mg/L as $CaCO_3$)	0	200	24	20
Chloride (mg/L)	15	40	3	5
Sulfate (mg/L)	5	18	1	7
Total chlorine (mg/L as Cl_2)	0-1.5	3.0	1.0	0.8
Phosphate (mg/L as PO_4)	0-2.0	0	0	0
Conductivity (µmhos/cm)		420	60	83

Experiments were conducted using chlorine-containing water at several sites. These included tap water at the laboratory of University of Washington in Seattle, the water laboratory of HDR Inc. (Bellevue, WA), San Francisco Water Utility Lab in Millbrae (San Francisco, CA), and a water quality station located at the Kellar Water Treatment Plant (Pinellas County, FL). Relevant water quality parameters are listed in Table 5.3.

Metal Analyses

Analysis for metals throughout this study was carried out using a inductively-coupled plasma atomic emission (ICP-AE) spectrometer with sequential light detection. A cyclonic spray nebulizer was employed for sample atomization. The samples were run using a random-access autosampler. Prior to analysis, all metal-containing samples were acidified with concentrated nitric acid (0.1 mL per 5 mL sample). The calibration of the instrument was conducted using standard solutions containing 0 μg/L and 1000 μg/L copper. ICP-grade standards were used for calibration. The method employed in the ICP measurements determined an average of three repeated measurements of ICP emission intensity. All acidified samples were stored in standard plastic test tubes with screw caps until use.

Dissolved Inorganic Carbon

A carbon analyzer was used for dissolved inorganic carbon (DIC) analyses. The instrument was calibrated using solutions of sodium biphthalate and ultra-pure water boiled with an excess of sodium persulfate as described in the instrument's manual and in Standard Methods (19th Edition 1995). The calibration was checked daily. The standard deviation of the measurements was within 10% and the reproducibility within 5%.

Chemicals and Water

Preparation of all containers followed the recommendations of Standard Methods (19th Edition 1995). Reagent water residual DOC was less than 0.1 mg/L and resistance was 18.2 MΩ/cm. All chemicals were American Chemical Society reagent grade. Sodium bicarbonate, sodium chloride, sodium sulfate and sodium phosphate used to prepare the synthetic water quality were purchased stock solutions.

Phosphate buffer stock was prepared by dissolving 68.1 g KH_2PO_4 and 11.7 g NaOH per one-liter water. For a working solution, 50 mL of this stock was added to the reagant to be buffered, before dilution to one-liter in total volume. The resultant concentration of this buffer was 0.25 M KH_2PO_4.

Free chlorine stock solution was prepared by dilution of a reagent-grade sodium hypochlorite solution (5.6% available chlorine) with Milli-Q water. The solution was standardized by the N, N-diethyl-p-phenylenediamine colorimetric method in accord with Standard Method 4500-Cl G (Standard Methods 19th Edition 1995), calibration curves were obtained using a potassium permanganate solution, whose absorbance was measured at 515 nm. Standardized solutions of chlorine were stored at 4°C in glass bottles wrapped in aluminum foil. Concentrations of free chlorine were measured using the N, N-diethyl-p-phenylenediamine colorimetric method as described in Standard Methods 4500-Cl G (1995). In these measurements, the absorbance of solutions containing the N, N-diethyl-p-phenylenediamine active chlorine indicator was measured at 530 nm using a 3 cm optical cell. A Hach DR-4000 spectrophotometer was employed in these analyses.

The pH of waters was measured with a pH meter equipped with a combination electrode. The pH meter was calibrated using pH 4.0, 7.0 and 10.0 standard buffer solutions. When necessary, the pH was adjusted with sodium hydroxide or sulfuric acid solutions.

LABORATORY EXPERIMENTAL FINDINGS

In this section the general findings concerning E_{corr} and metal release patterns, their correlation and the effects of metal surface aging are presented. Also, the effect of water quality parameters (pH, alkalinity, phosphate and chlorine concentration) on the relationship between copper release and E_{corr} values are presented.

Typical Behavior of the Corrosion Potential During Stagnation in CPSF Systems

Data presented in Figure 5.10 to Figure 5.12 show the typical behavior of E_{corr} observed in the experiments using synthetic water and Pinellas County water (without chlorine). The results were obtained with two cell designs: CPSF super long coil copper cell (Figure 5.10) and CPSF NRC (Figure 5.11 and Figure 5.12).

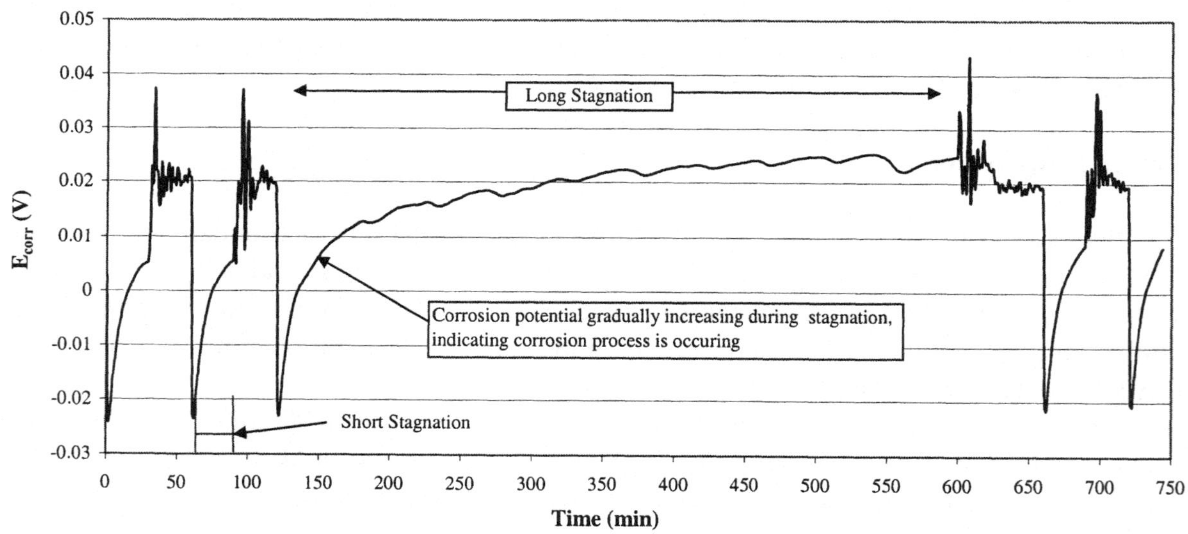

Figure 5.10 E_{corr} profile for the CPSF super long coil copper cell exposed to synthetic water (pH 7.5, alkalinity 15mg/L as $CaCO_3$, day 1 of operation)

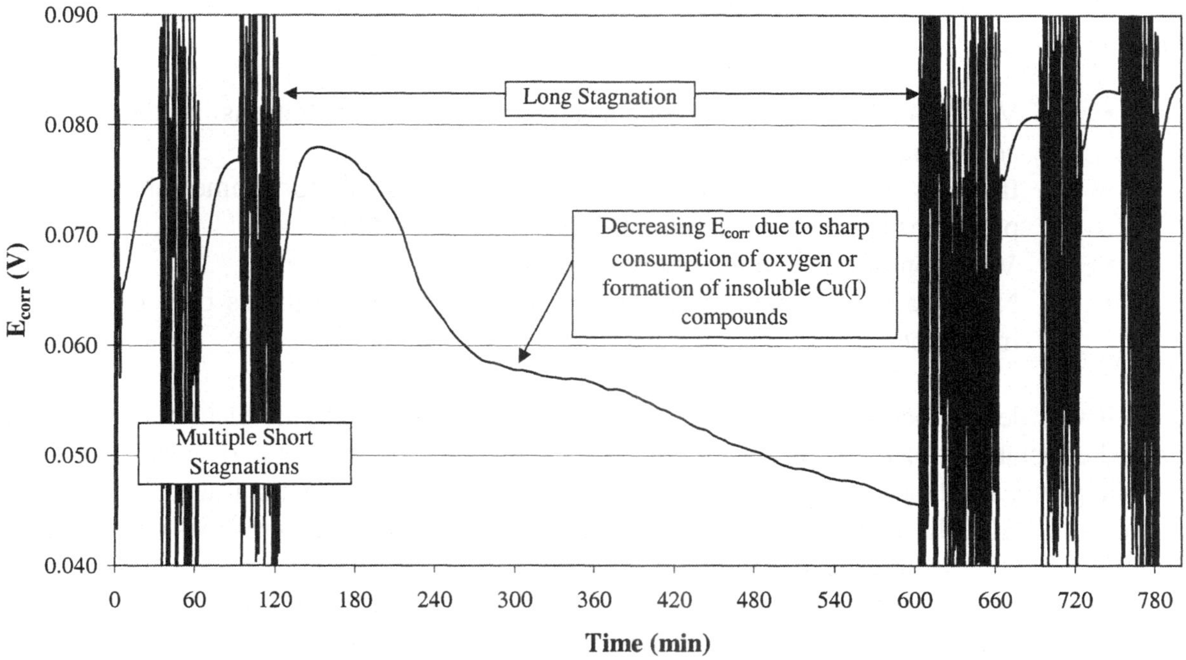

Figure 5.11 E_{corr} profile for CPSF NRC cell (C-2) exposed to synthetic water at pH 7.5 (day 1 of operation)

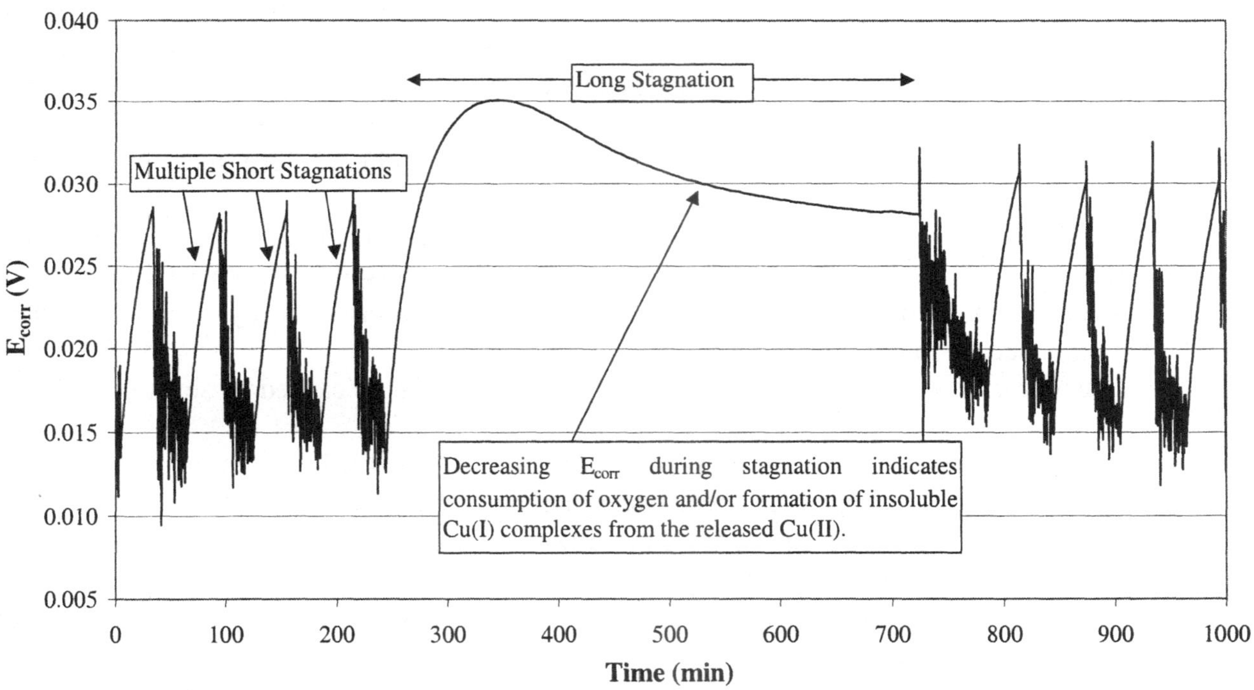

Figure 5.12 Typical E_{corr} profile for CPSF NRC cell (C-1) exposed to unchlorinated Pinellas County water (pH 7.5, Alk. 200 mg/L as $CaCO_3$, day 1 of operation)

The first important feature is the behavior of E_{corr} and the notable difference between episodes of stagnation and flow; E_{corr} exhibits instability during flow episodes. Although only the data obtained for stagnation episodes is discussed in this report, it is relevant to mention the major reasons for E_{corr} instability. There are four main causes of E_{corr} instability during flow episodes:

- Variations of dissolved copper and oxygen concentrations at the surface of corroding metal (due to turbulence);
- Intrinsic electrochemical noise, which is typical for some electrochemical processes (e.g., Leban et. al. 2001, Schmitt et. al. 2001);
- Variation of the junction potential; and
- Noise caused by external electromagnetic fields (contributes due to a very high impedance of the reference electrode).

It was determined during the project that noise caused by external electromagnetic fields caused the instability of E_{corr} during flow episodes. Specifically, the cause of instability was the electromagnetic field caused by operations of the electric motor of a peristaltic pump positioned close to the CPSF cell. However, the presence of E_{corr} instability during episodes of flow, did not affect the data obtained during stagnation episodes.

Contrary to behavior of E_{corr} during flow episodes, stagnation periods were almost always noise-free. As predicted by the CPSF theory, in all unchlorinated experiments, E_{corr} gradually increased following the start of stagnation, which indicated that corrosion reactions were in progress and copper concentration in the CPSF cell was increasing. As described in the CPSF theory, and verified in the experiments, when the dissolved oxygen concentration at the metal surface is nearly constant, a gradual increase of E_{corr} associated with copper release is expected.

For example, in Figure 5.10 (CPSF super long coil copper cell exposed to synthetic water quality, with pH 7.5 and alkalinity 15 mg/L), an increase of about 45 mV can be observed during the 8-hour long stagnation period. During a much shorter period (approximately 30 minutes) immediately after the start of stagnation, the rate of E_{corr} change was more rapid and its magnitude was about 25 mV.

In the CPSF NRC experiment, shown in Figure 5.11 (synthetic water), there was a decrease of E_{corr} values in the later part of a long stagnation episode. Another CPSF NRC cell (C-1) exposed to non-chlorinated water from Pinellas County, Florida, also yielded a similar E_{corr} profile with a decreasing limb (Figure 5.12).

The decrease of E_{corr} during long stagnation episodes can be caused by several factors:

- A localized decrease in the concentration of dissolved oxygen near the surface of corroding copper can occur faster than predicted by the CPSF theory.
- Cu^{2+} ions produced at the electrode surface can interact with copper (0) to form Cu^{+} species, which then precipitate on the corroding metal surface (e.g., Merkel et. al. 2002, Ferguson, Franque and Shock 1996). This mechanism was not accounted for in the general CPSF theory.
- The precipitation of solid phases at the surface can lead to changes of E_{corr} that were not accounted for in the current version of the CPSF theory.

Changes of E_{corr} vs. Time Profiles in Week-Long Experiments With Recirculated Water

For corrosion experiments performed in the laboratory of University of Washington, synthetic water was recirculated and exchanged once per week. Immediately after any water exchange, the copper concentration was minimal. Due to the corrosion processes in the CPSF cell, the concentration of copper in the reservoir increased between water exchanges. This gradual increase of Cu_{total} caused consistent changes in the behavior of E_{corr} (Figure 5.13 and Figure 5.14). As predicted by CPSF theory, the value of E_{corr} immediately after a water exchange had noticeably more negative (cathodic) potentials than at any time later in the run. As the experiment progressed and the concentration of copper increased, E_{corr} values at the start of each stagnation episode gradually increased as well. The shift of E_{corr} values was evident both at the beginning of stagnation and during longer periods of stagnation. These changes of the position of E_{corr} were especially prominent for the first 2 days after each water exchange. These trends were evident for exposure to any water quality.

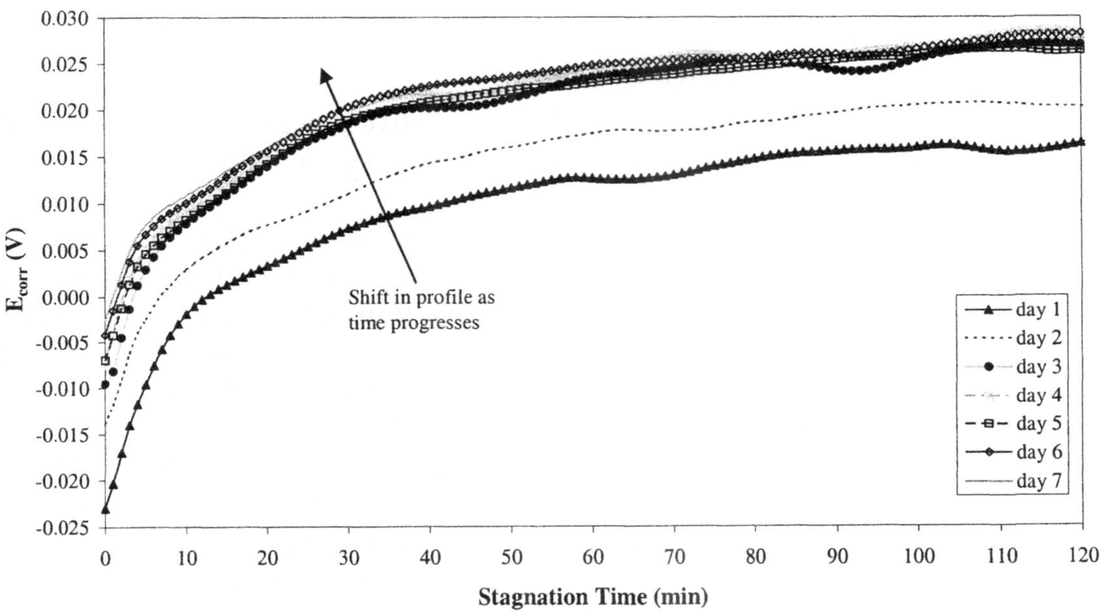

Figure 5.13 Comparison of E_{corr} profiles for seven consecutive days of exposure to synthetic water. 120 minutes stagnation time, pH 7.5, CPSF super long coil copper cell

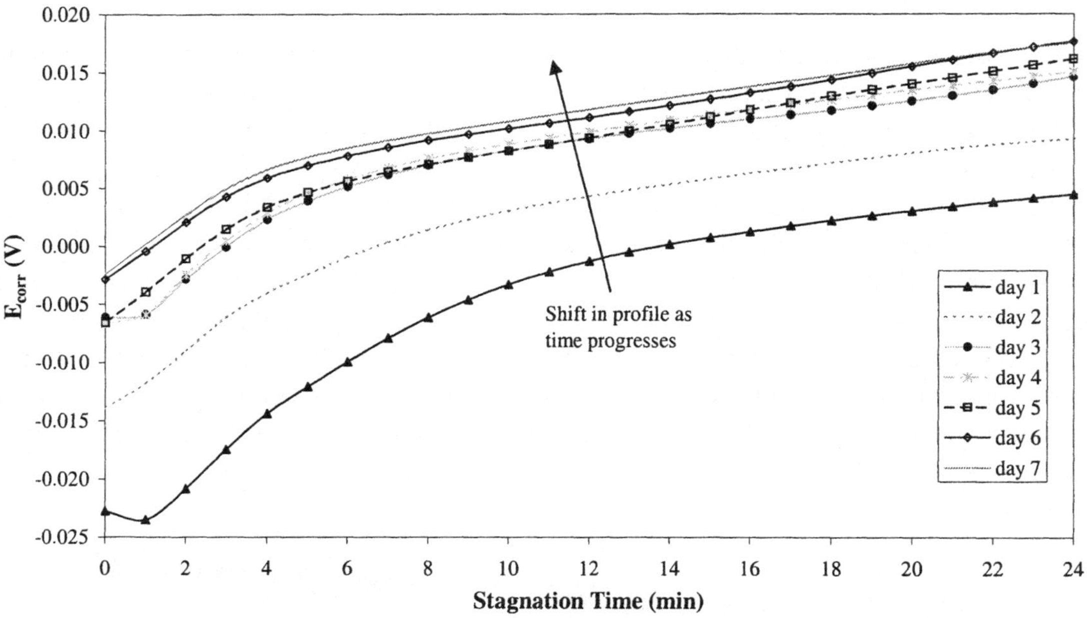

Figure 5.14 Comparison of E_{corr} profiles for several consecutive days of exposure to synthetic water, 25 minutes stagnation time, pH 7.5, CPSF super long coil copper cell

As the surface aged, the position of E_{corr} at the beginning of each week-long experiment, tended to shift to a more positive value than in the previous experiment (Figure 5.15). The trend is most likely due to a decrease in the intrinsic current of copper oxidation (i_{Cu}) for aged surfaces. This can be illustrated semi-qualitatively with the aid of an Evans diagram showing the electrochemical currents and potential of a corroding system. The point where anodic current of copper oxidation and cathodic current of oxygen reduction intersect, defines the experimentally measurable E_{corr} and i_{corr} of the system (Figure 5.16a). As the surface ages, anodic current of copper oxidation (i_{Cu}) tends to shift to a lower value while the current of oxygen reduction remains largely unaffected (Merkel et. al. 2002). As a result of this effect, the relative positions of the oxidation and reduction branches of the relevant electrochemical currents change thus causing an anodic shift of the intersection point (E_{corr}) in the Evans diagram (Figure 5.16b). Thus, the anodic shift of E_{corr} values corresponds to a decrease of the intrinsic copper oxidation current for aged copper surface.

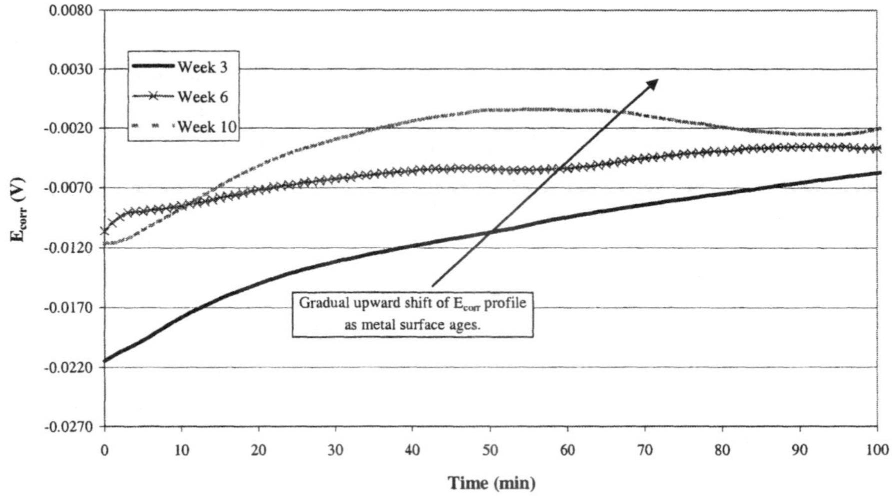

Figure 5.15 Comparison of E_{corr} profiles for several weeks of exposure to synthetic water. CPSF super long coil copper cell, pH 7.5

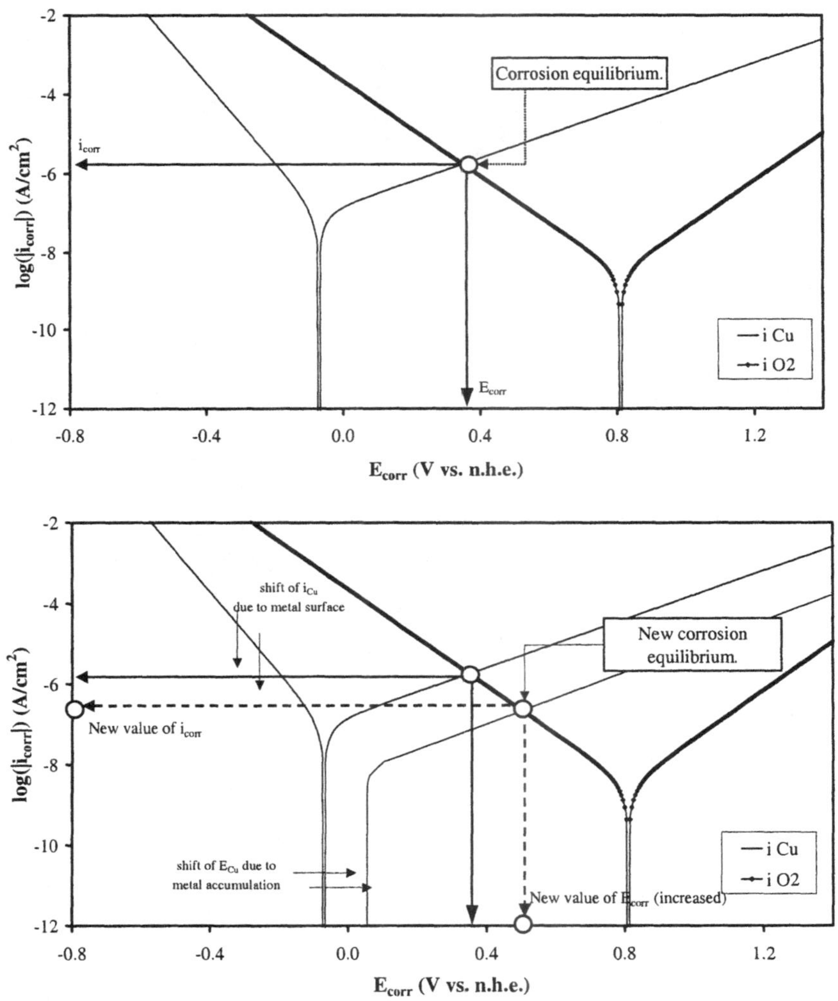

Figure 5.16 a and b Evans diagrams and demonstration of an anodic shift of E_{corr} associated with a decrease of electrochemical current of copper oxidation (i_{Cu}).

Typical Metal Release Patterns

The total soluble copper concentration found in the recirculated water slowly increased with time. Figure 5.17 shows the results of experiments with the CPSF super long coil copper cell at pH 7.5 and alkalinity 15 mg/L. In this figure, it can be seen that the rate of metal release for more freshly-exposed surfaces is notably higher than for metals which have been exposed to water for longer times.

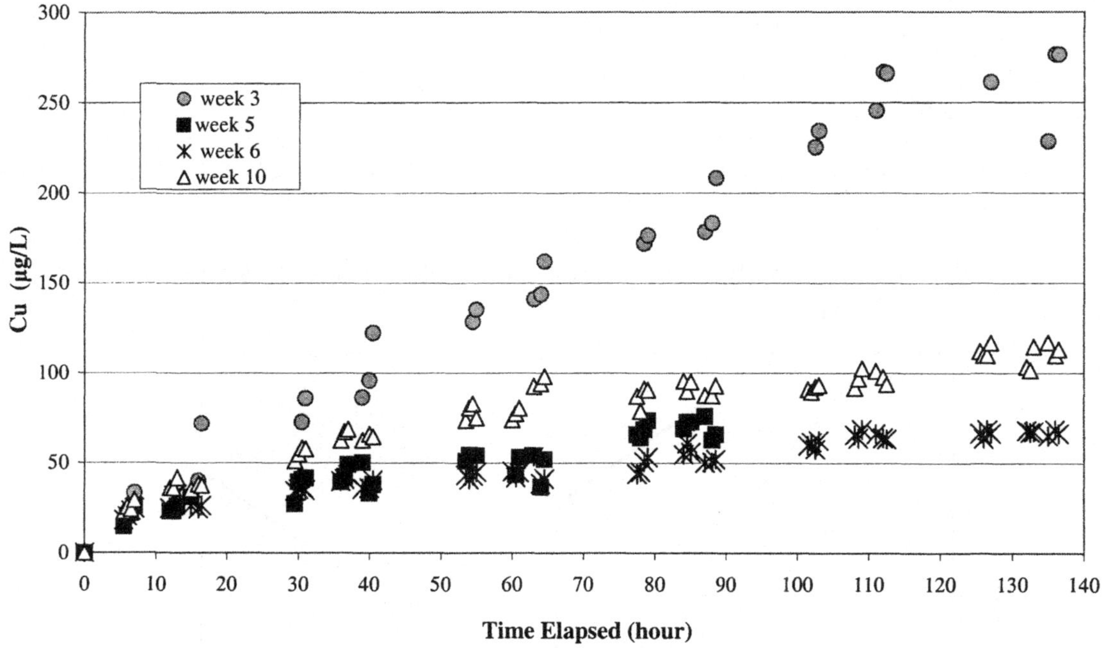

Figure 5.17 Effects of aging on metal release. Synthetic water, alkalinity 15 mg/L, pH 7.5, CPSF super long coil copper cell.

As expected, water quality parameters such as pH and alkalinity affected the rate of metal release. An increase in pH from 6.5 to 8.5 caused the concentrations of copper in the recirculated water to decrease several-fold, most prominently in the beginning of each week-long experiment (Figure 5.18). Increasing the alkalinity of the synthetic water from 15 mg/L to 50 mg/L and 100 mg/L resulted in a decrease of the copper concentration (Figure 5.19)

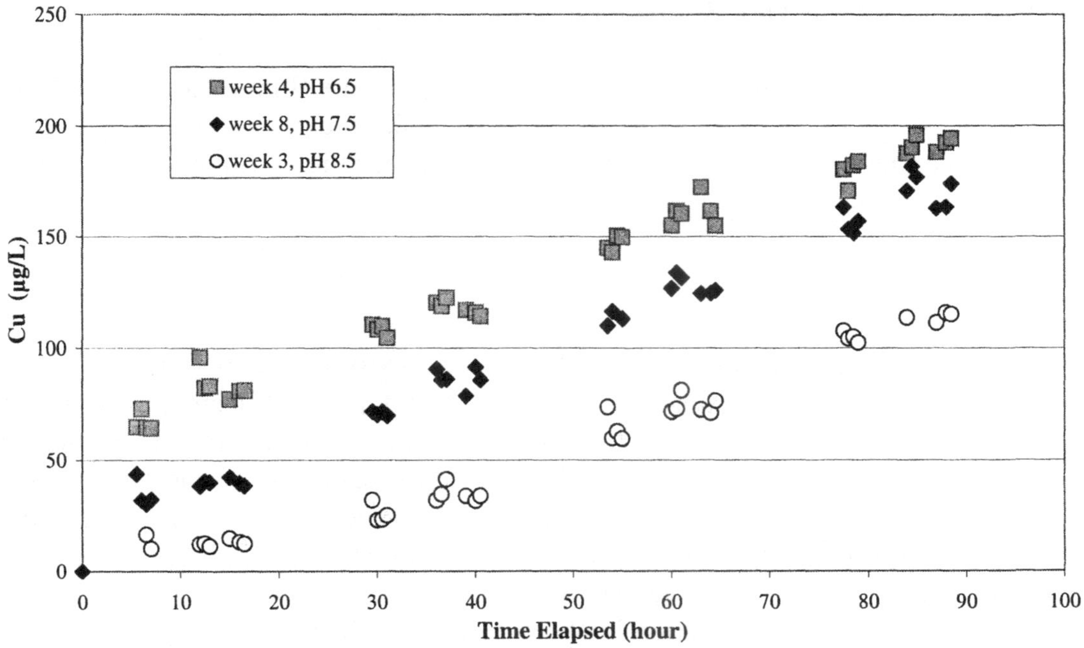

Figure 5.18 Effects of pH on metal release. Synthetic water, alkalinity 15 mg/L, CPSF NRC cell (C-2).

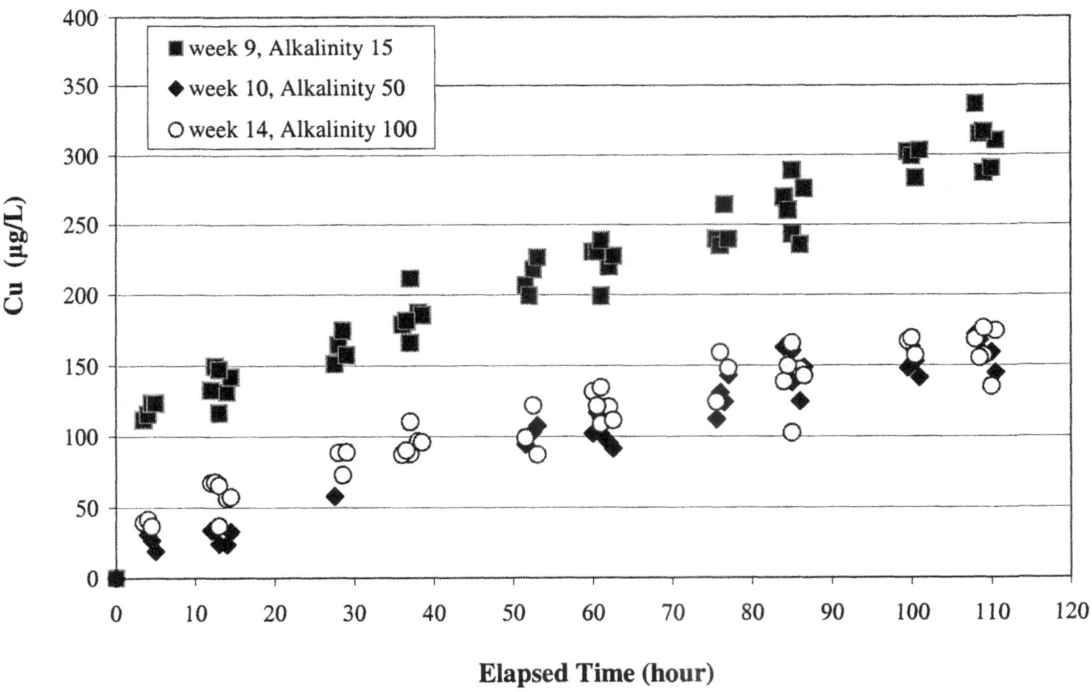

Figure 5.19 Effects of alkalinity on metal release. Synthetic water, pH 7.5, CPSF NRC cell (C-3).

Correlation Between E_{corr} and Total Copper Concentrations

The analysis of the corrosion processes discussed in Appendix B predicted that as long as the concentration of oxygen at the metal surface is quasi-stable, the corrosion potential during stagnation episodes can be described by the simplified master equation (Equation 5.1) that can be simplified as shown below:

$$E_{corr} = E_0 + \gamma \log(Cu_{total}) \qquad (5.1)$$

where E_0 is the value of E_{corr} at standard conditions (the activity of total copper is one) and γ is the slope of E_{corr} vs. $\log(Cu_{total})$ correlation. The values of both E_0 and γ depend on water quality parameters (pH, alkalinity) and intrinsic electrochemical parameters of the system. At the same time, Equation 5.1 predicts that E_{corr} values should be linearly proportional to the potential of an ion-selective electrode (ISE) specific to Cu^{2+} ions in the same water. This is because, as long as there is no precipitation of copper-containing solids, the concentrations of Cu_{total} and $[Cu^{2+}]_{free}$ are linearly correlated. Typical data obtained using CPSF cells show the validity of these predictions. A very strong linear correlation between E_{corr} values measured at the beginning of any stagnation episode and the logarithm of the concentration of total copper in the water residing in the CPSF cell was observed (Figure 5.20). At the same time, the behavior of the potential of a Cu^{2+}-selective ISE (placed in the reservoir containing the recirculated water flowing through the CPSF cell) was virtually identical with that of E_{corr} (Figure 5.21).

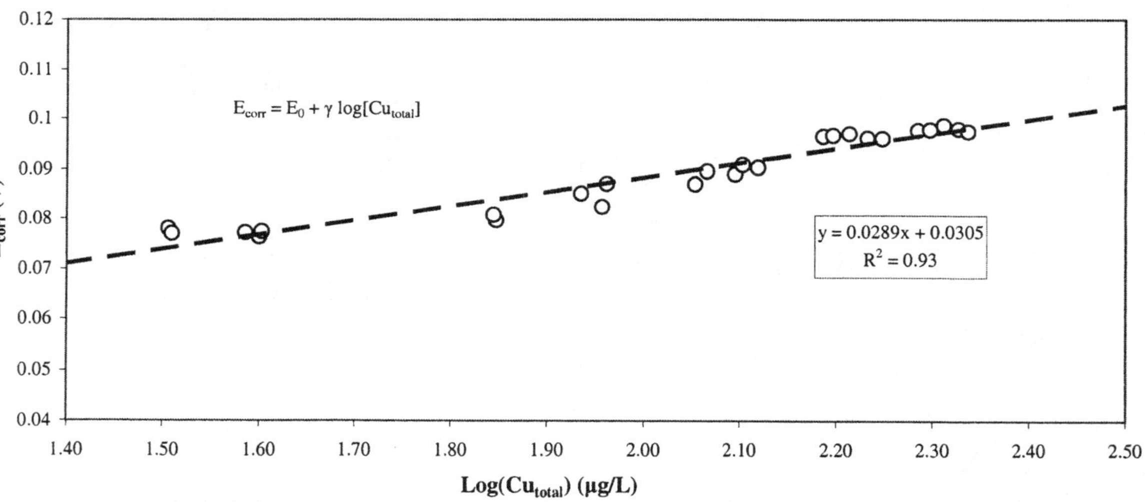

Figure 5.20 Correlation between E_{corr} and total soluble copper concentration at the beginning of stagnation. Week 4 of exposure to synthetic water, pH 7.5, alkalinity 15 mg/L, CPSF NRC cell (C-2).

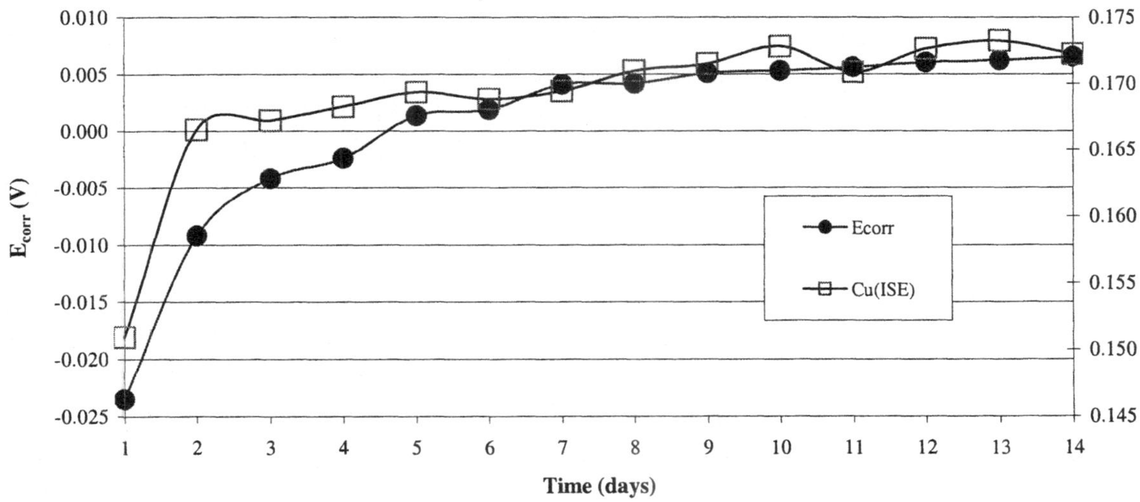

Figure 5.21 Comparison of the behavior of E_{corr} measured in the CPSF cell and potential of Cu^{2+}-selective ISE in the reservoir containing recirculated water. Synthetic water, pH 7.5, alkalinity 15 mg/L.

The aging of the surface and the associated changes of its electrochemical properties did not affect the basic validity and functional type of E_{corr} vs. Cu_{total} relationships. The relevant data obtained for the CPSF super long coil copper cell are presented in Figure 5.22. That cell was exposed to the synthetic water quality (pH 7.5, alkalinity 15 mg/L) for several months. The data shows that at any moment in time there is a strong linearity between the values of E_{corr} and the logarithm of the total copper concentrations at the beginning of any stagnation episode.

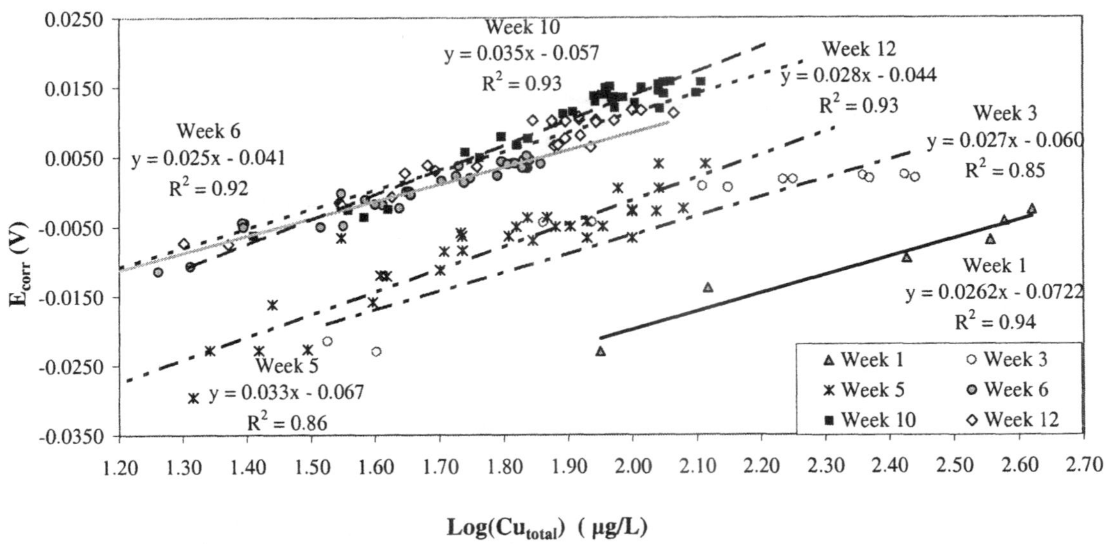

Figure 5.22 Effects of aging on the correlations between E_{corr} and soluble Cu concentration. CPSF super long coil copper cell, synthetic water quality, pH 7.5, alkalinity 15 mg/L.

The other trend seen in Figure 5.22 is that the data points corresponding to E_{corr} and Cu_{total} values gradually shift towards higher E_{corr} as the metal surface ages, while the rate of metal release tends to decrease. This gradual shift appears to have stabilized for exposure on Week 6 and later. This indicates that by week 6 the metal surface had undergone the initial transient phase of corrosion and had achieved a quasi-steady state. The mechanism that leads to this effect is a gradual decrease of the intrinsic current of copper oxidation (i_{Cu}), while the oxygen reduction current remained largely unaffected.

To demonstrate that the aging of the surface appears to affect the current of copper oxidation but does not affect other electrochemical parameters (e.g., Tafel slopes of the reactions), it was instructive to examine the values of the slope of the linear best-fit correlations between E_{corr} and $\log(Cu_{total})$ data (Figure 5.23). The linear relationship between E_{corr} and $\log(Cu_{total})$ is defined only by a combination of the Tafel slopes of the respective electrochemical reactions of copper oxidation and oxygen reduction (Equation 5.17). Examination of the experimental data shows that the slopes of the best-fit correlations exhibit some variability for data obtained for weeks 1 through 12 of exposure of the CPSF super long coil copper cell to the synthetic water quality. However, variability is likely to be caused only by the normal scatter of pertinent experimental data and does not seem to indicate any consistent and significant changes of the intrinsic Tafel coefficients of the coupled electrochemical reactions. It can be concluded that despite the change of the corrosion rates associated with the aging of corroding surfaces, the relationship between E_{corr} and $\log(Cu_{total})$ remains linear. This statement is applicable presuming a completely uniform mixing of water inside the CPSF cell and the absence of chlorine species. The presence of chlorine species may dramatically alter the functional form of the relationship that correlates E_{corr} to $\log(Cu_{total})$ data.

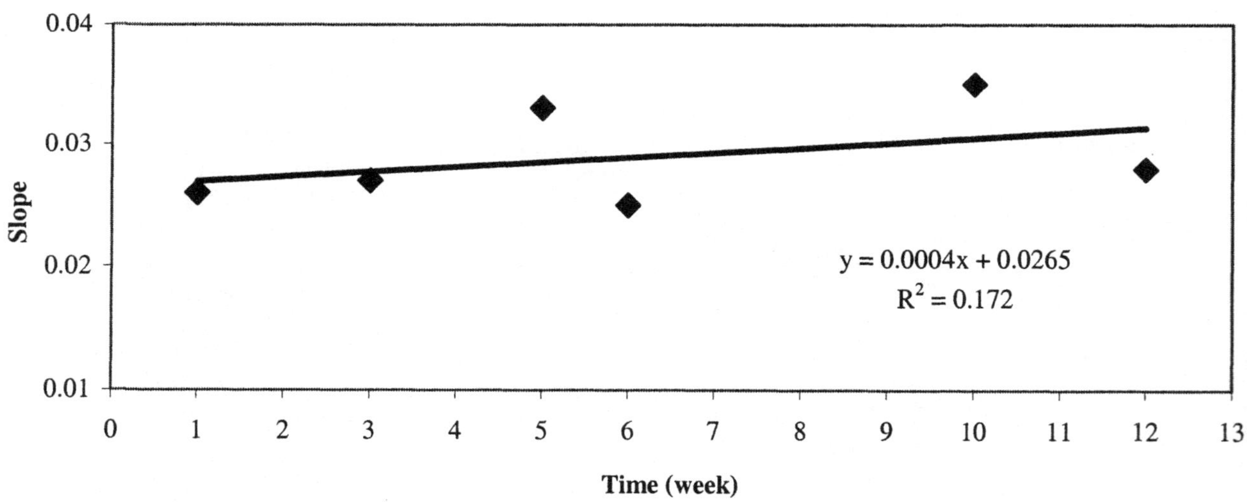

Figure 5.23 Values of the Tafel slopes of E_{corr} vs. $\log(Cu_{total})$ correlations as a function of exposure duration. Synthetic water quality, pH 7.5, alkalinity 15 mg/L, CPSF super long coil copper corrosion cell.

Effects of Water Quality Parameter Variations on E_{corr} vs. Cu_{total} Correlations

Almost universally, both corrosion rates and metal release increase significantly as pH decreases. Due to the importance of pH in corrosion control, laboratory experiments were carried out to examine whether pH variations would affect the functional type and strength of E_{corr} vs. $\log(Cu_{total})$ correlations. In these experiments, the synthetic water quality was used, pH was varied between 6.5 to 8.5, while the alkalinity was kept constant at 15 mg/L. During these tests, water was recirculated through the system and a 20-liter tank.

Representative results obtained in these experiments are presented in Figure 5.24. As pH was lowered, higher rates of copper release were observed and the values of E_{corr} increased. Overall, the correlations between E_{corr} and $\log(Cu_{total})$ remained highly linear at all pHs.

Analysis of the data shown in Figure 5.24 also appears to indicate that all the results obtained for varying pHs can be considered as one coherent dataset. Based solely on the data shown in Figure 5.24, it cannot be concluded that variations of pH do not affect the formal coefficients that characterize E_{corr} vs. $\log(Cu_{total})$ correlations. It is expected that prolonged exposures of copper to water with varying pHs will be accompanied by consistent changes of the positions of E_{corr} vs. $\log(Cu_{total})$ relationship. However, this was not verified with the performed experiments.

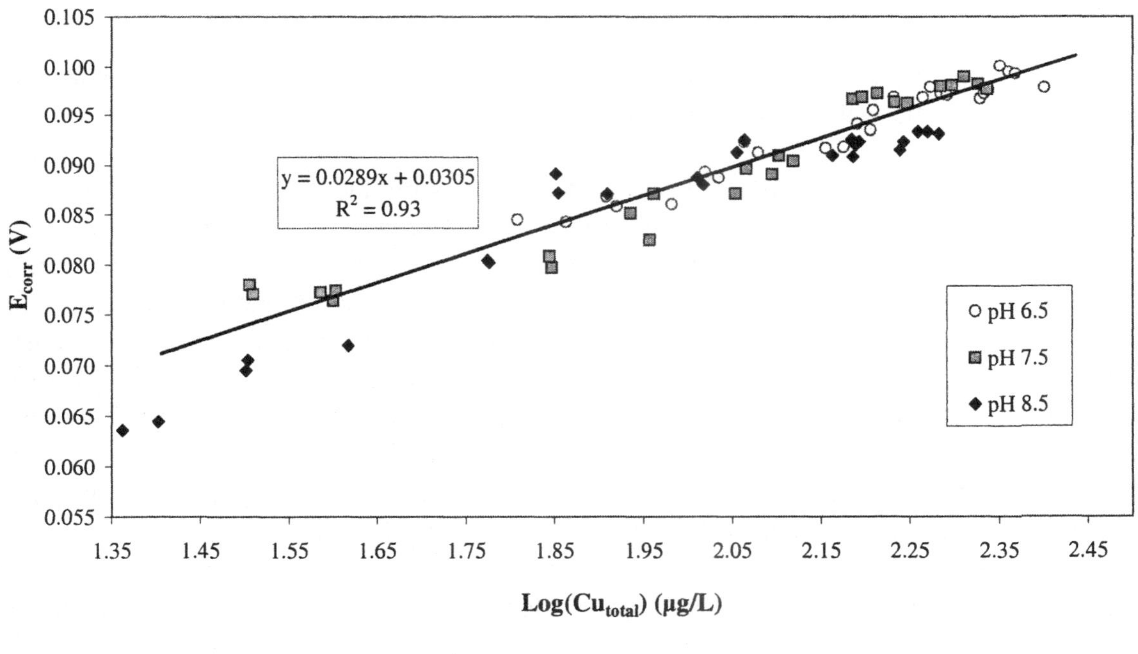

Figure 5.24 Effects of pH on correlations between E_{corr} and metal release. Synthetic water, alkalinity 15 mg/L. CPSF NRC cell (C-2).

It can be concluded from the research that E_{corr} and $\log(Cu_{total})$ data are strongly and linearly correlated at all tested pHs. The same conclusion can be made about the effects of the alkalinity. E_{corr} and $\log(Cu_{total})$ data obtained for the synthetic water quality at pH 7.5 with alkalinities of both 15 mg/L and 50 mg/L are compared in Figure 5.25. The data show that consistently lower E_{corr} values were observed at increased alkalinities. However, for both alkalinities there is a strong linear correlation between E_{corr} and logarithms of total soluble copper concentration.

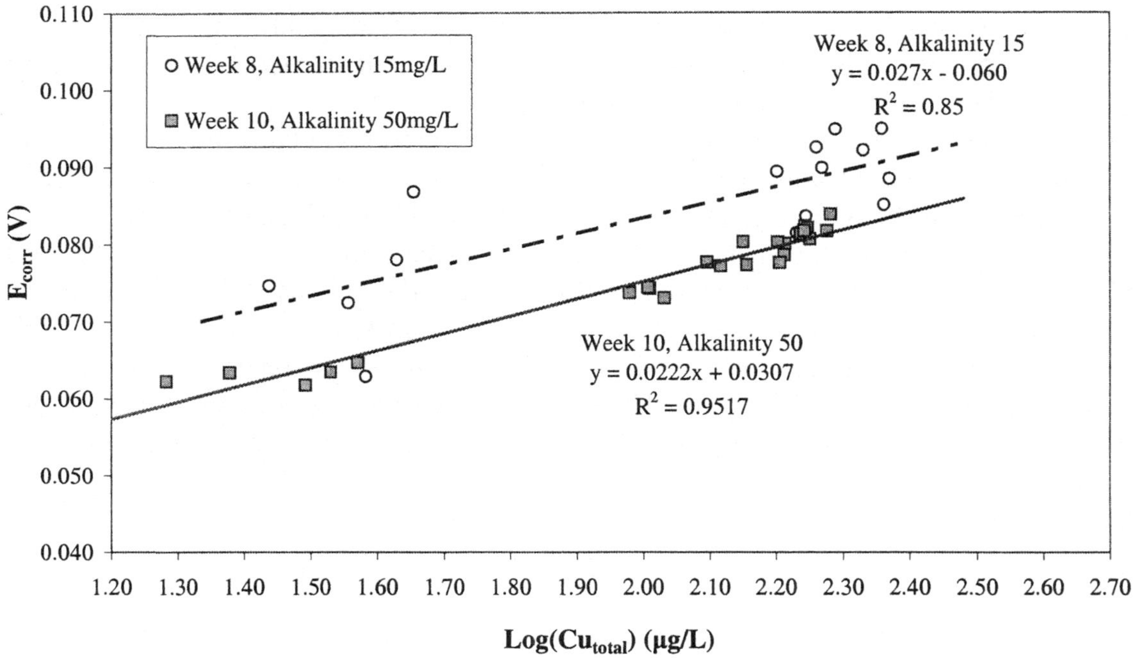

Figure 5.25 Effect of variations of alkalinity on the relationship between E_{corr} and total copper concentrations. Synthetic water, pH 7.5, CPSF NRC cell (C-3).

The observed cathodic shift of E_{corr} is associated with the formation of fairly strong complexes between Cu^{2+} and carbonate ion and ensuing decrease of the concentration of free copper ($[Cu^{2+}]$). However, detailed analysis of the actual E_{corr} values is necessary to determine whether the shift of E_{corr} is caused solely by the formation of the carbonate complexes of Cu^{2+}. Potentially the results may also be affected by changes of the intrinsic copper oxidation current caused by alterations to the surface state of the corroding metal in the presence of increased carbonate. Further calculations and long-term experiments are necessary to examine this issue in more detail. However, based on the data presented above, it was concluded that variations of both pH and alkalinity of drinking water do not affect the applicability of the most important result of the CPSF theory; the predicted linearity of E_{corr} and $\log(Cu_{total})$.

The characteristics of E_{corr} and metal release were also compared for the synthetic water quality and water shipped from Pinellas County, Florida. The major difference between their chemistries was that the water from Pinellas County had a much higher alkalinity (approximately 200 mg/L as $CaCO_3$) and hardness compared to those of the baseline synthetic water. Based on the experiments with varying alkalinities, it was anticipated that the values of E_{corr} in Pinellas County water would exhibit a cathodic shift but otherwise the correlation between E_{corr} and $\log(Cu_{total})$ would retain the same functional type. This indeed was seen in the experiments (Figure 5.26).

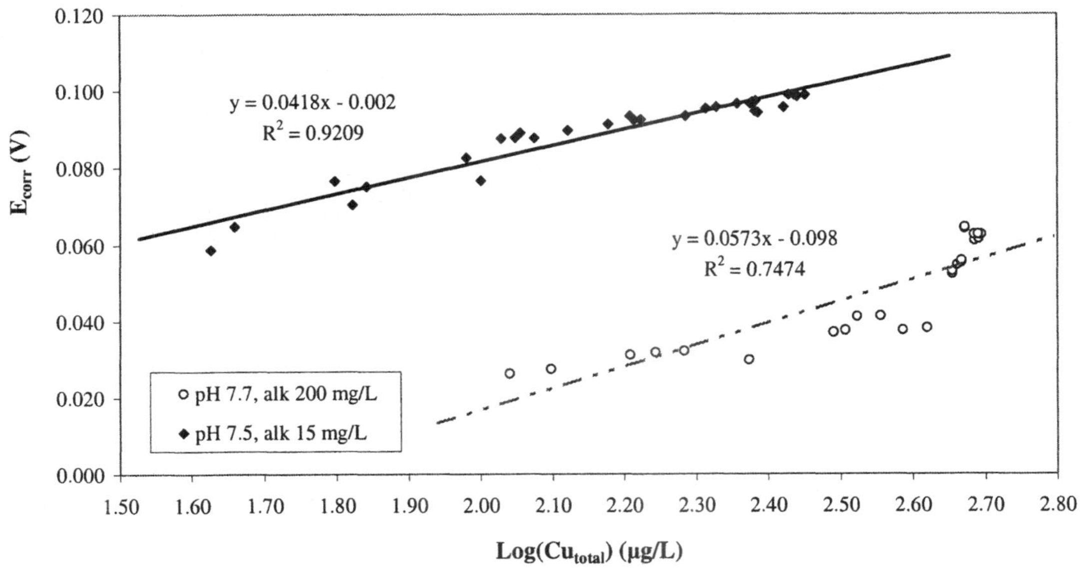

Figure 5.26 Comparison of E_{corr} and Cu_{total} data for the synthetic water quality (pH 7.5 alk. 15 mg/L) and Pinellas County, FL water (pH 7.7, alk. 200 mg/L).

Some significant differences were also identified. In general, the rate of metal release and the slope of the E_{corr} vs. log(Cu_{total}) correlation were somewhat higher compared to those in the synthetic water. It was also observed that in Pinellas County water, at the beginning of exposure, the values of E_{corr} tended to increase monotonically during stagnation. After some level of copper had accumulated in the water being recirculated, as was the case during Day 2 and later in any week-long lab experiment, the change of E_{corr} over time became non-monotonic and exhibited a maximum observed at varying times after the start of stagnation (Figure 5.27). At the same time, the rate of copper release from the cell gradually decreased as the week-long experiment progressed. This observation corresponded to a non-monotonic behavior of E_{corr} previously discussed.

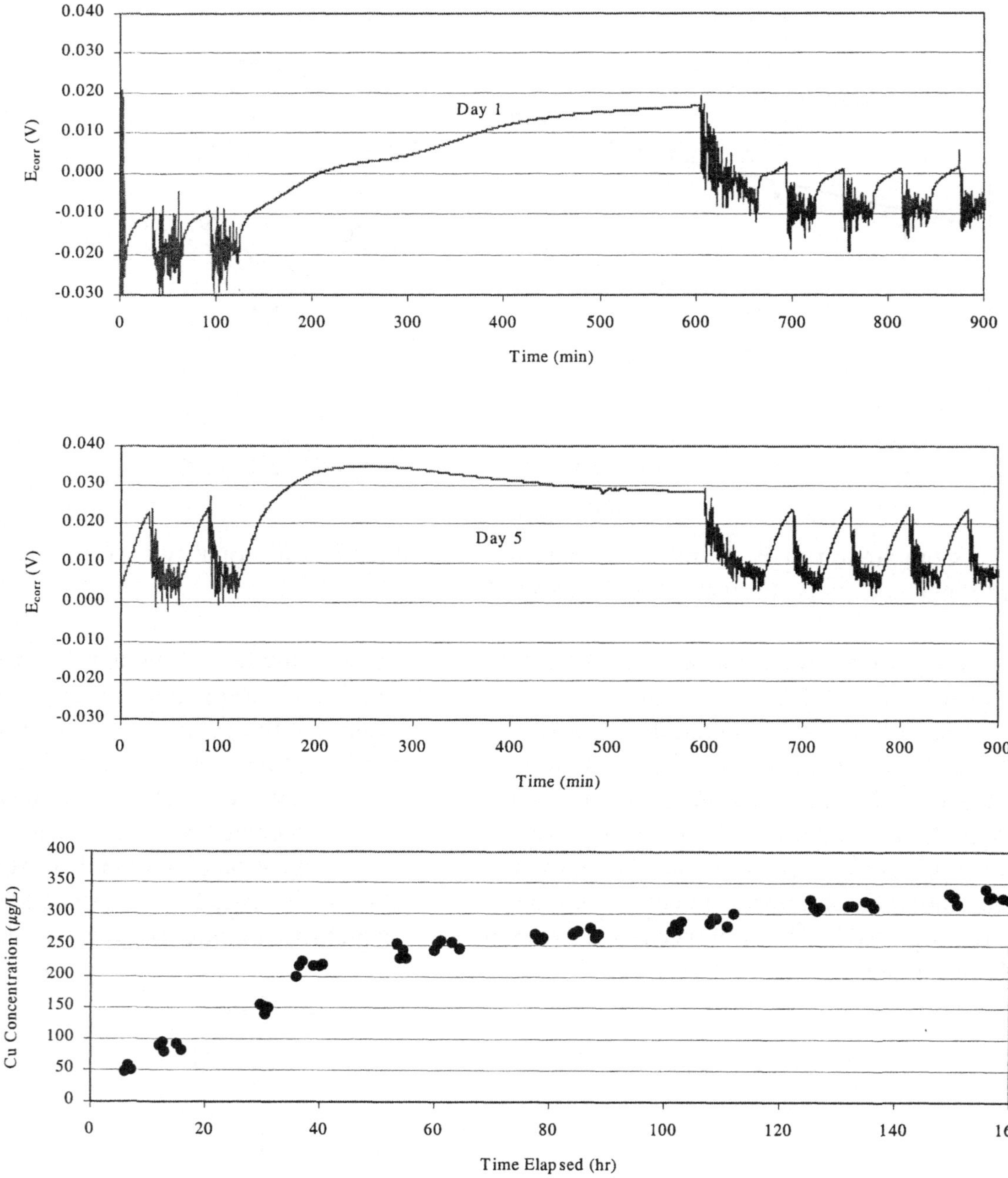

Figure 5.27 Changes of E_{corr} profile during stagnation and metal release during a week-long experiment with Pinellas County water. Second week of exposure, CPSF NRC cell (C-1).

It was hypothesized that in Pinellas County water, and in other water qualities, the non-monotonic behavior of E_{corr} over time during stagnation episodes could be considered to be a manifestation of two main processes. First, it was examined whether non-monotonic behavior could be related to a rapid decrease in the concentration of oxygen at the corroding surface. Experiments performed under conditions in which the rate of corrosion was expected to decrease but a non-monotonic behavior of E_{corr} was prominent, gave convincing evidence that the depletion of oxygen at the surface is not likely to be the main cause of the non-monotonic behavior of E_{corr}. Second, it was examined whether this type of behavior of E_{corr} could be caused by a rapid disappearance of Cu^{2+} ions due to the formation of solid phases (e.g., malachite or copper phosphates) or reactions of Cu^{2+} with Cu(0) leading to the formation of low-soluble species of Cu(I). Because the non-monotonic behavior of E_{corr} during stagnation episodes became progressively prominent with the increase of total copper concentration in the solution and with aged surfaces, it can be concluded that rapid formation of insoluble copper phases is the most likely cause of non-monotonic behavior. Preliminary modeling of the E_{corr} behavior during stagnation did not predict the non-monotonic behavior of the corrosion potential in the case of formation of solid phases. This indicates that the model needs further development.

Effects of Phosphate on the Corrosion Potential and Metal Release

Following experiments with varying pH and alkalinities, the effects of phosphate on metal release and E_{corr} were examined. In this phase of work, CPSF NRC cell (C-3) previously exposed to synthetic water (pH 7.5, alkalinity 15 mg/L as $CaCO_3$) was used. The concentrations of orthophosphate were varied from 0 to 2 mg/L (as PO_4). The cell was exposed to the synthetic water with a target phosphate concentration for a week, after which its concentration was increased to the next level.

The presence of phosphate caused significant changes in the rates of metal release and behavior of E_{corr}. Copper release decreased noticeably even at the lowest phosphate dose (0.5 mg/L). At the highest dose of phosphate (2 mg/L as $CaCO_3$), the decrease was several-fold (Figure 5.28). Simultaneously, the values of E_{corr} also decreased (Figure 5.29). This phenomenon was detected almost immediately following the addition of phosphate to the water.

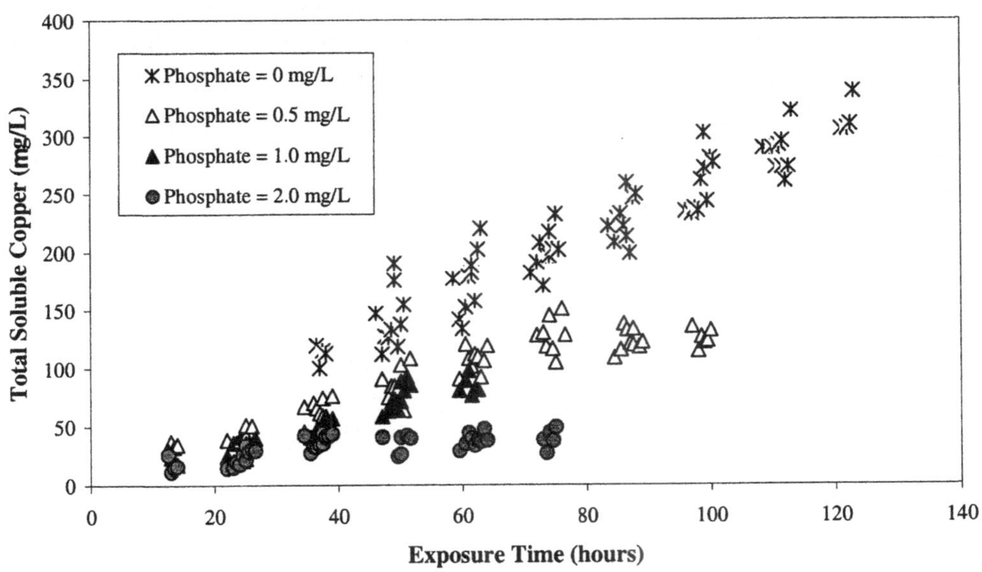

Figure 5.28 Effects of phosphate concentration on copper uptake from the cell and accumulation in the water tank. Week-long experiments, synthetic water quality (pH 7.5, alk. 15 mg/L), CPSF NRC cell (C-3).

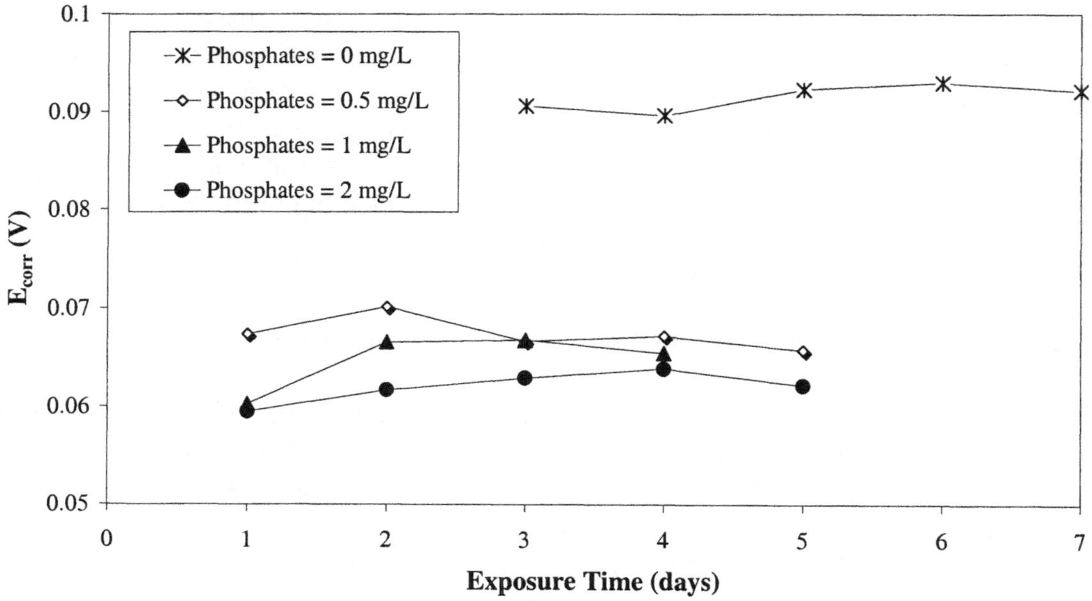

Figure 5.29 Effects of phosphate concentration on the value of corrosion potential. Week-long experiments, synthetic water quality (pH 7.5, alk. 15 mg/L as $CaCO_3$), CPSF NRC cell (C-3).

In addition to the general decrease of E_{corr}, its rate of change during stagnation also underwent changes. Instead of exhibiting a slow increase typical for fresh copper surfaces exposed to the synthetic water quality (without corrosion inhibitors and chlorine), the value of E_{corr} largely stabilized at stagnation times greater than approximately 20 minutes (Figure 5.30). Measurements of metal release in these conditions also indicated that no significant increment of copper concentration in the CPSF cell was observed for stagnation times exceeding that typical for E_{corr} stabilization. It was also evident that increasing phosphate concentration to more than 1.0 mg/L did not cause substantial changes of E_{corr}.

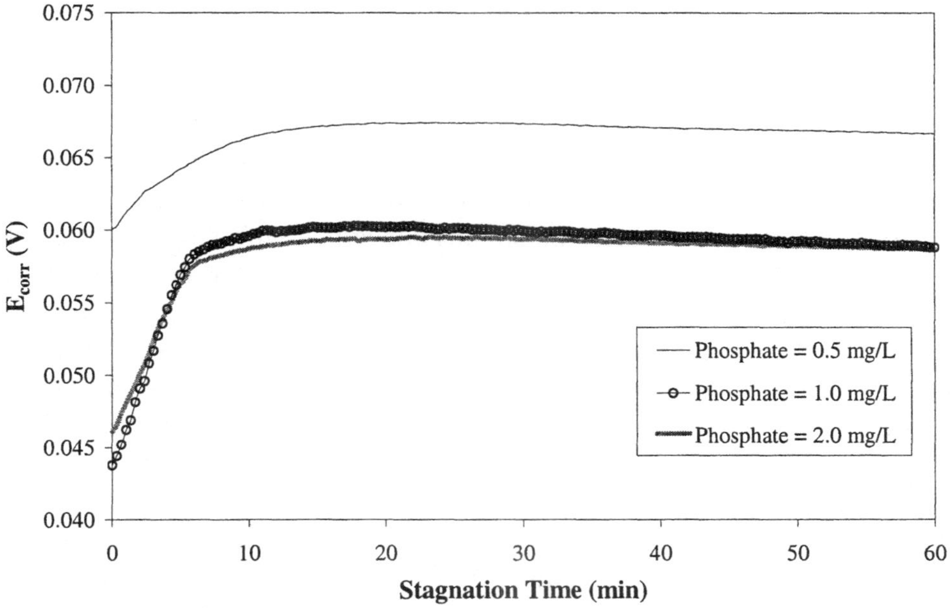

Figure 5.30 Effect of phosphate concentration on E_{corr} vs. time profiles for copper. Synthetic water quality, pH 7.5, alk. 15 mg/L, CPSF NRC cell (C-3).

The correlations between E_{corr} and copper concentrations were highly sensitive to the presence of phosphate (Figure 5.31). The most prominent shift of E_{corr} occurred when 0.5 mg/L phosphate was added to the synthetic water quality. Further increase of its concentration did not seem to affect the position of the E_{corr} and $\log(Cu_{total})$ relationship. There is no doubt that the concept of linear relationship between E_{corr} and $\log(Cu_{total})$, at the beginning of stagnation episodes, is as applicable for corrosion in the presence of phosphate as it is in all other cases in the absence of chlorine.

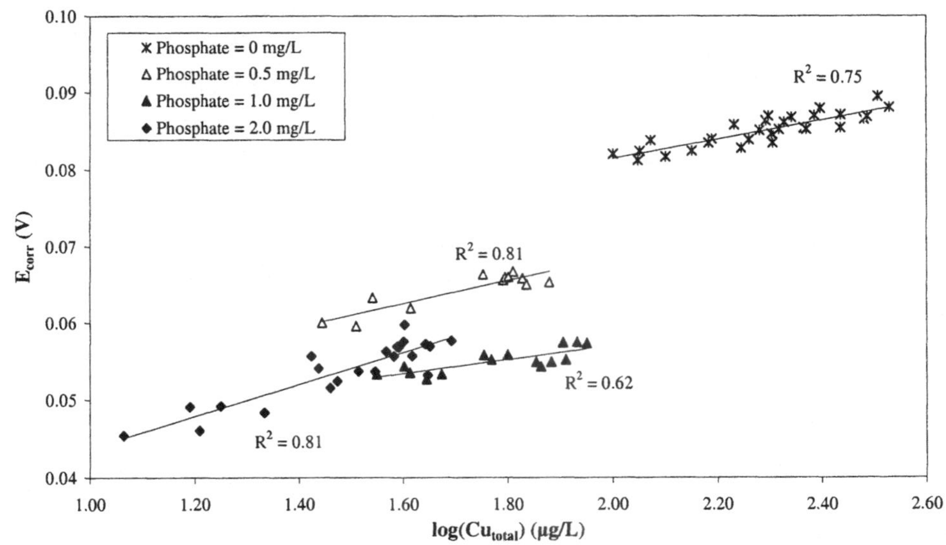

Figure 5.31 Effect of phosphate concentration on correlations between E_{corr} and Cu_{total} data. Synthetic water quality (pH 7.5, alk. 15 mg/L as $CaCO_3$) CPSF NRC cell (C-3).

Effect of Chlorine on the Behavior of E_{corr} and Correlations Between Total Copper and E_{corr} Values

The initial approach utilized in the development of the CPSF method implied that chlorine species, due to their low molar concentration compared to that of oxygen, are unlikely to be "major" players that define the rate of metal release. This approach is largely based on existing literature and results of corrosion studies concerned with elucidation of effects of water quality parameters (pH, alkalinity, phosphate) on the rate of corrosion and metal release. Thus, the discussion has so far been concerned with the results generated in the absence of chlorine.

Data obtained immediately following the start of the field experiments indicated that chlorine species, due to their distinctly different electrochemical properties (most importantly, higher standard electrochemical potential of the $HOCl/Cl^-$ couple) cause pronounced changes of E_{corr} and patterns of its behavior during stagnation episodes. This was further validated by analysis of the data obtained during the preliminary phase of the CPSF concept development.

For laboratory experiments designed to explore the effects of chlorine, synthetic water was used with a pH of 7.5 and an alkalinity of 15 mg/L. No phosphate and/or ammonia were added. In all cases, the concentration of chlorine was increased gradually from 0 mg/L to as high as 1.5 mg/L. For each concentration of chlorine, E_{corr} measurements were carried out for three days. In addition to CPSF measurements, the rate of copper release was also measured for stagnation times varying between 1 to 60 minutes. The measurements of copper release were carried out in flow-through mode and the water was not recirculated. All experiments on the effects of chlorine were carried out using CPSF NRC cell (C-2).

The measurements showed that the presence of even small amounts of chlorine caused considerable changes in the behavior of E_{corr} and its correlation with copper release. The profiles of E_{corr} over time for chlorine concentrations from 0 mg/L to 0.15 mg/L are shown in Figure 5.32, while for higher chlorine doses they are shown in Figure 5.33. Inspection of these figures indicates that in the presence of even 0.1 mg/L of chlorine, the typical trend of E_{corr} behavior observed for all water qualities in the absence of chlorine, is reversed. In the absence of chlorine, the value of E_{corr} tends to increase immediately after the start of a stagnation episode. This is caused by the accumulation of copper (II) species inside the CPSF cell (provided that the oxygen concentration at the surface remains largely unaffected). In contrast to that, in the presence of chlorine, the value of E_{corr} decreases immediately after the start of stagnation. This trend becomes increasingly prominent with the increase of free chlorine concentrations and can be seen by comparing the E_{corr} data shown in Figure 5.32 and Figure 5.33. At the same time, the value of E_{corr} behaves non-monotonically; it increases for low chlorine doses but decreases for chlorine doses greater than 0.25 mg/L.

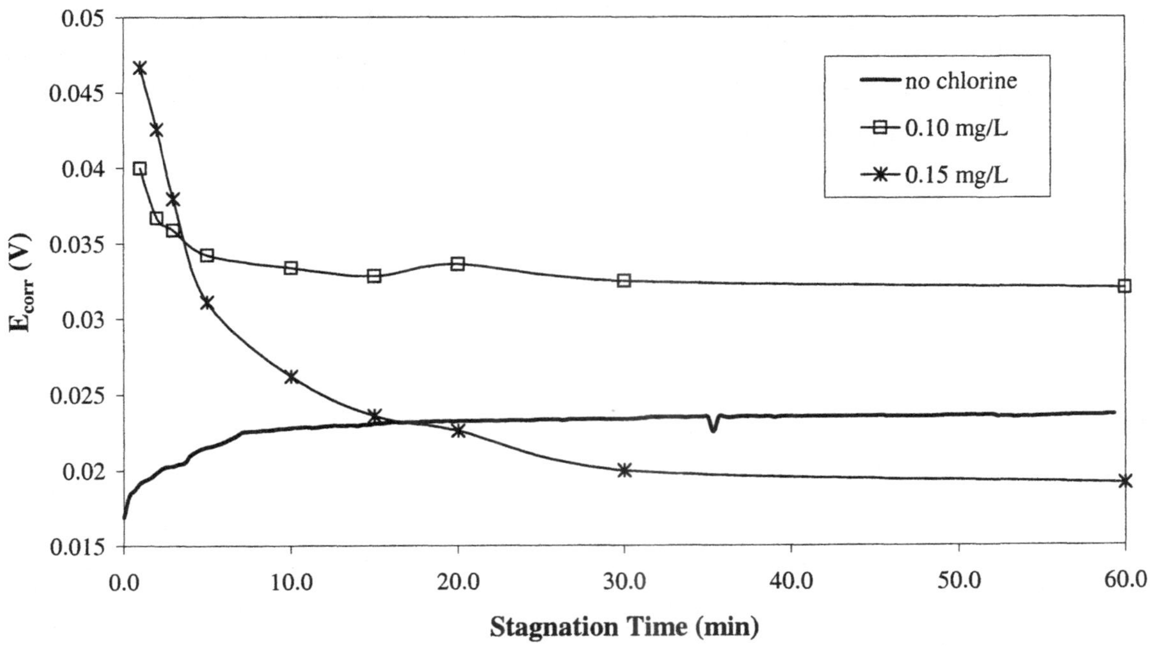

Figure 5.32 Dependence of E_{corr} vs. time in the absence of chlorine and at free chlorine concentrations 0.10 mg/L and 0.15 mg/L.

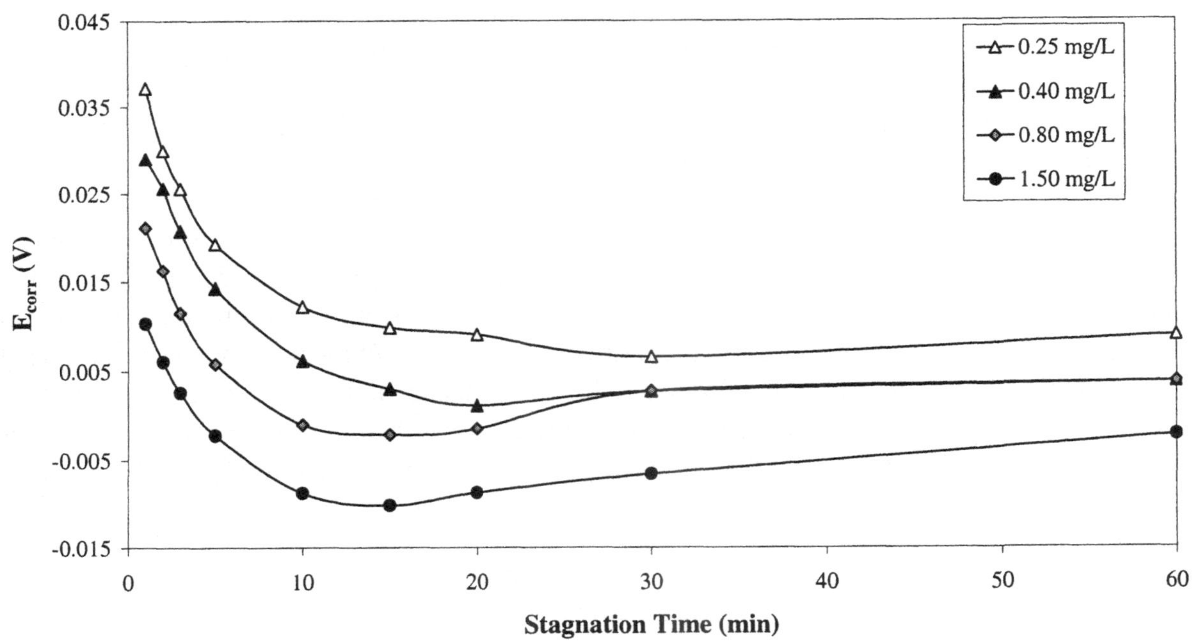

Figure 5.33 Dependence of E_{corr} vs. time at free chlorine concentrations varying from 0.25 mg/L to 1.50 mg/L.

The rate of change of E_{corr} immediately after the start of stagnation is complex and depends on the chlorine concentration. The magnitude of change of E_{corr} (within the first 5 minutes of stagnation) reached a maximum when the chlorine concentration was 0.25 mg/L and decreased somewhat at higher free chlorine doses (Figure 5.34). In the absence of chlorine, the magnitude of change of E_{corr} during the first 5 minutes of stagnation was minimal. Additionally, the sign of E_{corr} increased with time when chlorine was not present, while it decreased with time for all tested concentrations of chlorine.

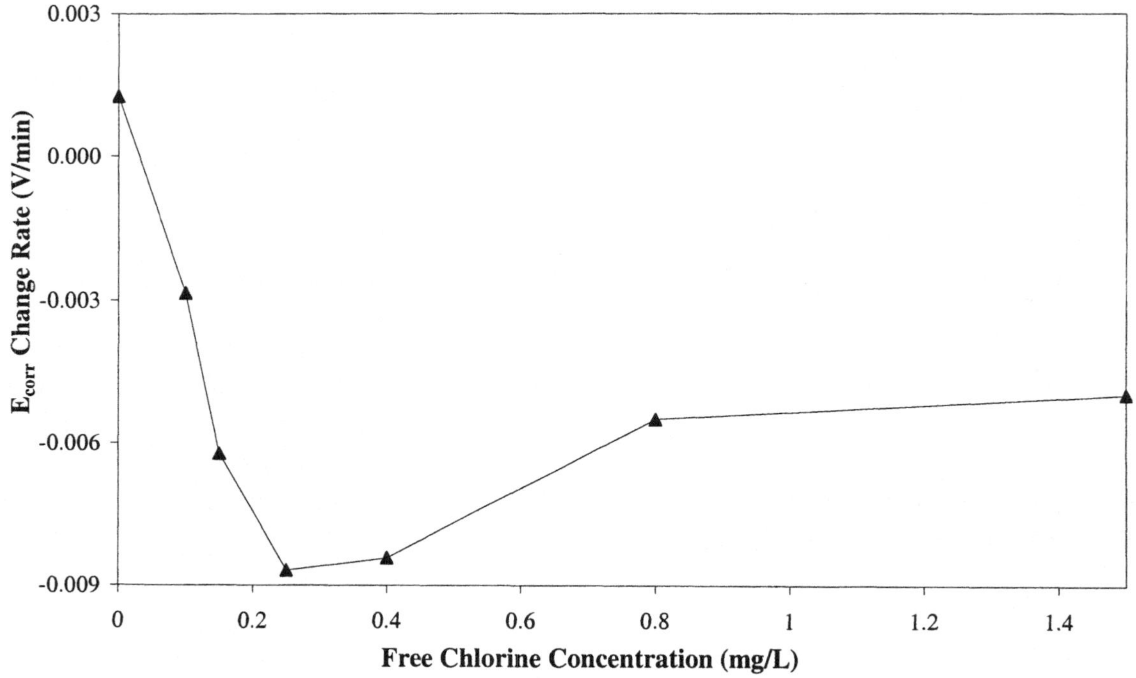

Figure 5.34 Effects of chlorine concentrations of the rate of E_{corr} change within first 5 minutes of stagnation. Laboratory CPSF experiments.

Similar to the phenomena observed in the behavior of E_{corr}, the rate of copper release in the CPSF cell was also notably affected by free chlorine (Figure 5.35). In all cases, the concentration of copper increased with time. The copper concentrations in samples withdrawn from the corrosion cell were almost three times higher (at a chlorine dose of 1.5 mg/L) than samples taken in the absence of chlorine. The dependence of the rate of copper release and chlorine concentration was somewhat complex; it had a local maximum at a Cl_2 concentration of 0.4 mg/L, was somewhat less at 0.8 mg/L and again higher at Cl_2 concentration of 1.5 mg/L.

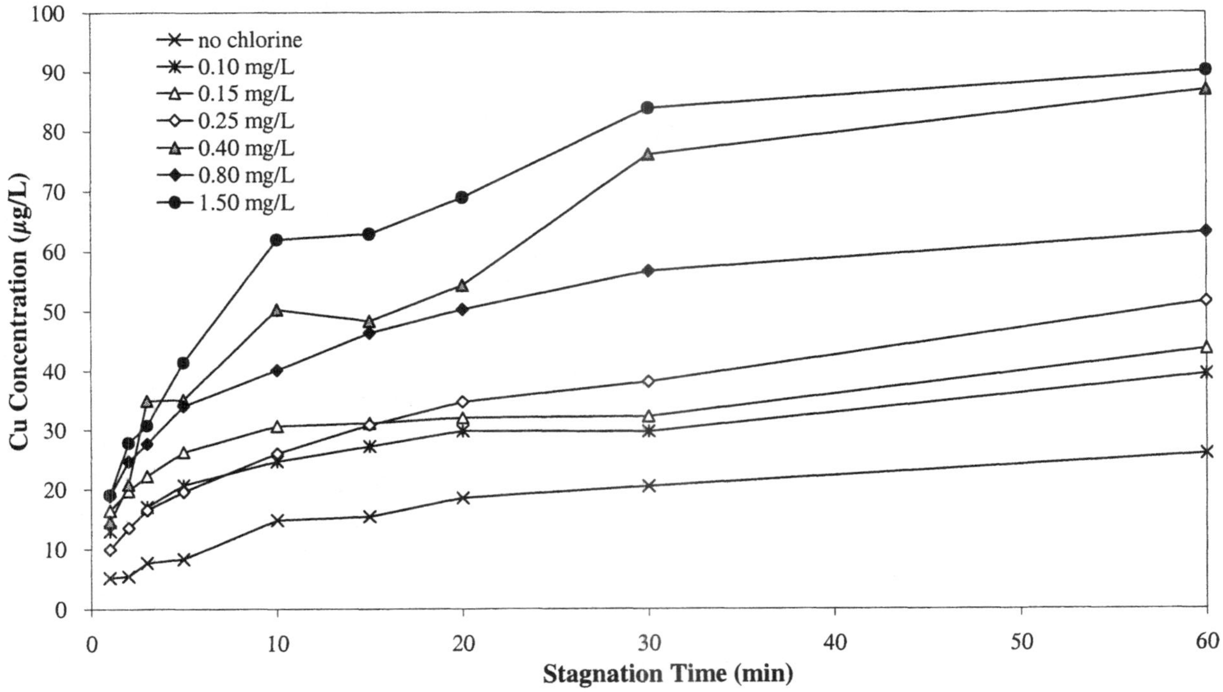

Figure 5.35 Copper release vs. stagnation time at varying free chlorine doses.

Despite the significant effects on the behavior of E_{corr} that are associated with the presence of chlorine, a strong linear correlation between E_{corr} and $\log(Cu_{total})$ was observed for samples withdrawn from the CPSF cell at varying stagnation times (Figure 5.36 to Figure 5.38). The relevant regression coefficient (R^2) values varied 0.74 to 0.92 (Table 5.4).

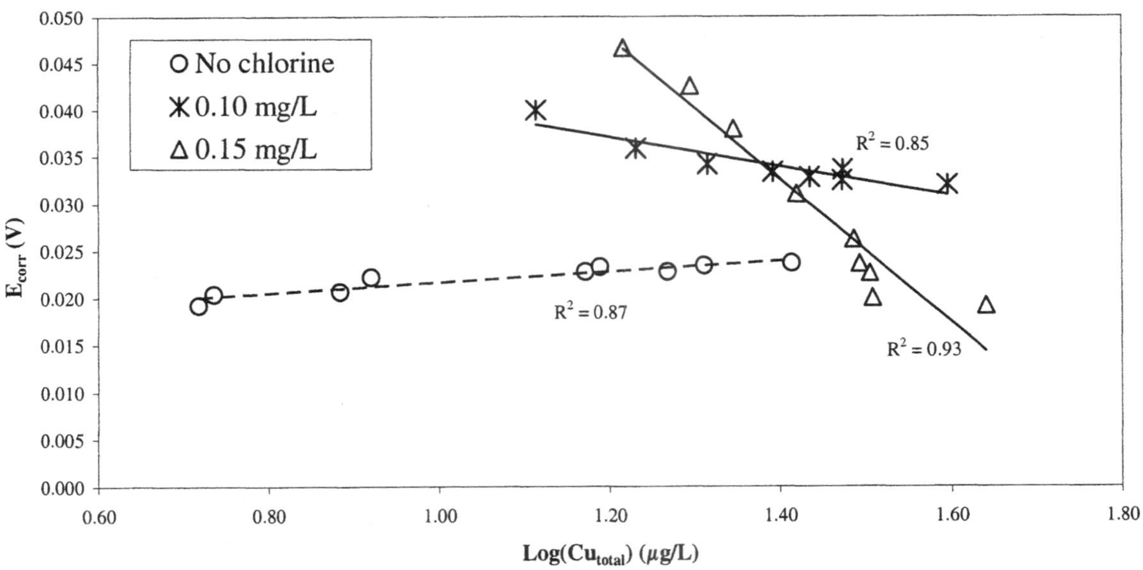

Figure 5.36 E_{corr} and Cu_{total} relationship in the absence of chlorine or presence of low chlorine concentrations. From laboratory CPSF experiments.

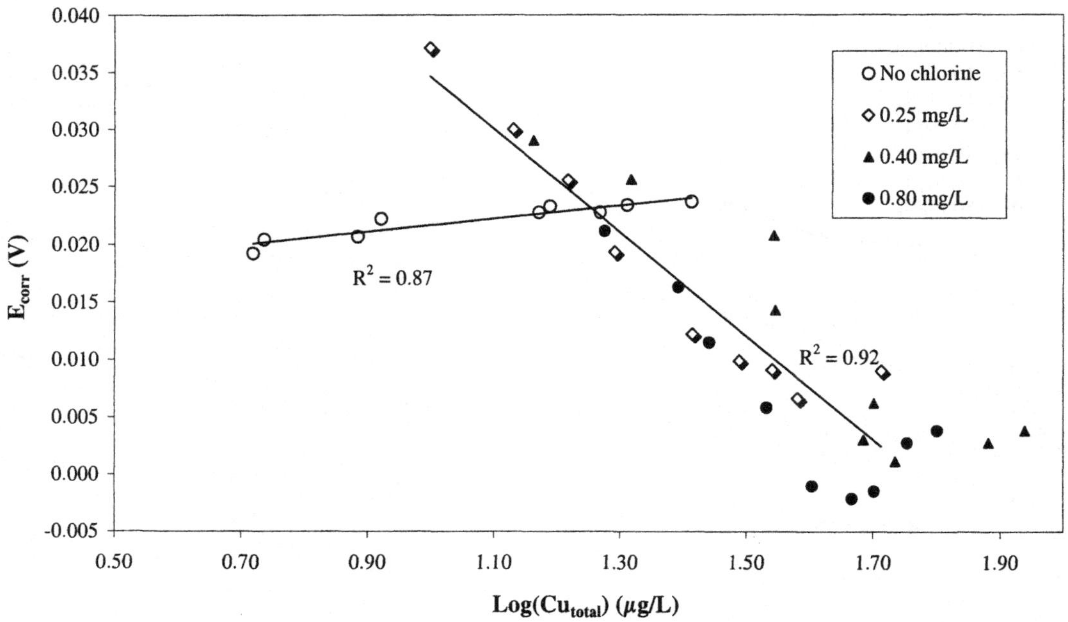

Figure 5.37 Relationship between E_{corr} and Cu_{total} values in the absence of chlorine and chlorine concentrations between 0.25 and 0.8 mg/L. Laboratory CPSF experiments.

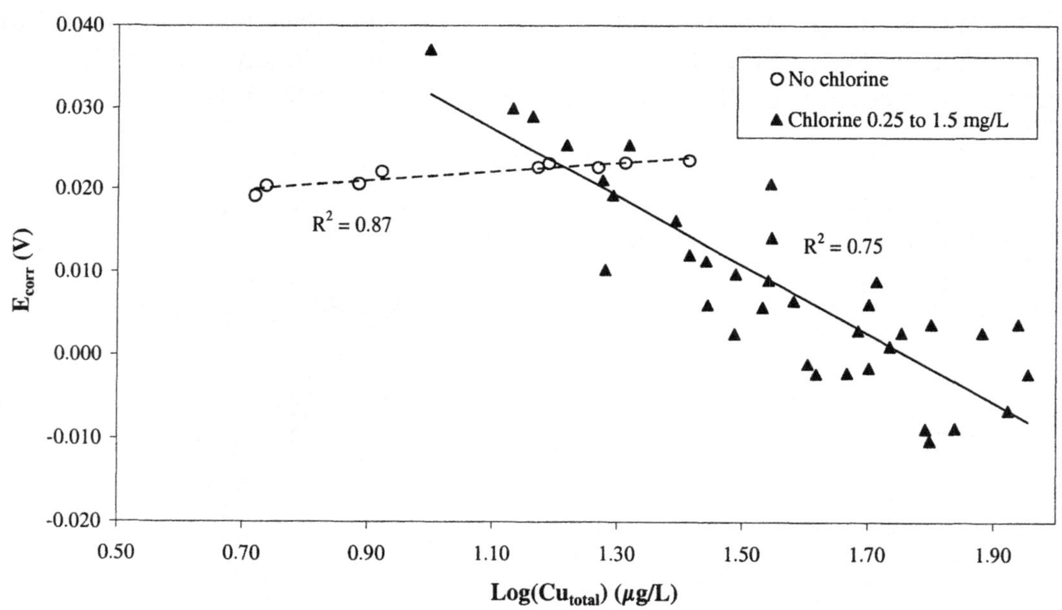

Figure 5.38 Cumulative relationship between E_{corr} and Cu_{total} values for all chlorine concentrations ranging from 0.25 to 1.5 mg/L. Laboratory CPSF experiments

Table 5.4
Linear fit parameters for E_{corr} and Cu_{total} data obtained for CPSF experiments at varying free chlorine concentrations

Chlorine concentration	Factor A	Factor B	R^2
No chlorine	153.5	-2.3	0.87
0.10 mg/L	-54.8	3.3	0.85
0.15 mg/L	-12.3	1.8	0.93
0.25 mg/L	-20.3	1.7	0.92
0.40 mg/L	-21.3	1.9	0.85
0.80 mg/L	-18.3	1.7	0.74
1.50 mg/L	-27.9	1.6	0.74
All data for chlorine 0.15 to 1.50 mg/L	-13.0	1.7	0.65
All data for chlorine 0.25 to 1.50 mg/L	-18.1	1.7	0.75
All data for chlorine 0.25 to 0.80 mg/L	-19.8	1.8	0.80

The nature of this relationship was different in the absence and presence of chlorine. In the absence of chlorine, an increase of E_{corr} was associated with an increase in copper concentration, as predicted by the general CPSF theory. In contrast copper concentration increased when E_{corr} decreased when chlorine was present.

The phenomenon of a consistent decrease of the corrosion potential, during stagnation times, in the presence of chlorine could be qualitatively reproduced in numerical calculations presuming that chlorine species at the electrode surface are significantly depleted soon after the start of stagnation. This indeed can easily happen because the initial molar concentration of chlorine was approximately 225 to 15 times less than that of oxygen (for Cl_2 concentrations 0.1 mg/L and 1.5 mg/L, respectively). The current status of the CPSF theory is not advanced enough to carry out quantitative modeling of E_{corr} behavior in the presence of both oxygen and chlorine. Therefore, in this report we will interpret the observed data on a phenomenological basis.

First, the analysis of correlations between copper concentrations found in the CPSF cell at varying duration of stagnation and at varying chlorine doses shows that the $\log(Cu_{total})$ values can be well correlated with the corresponding E_{corr} values using the following linear expression (Equation 5.2):

$$\log(Cu_{total}) = A\,E_{corr} + B \qquad (5.2)$$

This expression is similar to that observed in the presence of oxygen only, except that the direction of E_{corr} during stagnation of chlorinated water is opposite.

Experimental values for Equation 5.2 are given in Table 5.4; factors A and B, R^2 values varying chlorine doses, and linear fitting parameters for extended ranges of chlorine doses. Figure 5.39 graphically shows the values of factors A and B.

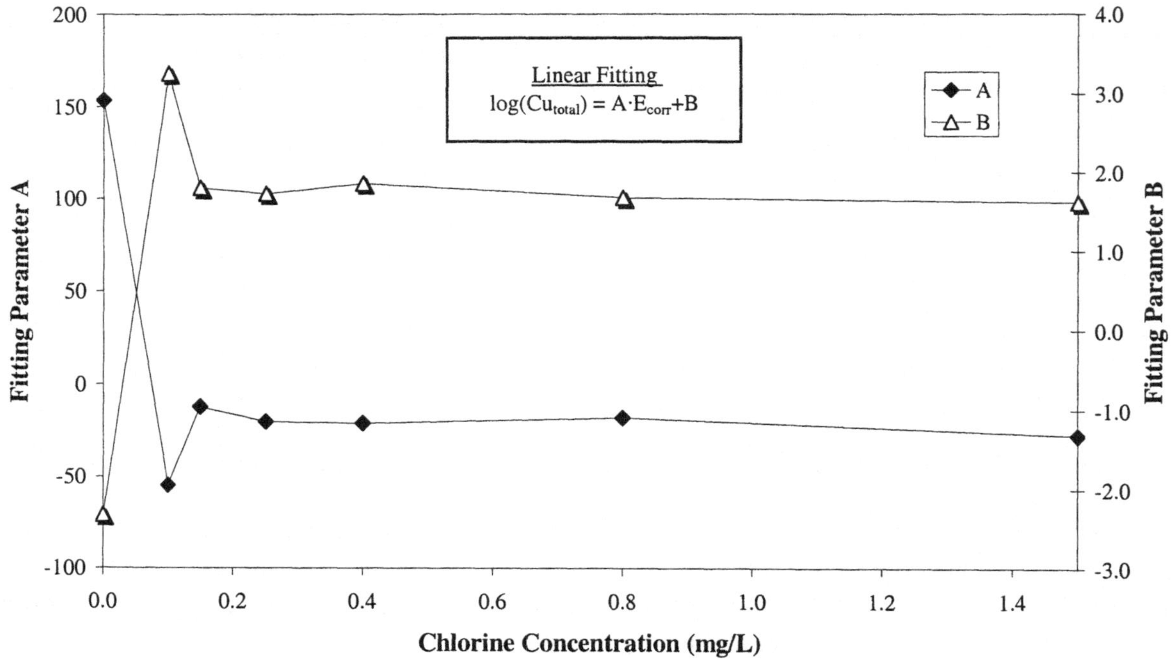

Figure 5.39 Dependence of linear fitting parameters vs. chlorine dose.

These calculations indicate that for the synthetic water quality used in the bench experiments the measured copper concentrations in the CPSF cell at varying stagnation times and chlorine concentrations (varying from 0.25 to 1.5 mg/L) can be correlated with the change of the corrosion potential using basically one best-fit expression. The applicability of this approach to field studies in which water qualities containing chlorine species were used is in the next section.

PREDICTIVE TECHNOLOGY FIELD STUDIES

Introduction

The field studies summarized in this chapter were designed to provided an alternative monitoring technique in a utility setting. The results of these studies were integral in determining the real-world applications and the development of final recommendations for this report.

The final version of the corrosion cell described in this chapter was field tested at two participating utilities and a local laboratory using tap water. The objective of the field studies was to test the viability of the technology in identifying features of the behavior of corrosion potential (E_{corr}) in drinking water containing residual chlorine, identify trends of metal release, and explore correlations between total metal concentration and E_{corr} values.

The following describes the field tests that occurred at each location, and presents the overall results and findings from this effort. Detailed descriptions and results from each of the three field studies are found in Appendix D.

Approach

Field tests of the Corrosion Potential Stagnation/Flow (CPSF) Narrow Rectangular Channel (NRC) corrosion cells were carried out at three sites; the San Francisco Water Utilities' Millbrae Water Lab, the Pinellas County Kellar Water Treatment Plant in Florida, and HDR Water Research Laboratory in Bellevue, WA. The relevant water quality parameters of each location are summarized in Table 5.5.

Table 5.5
Average water quality parameters for CPSF NRC cell* field tests

Parameter	Synthetic Water (Control)	Pinellas County Water	HDR Laboratory Tap Water	San Francisco Water
pH	6.5 – 8.5	7.7	6.5-8.5	8.6
Alkalinity (mg/L as $CaCO_3$)	15 - 100	200	16	20
Hardness (mg/L as $CaCO_3$)	0	200	24	20
Chloride (mg/L)	15	40	3	5
Sulfate (mg/L)	5	18	1	7
Total chlorine (mg/L as Cl_2)	0-1.5	3.0	1.0	0.8
Phosphate (mg/L as PO_4)	0-2.0	0	0	0
Conductivity (μmhos/cm)		420	60	83
Chlorine Residual (mg/L)	0	3.7 ± 0.2	1.0	0.9 ± 0.1

*Pinellas County water and select HDR laboratory experiments were performed using CPSF NRC cell C-1, while HDR Laboratory and San Francisco water experiments were performed using CPSF NRC cell C-4.

CPSF NRC cells, developed in the research laboratory of the University of Washington and detailed previously in this Chapter, were used at all three field testing locations. Cell C-4 was used in experiments carried out in San Francisco and in Bellevue, while Cell C-1 was used in Pinellas County and during select HDR laboratory experiments. Both cells were previously tested using the synthetic water quality and tap water at the University of Washington laboratory prior to their field deployment.

The experimental setup used for field E_{corr} measurements using the CPSF NRC cell is shown in Figure 5.40. At each test site, drinking water was allowed to flow continuously through the corrosion cell, at a flow rate of 1 L/min. The on/off condition of the flow was initiated using an automated controller. Electrochemical potential data were acquired using a digital multimeter, which was connected to a laptop computer to store the data.

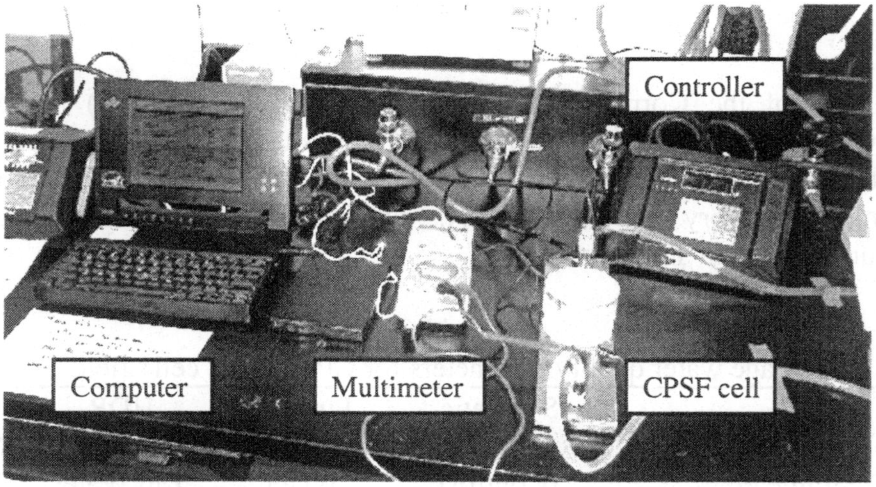

Figure 5.40 General view of the experimental setup for CPSF NRC cell field measurements.

The cell installation, data collection and data processing for these experiments were performed with the help of the personnel from San Francisco Water Utilities, Pinellas County and HDR Inc.

The E_{corr} data obtained during the first 20 to 60 minutes after the start of stagnation in chlorine-containing water qualities are most indicative and easy to interpret in the framework of the CPSF approach. As a result, data obtained during the first 20 to 60 minutes of corrosion reactions will be the focus of most discussions. Although the current level of the CPSF method development does not provide for quantifying the rate of metal release for any duration of stagnation, it is theorized that the rate of metal accumulation immediately after the start of corrosion is indicative of the corrosivity of that water. Properly calibrated CPSF data could potentially be used to predict metal concentrations at the end of stagnation periods longer than those to be used in this discussion.

Summary of Findings

E_{corr} Behavior

Three significant features were identified in field observations of E_{corr} during episodes of stagnation in chlorinated water:

- Following the initiation of a stagnation period, there is a rapid decrease (relaxation) in the value of E_{corr},
- As the stagnation time increases, so do copper concentrations, and
- There is a time at which, after a period of stagnation, E_{corr} will begin to slowly increase.

In the presence of residual chlorine, E_{corr} undergoes a rapid decrease after the initiation of stagnation. This was observed for all water qualities and is most likely associated with chlorine consumption in corrosion reactions, which causes the disappearance of chlorine at the metal surface. The relaxation of E_{corr} at the beginning of stagnation in HDR tap water (Day 15) is shown in Figure 5.41, and is a typical pattern for chlorinated water. The figure also shows that the relaxation of E_{corr} largely coincided with a rapid accumulation of copper in the cell. Copper concentrations in Figure 5.41 were sampled at the end of each stagnation period observed on Day 15.

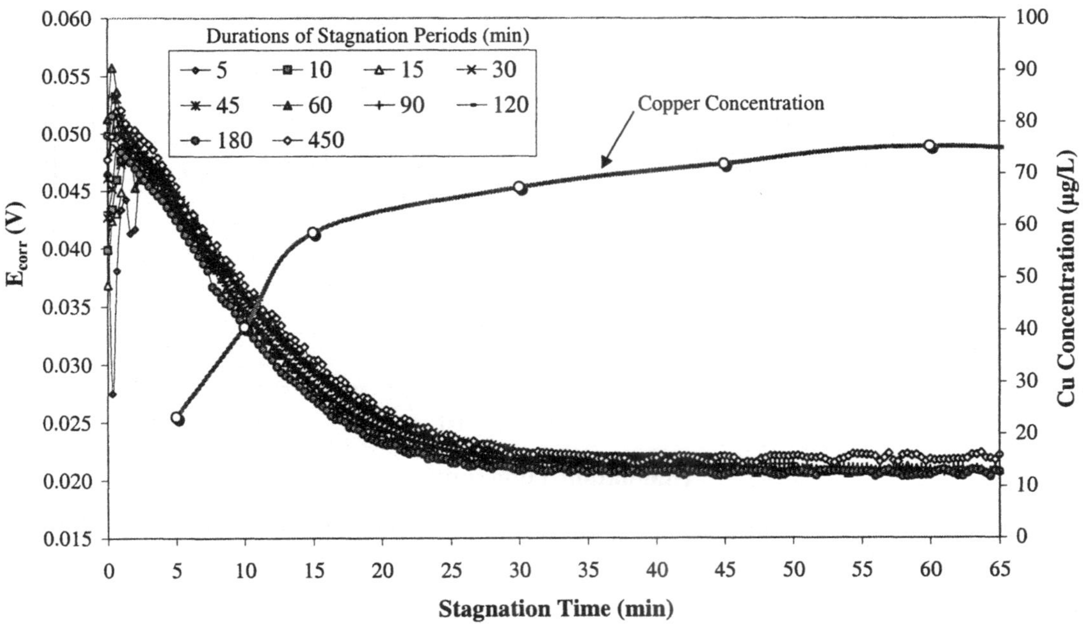

Figure 5.41 E_{corr} and metal release patterns for varying stagnation periods. Day 15, HDR water, CPSF NRC cell.

The relaxation of E_{corr} after the start of stagnation was also observed in San Francisco drinking water (Figure 5.42). The initial rapid decrease of E_{corr} lasted for about 60 to 90 minutes and was followed by a slow increase of E_{corr}, a feature which was seen in the data of other water qualities.

San Francisco water also had a relationship between the relaxation of E_{corr} and copper release in the CPSF cell (Figures 5.41 and 5.42) that was similar to that observed for HDR tap water. The copper concentrations in San Francisco water were considerably lower than for HDR tap water (and all other tested water qualities), which is in agreement with basic water quality parameters in those sources; HDR tap water had a lower alkalinity and pH compared to those of San Francisco water. Additionally, this could have been caused by the transient state of the surface and possible deficiencies in preparation of sample for metal analysis. Nonetheless, the position of the data and the similarity of the overall trends support the conclusion that the CPSF methods can be utilized to quantify, in situ, the rate of metal release, especially for short stagnation times.

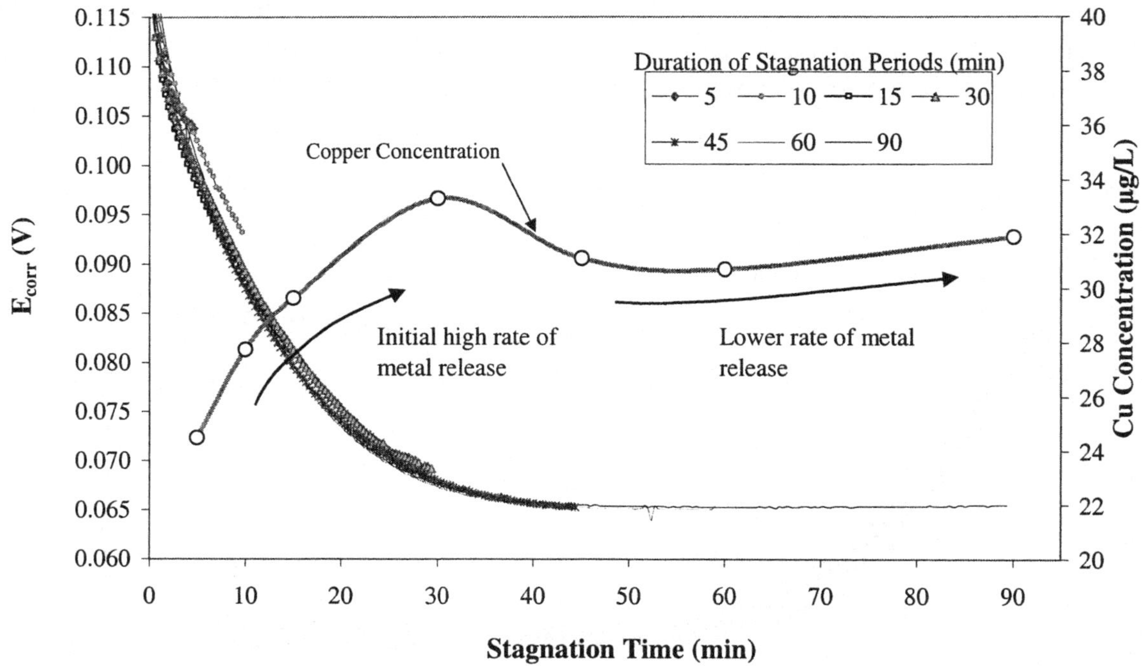

Figure 5.42 E_{corr} relaxation and metal release patterns for varying stagnation periods. Day 2, San Francisco water, CPSF NRC cell.

Pinellas County water had a much higher concentration of chlorine than the other water qualities tested. However, the major features seen in the behavior of E_{corr} during stagnation episodes in Pinellas County drinking water were still similar to those in other field tests. The value of E_{corr} rapidly decreased following the initiation of stagnation (Figure 5.43). Only for stagnation times longer than three hours was some increase of E_{corr} seen. The increase in E_{corr} was associated with the predominance of the oxygen-driven reactions (as opposed to mainly chlorine-driven copper oxidation at shorter times).

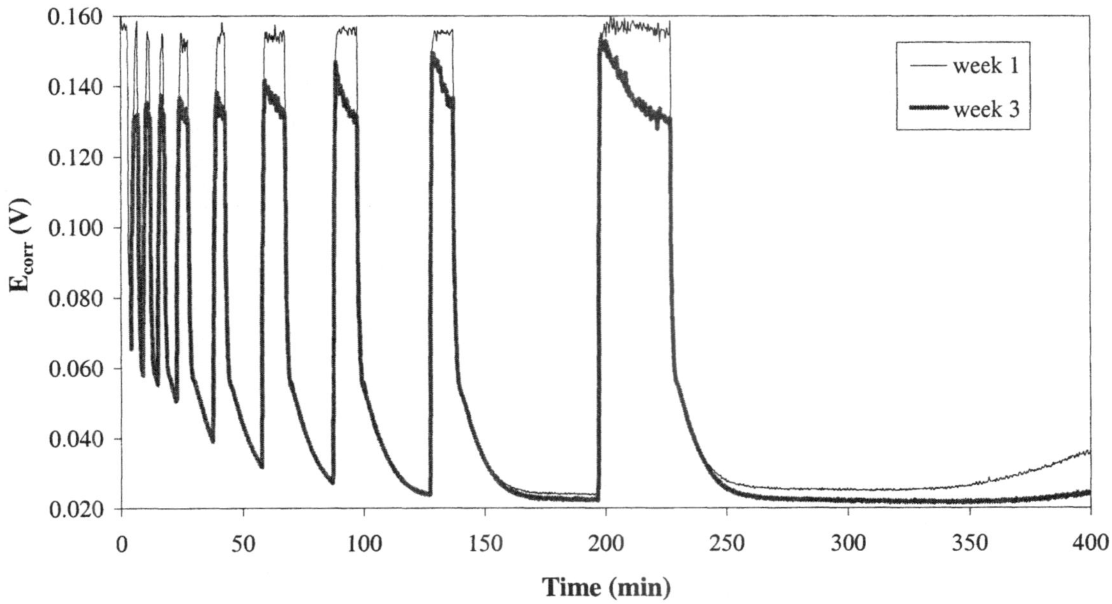

Figure 5.43 E_{corr} data profiles at varying stagnation times for week one and three of exposure. Pinellas County water, CPSF NRC cell.

Generally, Pinellas County water experiments were performed for stagnation times between one minute and 180 minutes. However, on the first day of testing, one experiment with a stagnation time of 480 minutes was performed. The first and last days were the only days on which samples were collected for stagnation times of less than 30 minutes. Direct measurements of copper release in samples with stagnation times of less than 30 minutes showed two main features (Figure 5.44): the aging of the surface for almost two months caused only minor changes in the rate of copper accumulation in the CPSF cell; and the kinetics of copper accumulation in the corrosion cell was somewhat unusual. A substantial concentration of copper was found in the cell for stagnation times as low as one minute. Copper corrosion increased with time but the comparison of copper concentrations determined for 30 and 480-minute stagnation times showed only a 25% increment. The rate of copper release observed in the CPSF cell was notably higher in Pinellas County water compared with that seen in the experiments carried out in the University of Washington laboratory (Figure 5.45).

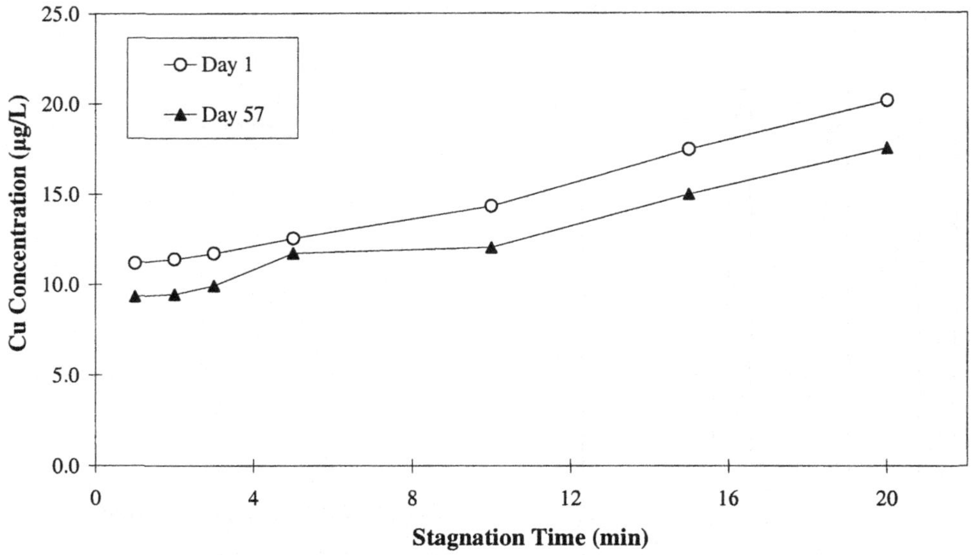

Figure 5.44 Comparison of copper accumulation on Day 1 and Day 57 of experiments. Pinellas County water, 1-liter samples, CPSF NRC cell.

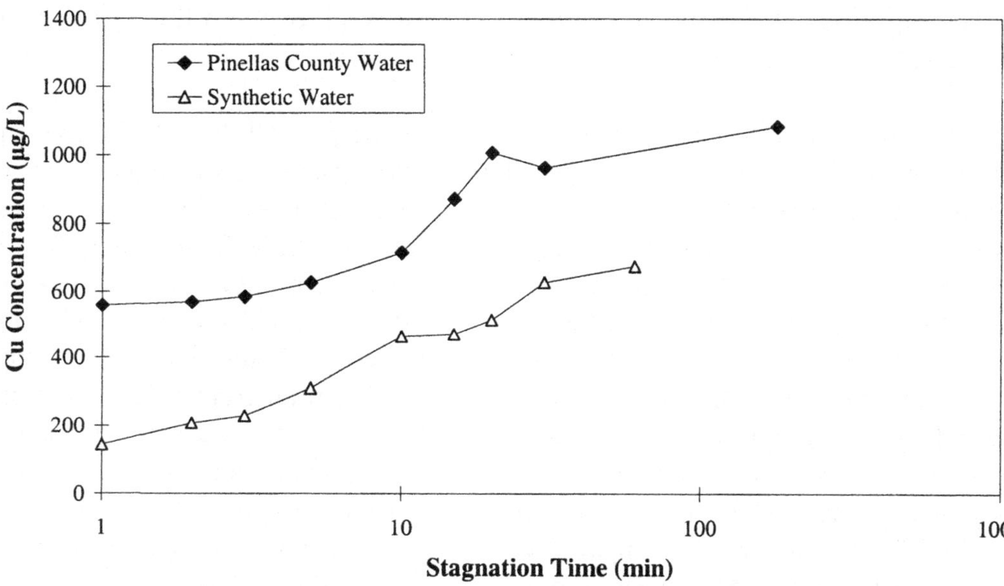

Figure 5.45 Comparison of metal release in Pinellas County water ([Cl] = 3.7 mg/L) and synthetic water quality ([Cl] = 1.5 mg/L). Metal concentrations shown represent total soluble metal inside the CPSF NRC cell.

The field tests indicated that the CPSF method can be used to monitor in situ metal release and corrosion in drinking water qualities. The major drawback of the approach (at its current state of development) is related to uncertainty in predicting metal release for long stagnation times. This limitation will have to be addressed in future research. The method did perform well for stagnation times shorter than 60 minutes.

The time at which E_{corr} reaches a minimum during stagnation episodes (and in the presence of chlorine) and, after which, E_{corr} values start to slowly increase, is called the relaxation time (t_{relax}). For HDR tap water, the value of t_{relax} was in the range of 25 to 40 minutes. For San Francisco water, t_{relax} was in the range of 60 to 90 minutes. As mentioned previously, t_{relax} was approximated at around 180 minutes for Pinellas County.

Although formal calculations to examine the effects of water quality and corrosion on the value of t_{relax} were not conducted, preliminary data suggests that t_{relax} is affected by chlorine concentration, pH, alkalinity, and other parameters. Additionally, t_{relax} appears to be inversely proportional to the rate of corrosion and t_{relax} increased for lower values of i_{corr}. These properties make t_{relax} useful in identifying the influence of aging on the corrosion processes.

Log(Cu_{total}) vs. E_{corr} Relationship

The functional type of correlations between E_{corr} and total soluble copper concentrations measured at varying stagnation times in the presence of chlorine species was similar to those determined in experiments with the synthetic water quality. This was demonstrated in experiments with HDR tap water (Figure 5.41) and San Francisco water (Figure 5.42). In those figures, the relaxation of E_{corr} and the accumulation of copper are obviously related.

When E_{corr} and log(Cu_{total}) are compared, the relationship is linear. The strength of the linear correlation was especially high for HDR tap water and Pinellas County water. The correlation between log(Cu_{total}) and E_{corr} for HDR tap water and Pinellas County water are shown in Figure 5.46 and Figure 5.47, respectively. For San Francisco water, the general trend was similar but the data of metal analyses exhibited more scatter than in the other cases.

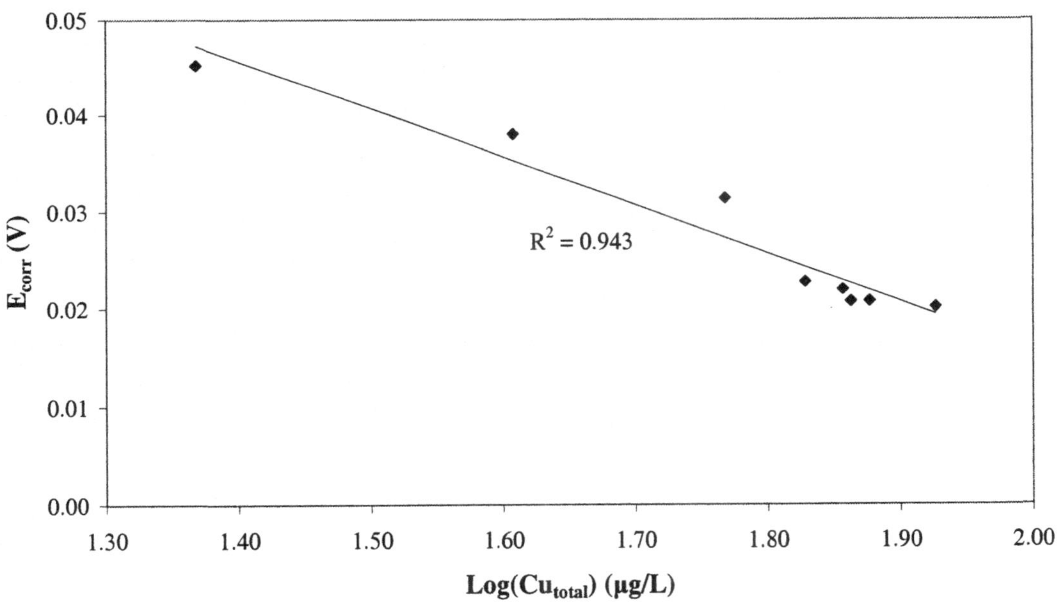

Figure 5.46 Correlation between E_{corr} and total copper concentration for stagnation times varying from 5 to 60 minutes. HDR laboratory tap water, 1-liter sample, CPSF NRC cell.

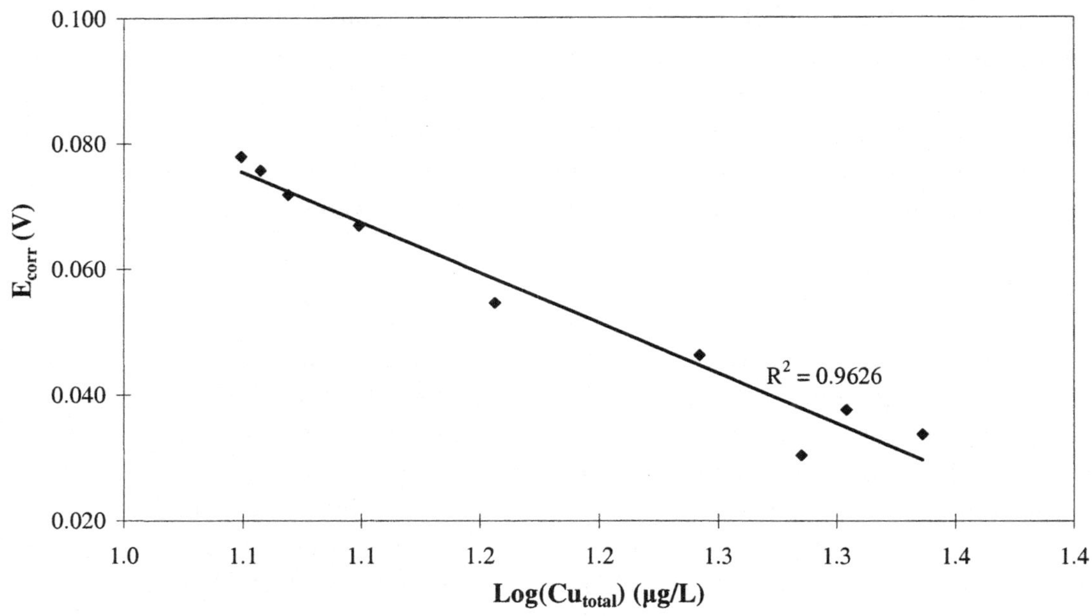

Figure 5.47 Correlation between E_{corr} and total copper concentrations. Pinellas County water, 1-liter sample, CPSF NRC cell.

Although there was a strong linear correlation, Pinellas County water did exhibit significant differences from HDR tap water and San Francisco water, due to the high concentration of chlorine. When the E_{corr} and log(Cu_{total}) correlation for Pinellas County water (total chlorine concentration of 3.7 mg/L) is compared to the correlation for synthetic drinking water (total chlorine concentration of 1.5 mg/L), there are three main differences (Figure 5.48). The values of E_{corr} and the copper concentration are both much higher in Pinellas County water, which is undoubtedly due to a higher chlorine concentration. Additionally, the slopes of the correlations between E_{corr} and log(Cu_{total}) are different.

Currently, the CPSF method is not exhaustive enough to explain the difference in the slopes of the lines. It could be caused by different chlorine concentrations, alkalinities, exposure times, or a combination of the above. Further experiments and numerical modeling are necessary to explore this issue.

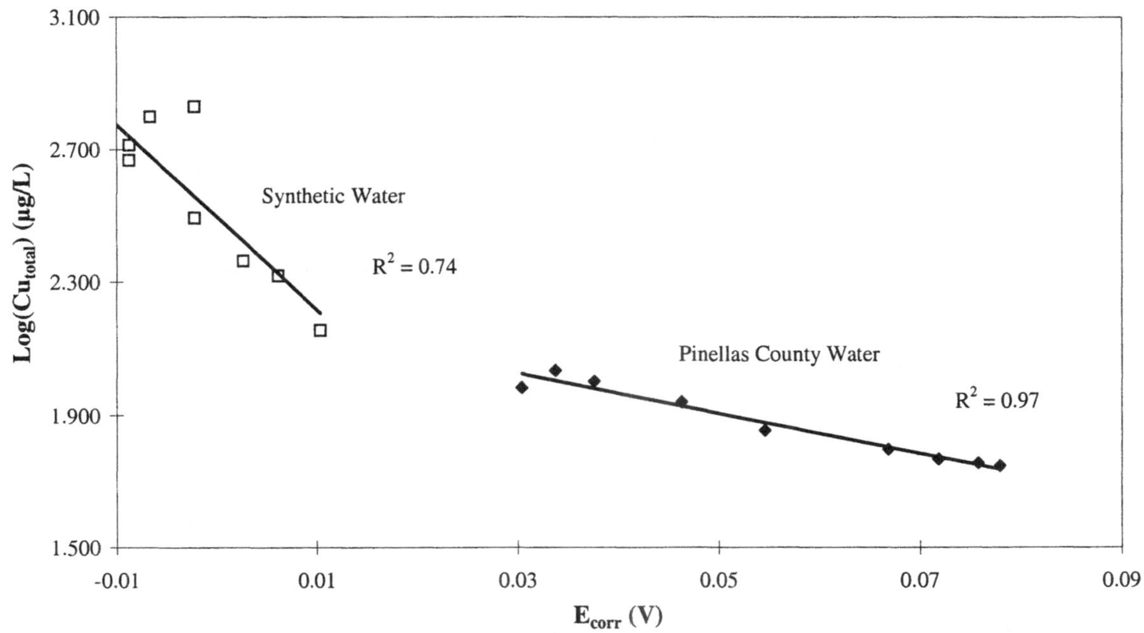

Figure 5.48 Correlation between the values of E_{corr} and copper concentrations with Pinellas County water ([Cl] = 3.7 mg/L) and synthetic water ([Cl] = 1.5 mg/L). CPSF NRC cell.

Surface Aging

The results pertaining to surface aging varied greatly between locations. For HDR tap water, the aging of the surface and effects of variations of water quality parameters could be traced based on E_{corr} and t_{relax} measurements. Effects of aging were not clear in experiments with San Francisco drinking water, while for Pinellas County water, the behavior of E_{corr} and metal release were stable regardless of surface age.

To identify effects of surface aging on E_{corr} and t_{relax} in CPSF experiments with HDR water, E_{corr} profiles were recorded at different days of experiments with HDR tap water (Figure 5.49). As the metal surface aged, E_{corr} values gradually increased. This was similar to the theoretical results described previously in this Chapter. The anodic shift of E_{corr} may be caused by a gradual decrease of the intrinsic copper oxidation current while the oxygen reduction current remains stable.

In the same series of experiments, with HDR tap water, the values of t_{relax} that were measured increased gradually as the surface aged (Figure 5.50). This indicated a gradual decrease in the i_{corr} of the surface.

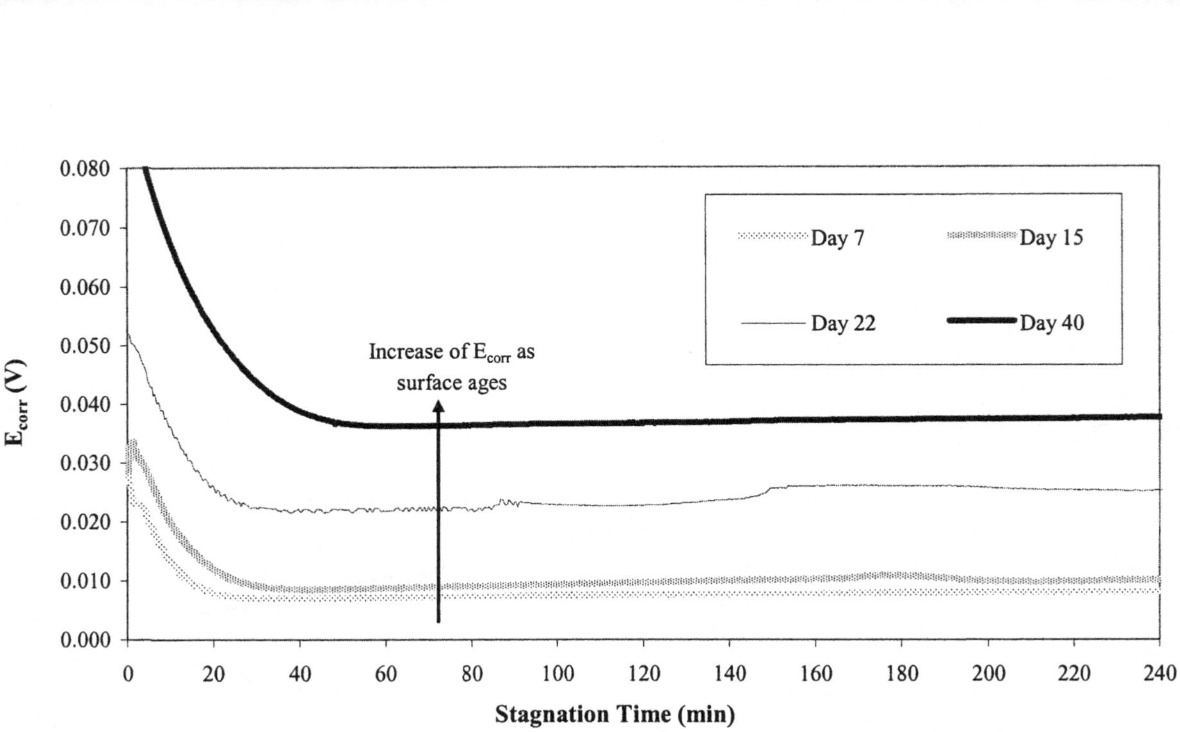

Figure 5.49 Effects of surface aging on E_{corr} profiles during stagnation. HDR water, CPSF NRC cell.

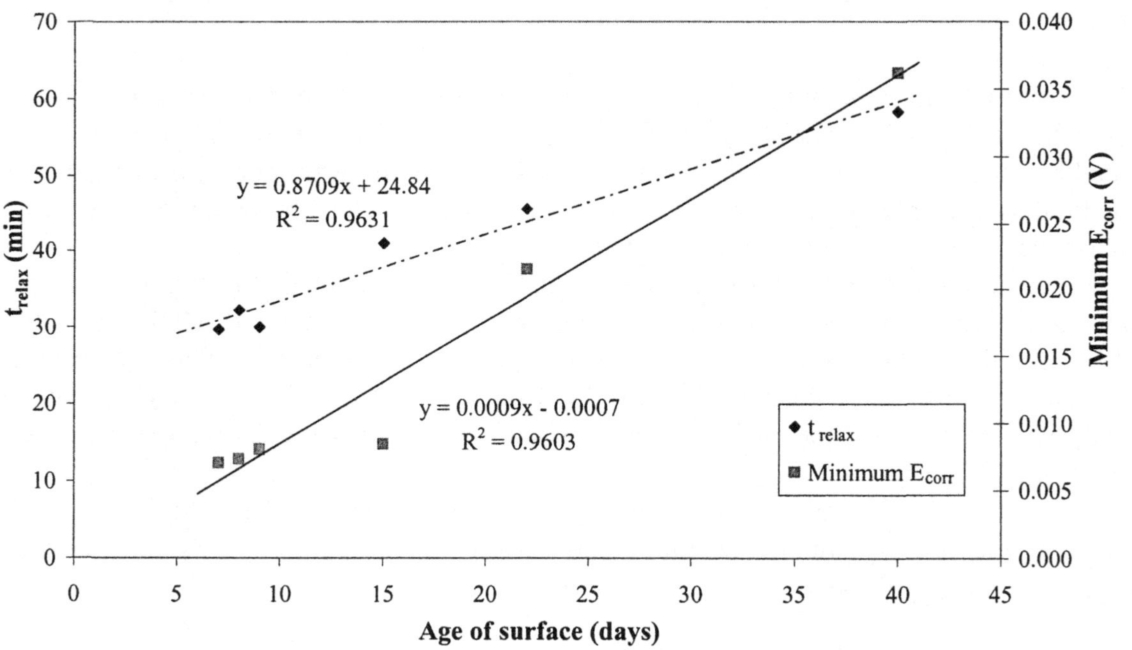

Figure 5.50 Effects of surface aging on minimal values of E_{corr} and relaxation time profiles during stagnation. HDR water, CPSF NRC cell.

For San Francisco drinking water, there were no consistent trends in the behavior of E_{corr} relaxation and copper release associated with the aging of the surface. Concentrations of copper in samples withdrawn from the CPSF cell at the end of various 8-hour stagnation periods also showed no apparent trends over time (Figure 5.51).

Throughout the field experiments in Pinellas County, the aging of the surface did not appear to affect either the behavior of E_{corr} or copper release. The slight increase of E_{corr} values at stagnation times longer than 3 hours was less prominent during exposure in Week 3 than during Week 1 (Figure 5.43). However, the initial rapid relaxation of E_{corr} during the initial phase of stagnation was virtually identical to that throughout the experiments (Figure 5.52).

Measurements of copper concentrations (withdrawn from the CPSF cell at the end of an eight-hour stagnation) showed that total soluble metal in those samples was largely stable after 10 days of exposure (Figure 5.53) and was close to 25 μg per 20 mL in the CPSF cell, which is calculated to be a concentration of 1250 μg/L. This could be estimated as 1.3 mg/L, which is the current action level for copper.

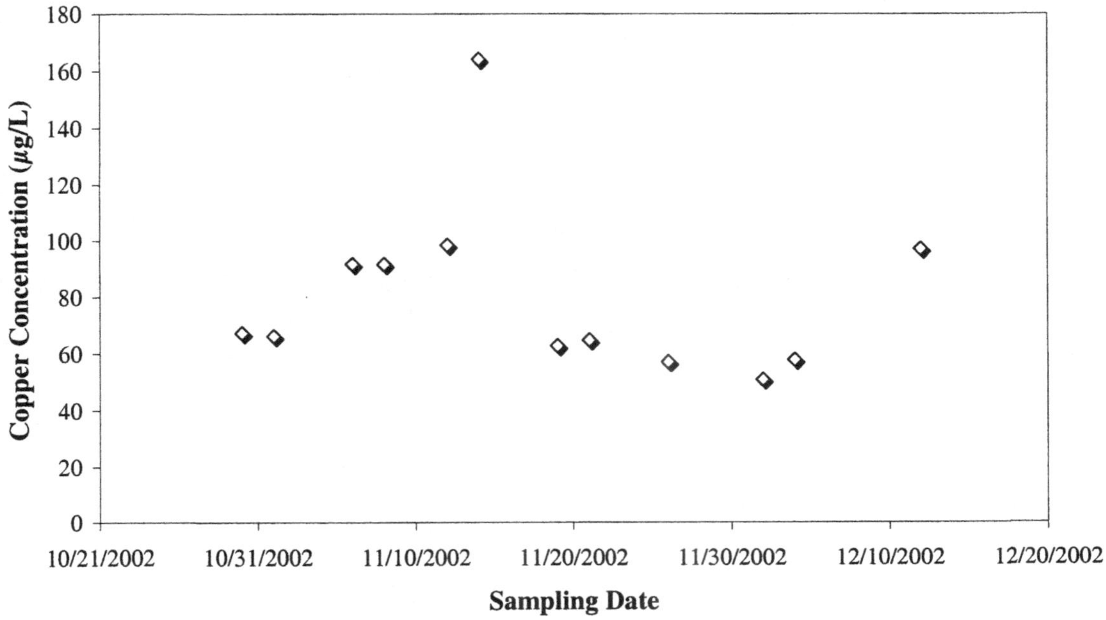

Figure 5.51 Copper concentrations in samples withdrawn at the end of 8 hours stagnation. San Francisco water, 1-liter sample, CPSF NRC cell.

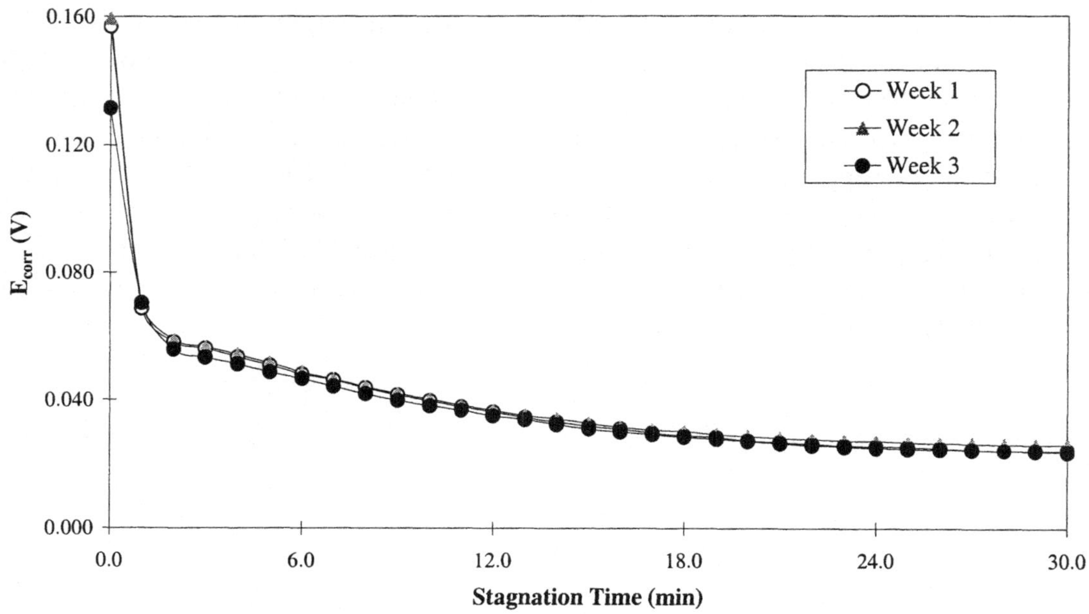

Figure 5.52 Comparison of the initial phase of E_{corr} vs. stagnation time profiles obtained during first three weeks of exposure. Pinellas County water, CPSF NRC cell.

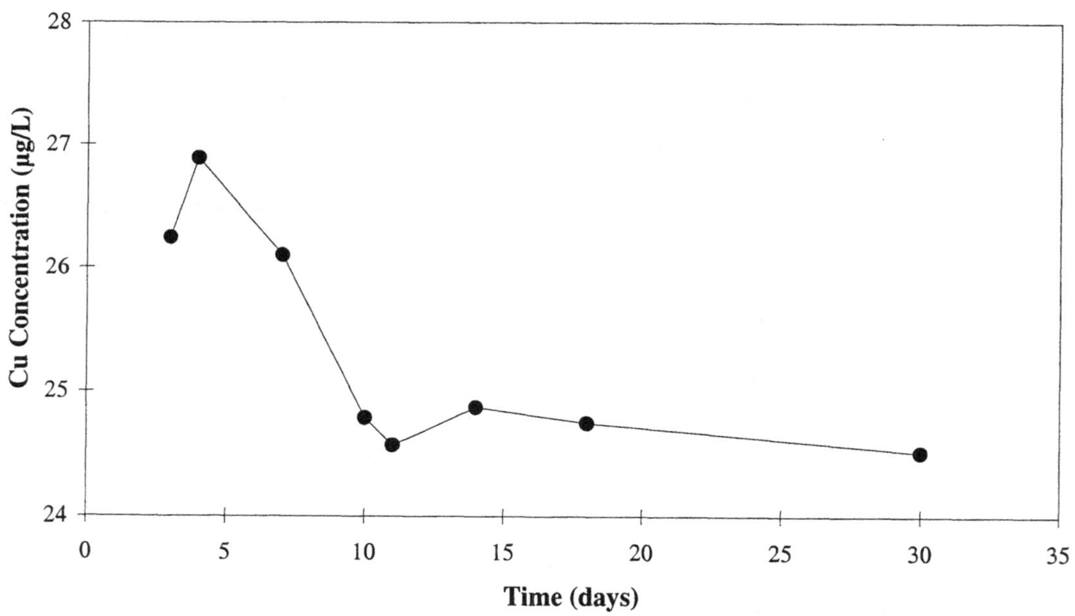

Figure 5.53 Concentrations of copper at the end of a 3-hour stagnation period. Pinellas County water, 1-liter sample, CPSF NRC cell.

During this study, the effects of two significant water quality parameters were studied: pH and chlorine concentration.

The pH of HDR laboratory tap water was manipulated so that the effects on t_{relax} could be observed. Measurements of the relaxation time from those experiments show that t_{relax} immediately decreased in value following the drop in pH from 7.5 to 6.5 (day 8), indicating an increased rate of corrosion (Figure 5.54). When the pH was increased from 6.5 to 7.5 (day 20), t_{relax} immediately increased. Prolonged aging of the copper surface at a pH of 7.5 caused the relaxation time to continue to increase gradually, although some variations were observed. When the pH was reduced again from 7.5 to 6.5 (day 80), the relaxation time immediately decreased again.

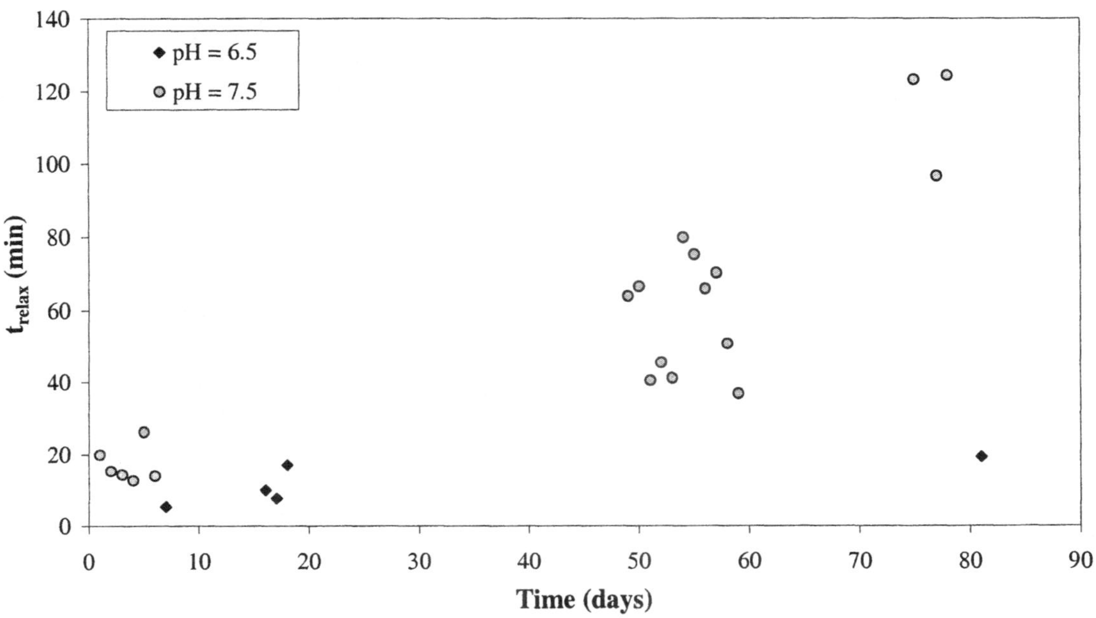

Figure 5.54 Changes of the relaxation time as a function of surface aging and pH variations. HDR water, CPSF NRC cell.

These results support the significance of measurement of t_{relax} in the presence of chlorine. The results show that t_{relax} data was useful for probing trends in the behavior of rate of corrosion associated with aging and variations of water quality parameters. However, further experiments are needed to examine and validate this approach in more detail.

In Pinellas County, high chlorine concentrations caused some differences in the results when compared with the other two water qualities. Although the majority of the chlorine was bound in the form of chloramine, the total chlorine concentration far exceeded concentrations in laboratory experiments or other field studies. Therefore, no data for comparable conditions were available. However, all major features seen in the behavior of E_{corr} during stagnation episodes in Pinellas drinking water were similar to those in the other field applications. The most significant differences noted were the increased values of E_{corr} and the shift in the slope of the $\log(Cu_{total})$ and E_{corr} relationship, both discussed previously and shown in Figure 5.48.

Although this study focused on the performance of the CPSF method for quantifying the corrosion of copper, the behavior of E_{corr} for other materials exposed to laboratory tap water had generally the same features as those described above (Figure 5.55 and Figure 5.56). The data shown in these figures were obtained using a Reiber corrosion cell, whose design and performance is different from the CPSF NRC cells developed in this study. Due to differences in the corrosion cells, the kinetics of E_{corr} relaxation determined with these alternative designs of the corrosion cells cannot be directly compared.

It is evident that all tested piping materials (copper, lead-containing brass, lead-tin solder and lead) exhibit similar trends during the relaxation of E_{corr} in stagnation episodes. The relaxation of E_{corr} is a manifestation of rapid corrosion processes that lead to chlorine consumption at the metal's surface simultaneous with the accumulation of metal in the cell. Although metal release was not measured in these experiments, the similarity between Figure 5.55 and Figure 5.56 and E_{corr} measurements obtained with the CPSF cell suggest that the CPSF method could potentially be applied to monitor the corrosion of all materials of interest for the drinking water industry. CPSF's actual usability is contingent on the further development and extensive tests of that method.

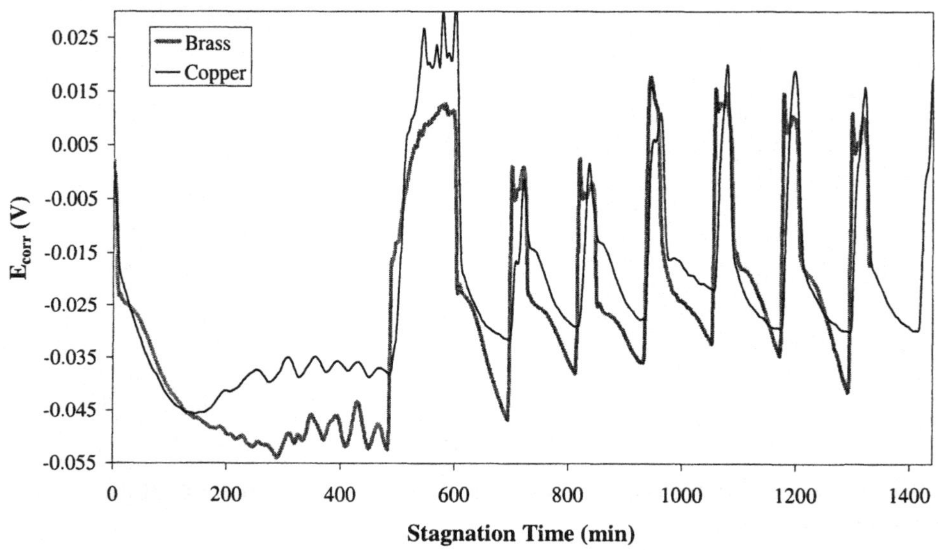

Figure 5.55 Comparison of E_{corr} vs. time profiles for copper and brass. HDR water, Reiber corrosion cell.

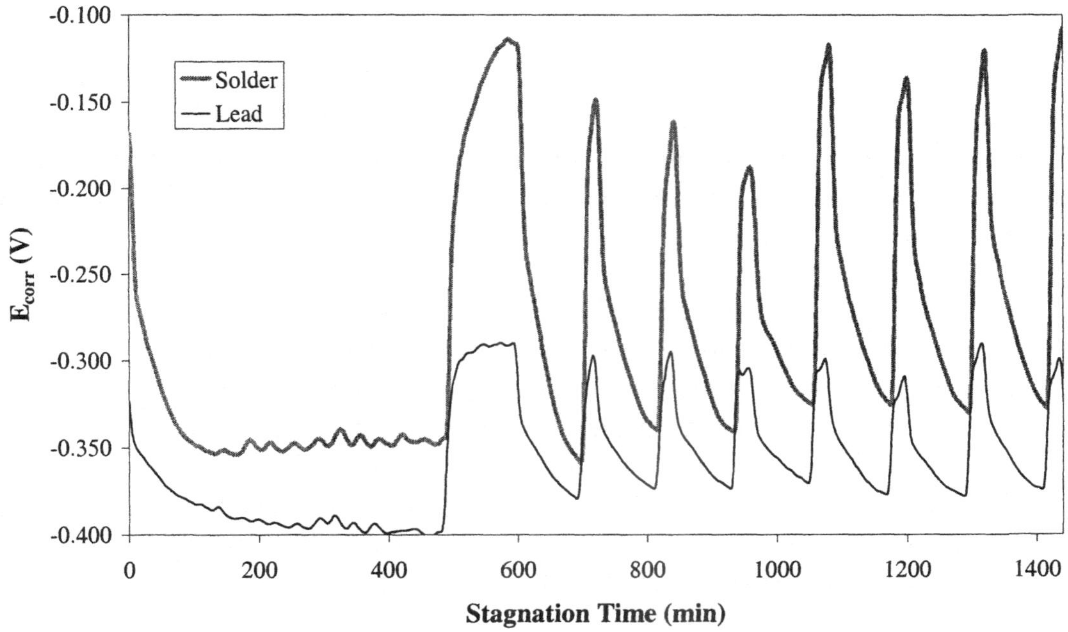

Figure 5.56 Comparison of E_{corr} vs. time profiles for lead/tin solder and lead. HDR water, Reiber corrosion cell.

Conclusions

All water quality parameters examined in the experiments (in the field tests, pH and chlorine concentrations) were observed to have affects on the values of E_{corr} and, therefore, on t_{relax}. It is significant that the relationship of the values of E_{corr} and the logarithm of the total soluble metal concentration in the CPSF cell are linear and strongly correlated in all field tests.

The nature of this relationship is different for unchlorinated and chlorinated water. In the presence of chlorine, the rapid consumption of chlorine at the surface appears to be the major cause of the decrease (rather than increase) of E_{corr} immediately after the onset of stagnation. The phenomenon was observed not only for copper, but also for leaded brass, lead/tin solder and lead. Because other metals showed a similar response, the CPSF method, albeit explored in detail for copper only, is likely to be applicable to monitor the corrosion of all metals of interest for the drinking water industry.

Examination of the CPSF method in the field conditions showed that the main results obtained in the studies employing the synthetic water quality were generally applicable for actual observations of corrosion.

The major limitation of the CPSF method (at its present state of development) is the fact that its performance has only been well explored for relatively short durations of stagnation (1 minute to 1 hour). Experiments show that the behavior of E_{corr} and metal release for stagnation times up to 8 hours is more complex and difficult to model. This limitation is not unique to the CPSF method. Other experimental and theoretical methods that have been employed to predict the kinetics of metal accumulation in drinking water exhibit similar difficulties. This limitation will have to be addressed in future research.

Further Study

The CPSF method is a conceptually novel approach to monitor corrosion and metal release in drinking water. This study provides a significant first step toward a field applicable device. Further experimental and theoretical studies are necessary to exhaustively test this concept and develop instrumentation for on-line monitoring of corrosion processes based on it. Some of these activities may include:

- Detailed numerical modeling of combined effects of chlorine and oxygen on the kinetics of E_{corr} relaxation for varying water quality parameters.
- Clarification of the effects of mass transfer limitations, precipitation of solid phases, and formation of species with alternative oxidation states (e.g., Cu(I), Pb(IV)) on the behavior of E_{corr} during stagnation episodes.
- Continuation of detailed experimental studies to ascertain the effects of water quality parameters (phosphate, chlorine and chloramine concentrations, conductivity, temperature) on the behavior of E_{corr} and metal release in short- and long-term stagnation episodes.
- Detailed experimental studies of E_{corr} and metal release correlations for brasses, lead/tin solder, and lead. These studies can be further expanded to examine the applicability of the CPSF method to monitor the corrosion of iron in drinking water.
- Examination of the performance of the CPSF method using tubular cells that more closely mimic the corrosion processes in drinking water distribution networks and household plumbing.
- Further expansion of field tests to examine the performance of the CPSF method with highly variable drinking water qualities and durations of exposures.

SUMMARY OF FINDINGS

Key findings derived from laboratory studies of corrosion monitoring using the CPSF approach are summarized as follows:

- The electrochemical potential (E_{corr}) of a corroding metal exhibits changes in a predictable and reproducible manner during stagnation episodes of varying duration. However, major features of E_{corr} behavior are different in the presence of oxygen only and in the presence of chlorine species.
- In the presence of oxygen only, there is a strong linear correlation between the logarithms of the total dissolved copper concentrations and E_{corr} values. Variations of pH, alkalinity and phosphate concentrations affect the position of E_{corr} vs. $\log(Cu_{total})$ relationships, but do not significantly affect their linearity and correlation.
- The relationship between E_{corr} and total soluble copper concentration is valid for metal surfaces of any age. However, (in the presence of oxygen only) the values of E_{corr} shift anodically as the metal surface ages. This is likely to be associated with

a decrease of the intrinsic anodic oxidation current (i_{Cu}), while the oxygen reduction current (i_{O_2}) remains largely unaffected.

- In the presence of oxygen, E_{corr} behaves non-monotonically during long stagnation episodes. This type of behavior can be caused by the depletion of oxygen at the surface, formation of Cu(I) species and precipitation of copper-containing solids on the metal surface. Further research is necessary to determine the contributions of these mechanisms.
- In the presence of chlorine, the decrease of E_{corr} after the start of stagnation episodes is accompanied by the accumulation of copper in the corrosion cell. The correlation between log(Cu_{total}) remains linear, but the value and sign of the slope log(Cu_{total}) vs. E_{corr} correlations are opposite compared to those found in the presence of oxygen only.
- In virtually all experiments that involved alterations of basic water chemistry (e,g., changes of pH, alkalinity, phosphate, chlorine concentration), changes of the behavior of E_{corr} could be observed immediately after the corroding surface was exposed to an altered water quality.

CHAPTER 6
CONCLUSIONS AND PROPOSED MONITORING STRATEGIES

INTRODUCTION

This project began with the objectives of:

- Developing a statistically valid alternative home sampling protocol to track lead and copper corrosion.
- Developing an on-line, accelerated testing methodology and equipment to track and predict impacts on lead and copper corrosion during treatment/source water changes.
- Performing case study investigations describing utility experience in achieving optimized corrosion control and the monitoring challenges following optimization.
- Assessing State regulator interpretation of the LCR *Minor Revisions* and their potential acceptance of alternative home sampling and predictive tools.

Three of these objectives were completed. Primacy agency interpretation of the LCR *Minor Revisions* was compiled nationwide showing regulator willingness to consider alternative protocols. Utility case studies documented that utilities could benefit from an alternative monitoring protocol that addressed sample variability, shrinking sample pools, and cost. An alternative tap monitoring was developed that can be presented to state regulators, and with approval, applied immediately by utilities. The development of an on-line monitoring tool faced many challenges from theoretical development, to corrosion cell design, to the effects of differing water chemistry, to extended data validity during extended stagnation times. While the completion of an on-line tool was not fully achieved, the findings of the project provided significant steps in the direction of future development.

Presented in this chapter is a summary of the Project's findings and the utility ready alternative in-home tap monitoring protocol. Each of the Project's findings are discussed separately.

ALTERNATIVE IN-HOME CUSTOMER TAP SAMPLING

The results and findings from the field studies that were completed, along with the historical information discussed in Chapter 4, provided the basis for the development of a customer tap sampling program for lead and copper levels. The alternative monitoring protocol was designed to provide regulators and utilities with an additional tool for verifying the impact of changes in treatment or source waters on corrosion control as measured by changes in tap lead and copper levels. The alternative tap monitoring protocol generates data that can be used to assess changes in tap lead and copper levels over time. The following is a discussion of the recommendations, data evaluation guidelines, the applicability of the 90th percentile statistics in evaluating changes in lead and copper levels, and the benefits and drawbacks of the alternative monitoring protocol.

The recommended alternative monitoring protocol addresses: (1) sample site selection, (2) sample size determination, (3) impact of sampling frequency, and (4) data collection methods. Table 6.1 provides a summary of the recommended alternative in-home tap monitoring protocol.

Table 6.1
Recommended alternative tap monitoring protocol summary

Metal	Lead	Copper
Number of sites	12-43	7-24 (per age category)
Number of samples/site	One per sampling period	
Evaluation period	Quarterly	
Site selection criteria	• Locate sites in areas of the distribution system that exhibit consistent water quality. • Select sites with similar physical conditions with respect to type and age of lead and/or copper source. • Select sites with consistent physical and age characteristics within each specific sample pool.	
	For Lead Sites: • Single family residences with lead solder installed prior to 1982 or lead service lines	For Copper Sites: • Single family residences with copper piping • Three age categories: >20 years ≥10 but ≤20 years < 10 years
Data collection and analysis criteria	• 1-Liter total. Split sampling with initial 125 mL followed by 875 mL. • Analyze total lead and copper levels • Minimum 6-hour stagnation time. All samples should have approximately similar stagnation times	

Sample Site Selection

The rankings of variability factors discussed in Chapter 4 indicated that the age of lead and/or copper source material and the consistency of the distributed water quality were important factors to be controlled when selecting monitoring sites for an alternative protocol. With respect to material age, literature sources indicated that newer copper piping will result in higher copper levels and newer faucets will cause elevated lead and copper levels in first draw samples. These facts were supported by results from limited field study monitoring from this project as well. However, the evaluation of the variability seen in copper levels from the field studies indicated consistent variation in copper levels across utilities, even though the data represented disparate water quality and sample site conditions.

Another interesting finding from the field studies was that intra-site sampling enabled highly variable sites to be identified. By conducting intra-site monitoring at potential sampling sites, the variability in lead and/or copper levels can be measured and the source of variability possibly determined. Decisions can then be made about which sites should be maintained in the sample pool for evaluating changes in lead and copper levels over time. Selecting sites for the sample pool that exhibited relatively higher average metals levels, but lower variability, would allow a greater level of accuracy to be obtained in the data, while still targeting higher metals levels sites.

When selecting sites for the alternative monitoring program, utilities should first determine what site characteristics and corresponding lead and copper levels are most important for them to evaluate; i.e. subsets of LCR monitoring, newer copper piping, and/or locations with newer faucets. Sample pools should be developed that contain consistent physical and age characteristics for the specific type of sites to be evaluated. If an evaluation of typical regulatory lead and copper levels is desired, then the sites selected for the monitoring program should be subsets of regulatory monitoring sites. If newer copper and lead sources are to be evaluated, then sites with new faucets and/or new copper plumbing should be targeted. If sites with a variety of ages and lead or copper sources are included in the sample pool, then variability in the results may be higher.

General recommendations for site selection for the customer tap monitoring protocol are as follows:

- Locate sites in areas of the distribution system that exhibit consistent water quality conditions and are not in dead-end/low flow areas.
- Screen proposed sites so that the sample pool contains sites with similar physical conditions with respect to type and age of lead and/or copper sources.
- Conduct initial intra-site monitoring to provide information on the variation in lead and copper at a particular site, and use this information to determine if a particular site should be included in the alternative monitoring program due to variability of the intra-site monitoring data.
- Choose sites that are subsets of the locations chosen for regulatory monitoring, if the desired objective is to predict and compare alternative monitoring results to regulatory results.
- Choose sites that represent newer faucet and/or piping conditions if the desired objective is to evaluate changes in lead and or copper levels from newer lead and copper sources in the systems.
- Choose sample sites that contain consistent physical and age characteristics within each specific sample pool.

Determining the Number of Sites

The number of sites sampled, combined with the frequency of their sampling should be adequate to allow a particular accuracy and confidence in estimating mean lead and copper levels in the system. Appendix A contains statistical background information on the relationship between sample size, accuracy, and confidence levels. The recommended tap monitoring protocol provides initial estimates for sample size based on results from the field study monitoring. Typical coefficients of variation for lead and copper level data from the alternative tap monitoring field studies for the inter-site and intra-site results were as shown in Table 6.2:

Table 6.2
Typical coefficients of variation for lead and copper level data from alternative tap monitoring

	Lead levels	Copper levels
CV for inter-site results	0.8	0.6
CV for intra-site results	0.4	0.2

Estimates of initial sample sizes can be made using Tables A.2 and A.3 in Appendix A for the CV values at 90% and 80% confidence. Tables 6.3 and 6.4 present these sample sizes, and as can be seen, accuracy levels better than 20% (i.e., 15%, 10%, or 5%) would result in very high sample numbers, particularly for lead levels where the variation is much higher.

For purposes of developing a reasonable monitoring program, the authors were cognizant of balancing the effort and cost of increased monitoring with the statistical accuracy and confidence required of the results. In order to establish reasonable sample sizes for customer tap monitoring, expectations about the desired accuracy and confidence in the data were somewhat reduced. Tables 6.5 and 6.6 display sample size estimates for lead and copper levels based on accuracy levels of 20% and 30% and confidence levels of 90% and 80%. Using these estimates, the minimum number of sample sites recommended for initial lead and copper level monitoring were determined. These initial sample numbers should be considered minimum recommendations, and utilities are encouraged to sample from more sites in order to provide greater statistical accuracy in the results.

By recommending fewer sampling sites than what is required for regulatory monitoring, more control may be gained over data collection and site characteristics, both of which may result in lower overall variability in the results generated. In addition, sampling from fewer sites addresses the possible reductions in targeted sampling sites that can occur with LCR monitoring, due to reductions in lead materials in the system and/or customer unwillingness to participate.

Table 6.3
Sample population for various coefficients of variation and accuracy levels (90% confidence level)

Accuracy level	Coefficient of variation (std deviation/mean)				
	0.2	0.4	0.6	0.8	1.0
5%	43	173	390	693	1082
10%	11	43	97	173	271
15%	5	19	43	77	120
20%	3	11	24	43	68
25%	2	7	16	28	43
30%	1	5	11	19	30
40%	1	3	6	11	17
50%	0	2	4	7	11

Table 6.4
Sample population for various coefficients of variation and accuracy levels
(80% confidence level)

Accuracy level	Coefficient of variation (std deviation/mean)				
	0.2	0.4	0.6	0.8	1.0
5%	26	105	237	420	657
10%	7	26	59	105	164
15%	3	12	26	47	73
20%	2	7	15	26	41
25%	1	4	9	17	26
30%	1	3	7	12	18
40%	0	2	4	7	10
50%	0	1	2	4	7

Table 6.5
Number of initial samples estimated for lead levels

Confidence level	Accuracy	No. sites (CV = 0.8)	No. samples/site (CV= 0.4)
90%	20%	43	11
90%	30%	19	5
80%	20%	26	7
80%	30%	12	3

Table 6.6
Number of initial samples estimated for copper levels

Confidence level	Accuracy	No. sites (CV = 0.6)	No. samples/site (CV= 0.2)
90%	20%	24	3
90%	30%	11	1
80%	20%	15	2
80%	30%	7	1

As discussed in the report, intra-site variability was lower than inter-site variability, both from the historical monitoring data and from results of the field monitoring studies. Based on the number of samples per site listed in Tables 6.5 and 6.6 and the fact that multiple samplings from individual sites would increase the overall monitoring burden, it is recommended that data collection sites be monitored only once during each sample round. Initial intra-site monitoring also provides a useful tool for determining which sites should be included in the monitoring program. As shown in the field study results, sites with unusually high variability can be evaluated and determinations made about the cause of the variability (faucet and/or piping replacement, unusually high stagnation periods, etc.) and whether the site should be included in the program.

The number of sites recommended for the alternative tap protocol are listed in Tables 6.5 and 6.6 for various accuracy and confidence levels. For lead level evaluations, the number of sites range from 12 to 43; and for copper level evaluations, the number of sites range from 7 to 24. Individual utility results with respect to lead and copper variability may differ from the estimates used to develop the sample sizes listed. These sample size recommendations are intended to be initial starting points, based on the results obtained from actual lead and copper monitoring evaluated for this project. They correspond to the highest accuracy levels (i.e. tightest band around the mean) that will result in realistic sample numbers, given the CV's calculated from actual lead and copper monitoring data. Utilities that have generated their own information on lead and copper variability can utilize the tables in Appendix A to develop individual starting points for sample size. Developing utility specific data will also provide information with which to assess the variability in the lead and copper levels and adjust the monitoring protocol accordingly.

Determination of Sampling Frequency

The frequency of data collection should be based on an understanding of the consistency in the quality of both the finished and distributed water, and the timeframe for any treatment or source water changes that may alter metals levels. Designed in conjunction with the number of sites, the frequency should be adequate to allow a particular accuracy and confidence in estimating mean lead and copper levels in the system. Therefore, recommendations must be site specific. Evaluation periods of 6 months to 1 year may be applicable in utility situations where distributed water quality is relatively stable and there is little seasonal variation. However, shorter evaluation periods would provide more timely data with which to make comparisons and determine if lead and/or copper levels have changed because of water quality changes in the system. The recommended evaluation (sampling) period for the field study data collection effort for this study was 10 weeks. As a starting point, it is recommended that the sampling frequency for alternative monitoring be quarterly. It would be preferable to have one year of monitoring completed prior to a source water or treatment change in order to have adequate data with which to evaluate potential differences in metals levels.

Data Collection and Analysis Procedures

It is recommended that consecutive 125 mL and 875 mL, first-draw tap water samples from the cold-water kitchen or bathroom tap be collected during any alternative tap monitoring protocol for lead and copper levels. This will allow the faucet contribution to lead and copper levels to be determined. To reduce the variability in the data, the stagnation times from all monitoring sites should be consistent (approximately 6 hours) and samples should be collected by either system personnel or residents who have been instructed on the sampling procedures required.

A large amount of the variability in household lead and copper levels may be attributed to the presence of particulate lead and copper in the water. The analysis of both total and dissolved lead and copper would provide data to determine the presence of particulate lead and copper, the impact they may have on total metals levels measured at the tap, and how the particulate levels may change with changes in source water and/or treatment.

Data Evaluation Guidelines

The purpose of an alternative tap monitoring program for lead and copper is to provide data with which to evaluate whether changes in treatment or source waters will cause changes in standing lead and copper levels at the tap. This evaluation can be used to compare alternative monitoring results with regulatory data and attempt to predict results from future regulatory monitoring, or to determine the impact that treatment changes may have on newer lead and copper sources in the system. In either case, data evaluation techniques can be utilized that compare the central tendency (mean or median) of the data sets and apply standard statistical tests to determine if significant differences are observed between the data sets (i.e., between two sets of alternative monitoring data, or between alternative monitoring data and regulatory data).

The alternative tap monitoring protocol provides utilities of all sizes with a tool requiring no additional equipment or training. The protocol provides optimized utilities with a more accurate representation of changes to lead and copper levels in a more timely and cost effective manner. In terms of regulator acceptance, the statistical basis is sound and focuses on monitoring activities in order to minimize misrepresentative data variability.

CPSF OLCC PREDICTIVE MONITORING

Applicability of On-line Corrosion Monitoring

The theory behind the CPSF approach to predictive monitoring was described in detail in this report and verification of the theory was successful when conducted under laboratory conditions. The development of a functional on-line corrosion cell proved to be challenging, however, the narrow rectangular channel (NRC) design functioned appropriately and had the operational capacity for carrying out CPSF experiments.

The CPSF theory and the NRC OLCC are a significant advancement toward a new approach to monitoring corrosion control in distribution systems. In addition to the theoretical and NRC OLCC development the project showed that:

- The electrochemical potential (E_{corr}) of a corroding metal exhibits changes in a predictable and reproducible manner during stagnation episodes of varying duration.
- In the presence of oxygen only, there is a strong linear correlation between the logarithms of the total dissolved copper concentrations and E_{corr} values.
- Variations of pH, alkalinity and phosphate concentrations affect the position of E_{corr} vs. $\log(Cu_{total})$ relationships, but do not significantly affect their linearity and correlation.
- The relationship between E_{corr} and total soluble copper concentration is valid for metal surfaces of any age.
- In virtually all experiments that involved alterations of basic water chemistry (e,g., changes of pH, alkalinity, phosphate, chlorine concentration), changes of the behavior of E_{corr} could be observed immediately after the corroding surface was exposed to an altered water quality.

However, at the current stage of development the technology is not ready for wide-scale deployment for utility use. The technology requires additional extensive field and laboratory development to address critical issues identified by this project as:

- The CPSF has only been well explored for relatively short durations of stagnation up to one hour.
- The major features of E_{corr} behavior are different in the presence of oxygen only and in the presence of chlorine species.
- In the presence of oxygen only the values of E_{corr} shift anodically as the metal surface ages. This is likely to be associated with a decrease of the intrinsic anodic oxidation current (i_{Cu}), while the oxygen reduction current (i_{O_2}) remains largely unaffected.
- E_{corr} behaves non-monotonically in the presence of oxygen during long stagnation episodes. This type of behavior can be caused by the depletion of oxygen at the surface, formation of Cu(I) species and precipitation of copper-containing solids on the metal surface. Further research is necessary to determine the contributions of these mechanisms.
- In the presence of chlorine, the decrease of E_{corr} after the start of stagnation episodes is accompanied by the accumulation of copper in the corrosion cell. The correlation between $\log(Cu_{total})$ remains linear, but the value and sign of the slope $\log(Cu_{total})$ vs. E_{corr} correlations are opposite compared to those found in the presence of oxygen only.

The future challenge will be to develop the technology that, in addition to greater ease of operation, provides data that can be readily and easily interpreted. Based on the level of sophistication involved in the CPSF theoretical development, the greatest challenge to medium and small utility applicability and regulator acceptance will be the simplification of data interpretation.

Equipment and Operational Expertise

During the course of this project the design of the OLCC underwent several unique stages. Each phase of design presented flaws that interfered with the collection of reliable data. The final OLCC design, the Narrow Rectangular Cell (NRC), provided a device that was easy to construct and use, easy to maintain, inexpensive, and provided reliable data. The NRC OLCC was shown in Figure 5.4 and 5.5, NRC specifications were summarized in Table 5.1. While additional work remains in the area of data interpretation and operation of the cell, the NRC OLCC is considered complete and functional for the purposes of conducting corrosion potential stagnation/flow experiments.

The NRC OLCC itself is simple in design and operation. The only element of the cell that requires regular maintenance is the probe measuring corrosion potential. Since the probe is a readily available probe from scientific supply houses the required maintenance is the same as any typical analytical probe. While the cell is the operational element, there are peripherals necessary to gather data such as the controller, multimeter and computer as previously shown in Figure 5.7.

Once assembled, the NRC OLCC operates, collects samples, and processes data automatically; operational expertise is minimal. The expertise required with the NRC OLCC is associated with the interpretation of generated data. Currently, the expertise required to interpret the data is likely limited to large utilities with laboratory staff familiar with electrochemical corrosion theory and full understanding of the current limitations of the CPSF method. For large utilities investigating corrosion control the NRC OLCC may provide a valuable tool into bench-scale investigations. In its current state of development it is not applicable in medium and small size utilities.

COMPARISON OF LCR PROTOCOL, ALTERNATIVE PROTOCOL, AND PREDICTIVE MONITORING

The existing LCR monitoring protocol, the alternative tap monitoring protocol, and the CPSF OLCC technology each present their own unique benefits, drawbacks, and trade-offs. In the case of the tap monitoring protocols, an increased and more regular monitoring program can provide data more reflective of corrosion control changes at a lower cost. Fewer customers are impacted, but those customers that do participate are more greatly impacted. The CPSF OLCC technology provides a significant reduction in monitoring costs, however at its current stage of development it is not ready for utility implementation or regulator approval. It remains to be shown that the technology will be widely applicable to medium and small size utilities.

Table 6.7 is a summary comparison of the LCR protocol and the two technologies developed by this project.

Table 6.7
Comparison and summary of alternative tap monitoring and CPSF OLCC monitoring approaches

Comparative Criteria	Lead and Copper Rule Protocol	Alternative Tap Monitoring Protocol	CPSF NRC OLCC Predictive Technology
Number of sites	100	12-43 Lead; 7-24 Copper	1
Number of samples per period	100	12-43 Lead; 7-24 Copper	N/A
Number of samples annually	100-200	48-172 Lead; 84-288 Copper	300-1000
Number of samples per site annually	1-2	4 Lead; 4 Copper	NA
Sample site requirements	Customer tap and stagnation period	Customer tap and stagnation period	Treatment plant and continuous flow
Additional costs beyond LCR monitoring	None	Yes	Yes
Impact of shrinking sample pool	Unable to identify required number of sample homes	Minor impact. Number of qualifying homes reduced by 75% over LCR	None
Customer participation	Yes	Reduced	None
Training requirement	None	None	Yes
Statistical soundness of data reflecting actual conditions	Reduced as sample pools shrink	Increased and more indicative of corrosion control changes	Under development
Additional equipment cost	None	$200 (chemicals)	$2,500[a]
Annual operational cost	$20,000 annually[b, e]	$10,000 annually[e]	$1,200[c]
Cost per sample[d]	$200	$100	$4.50[f]
Benefits	• LCR specified	• Reduced variability in data • Reduced customer management • Directly comparable data • Reduced cost • Non new monitoring or analytical protocols	• Minimal monitoring sites • Elimination of customer participation • Minimal staff requirement • Easy to operate • Low cost
Drawbacks	• Greater variability in data • Increased customer management • Limited comparable data	• Requires Primacy approval • Increased sample number	• Not yet sufficiently developed for field installation • Difficult to interpret data

a. Includes CPSF NRC, voltmeter, solenoid valve, and dedicated computer.
b. Assumes optimized system and one round (100 samples) of manual sampling.
c. Includes CPSF NRC maintenance and data management.
d. Assumes one sample per day and equipment costs annualized over five years for OLCC monitoring. Tap monitoring includes administrative, management, collection, and analytical costs.
e. Based on reported cost estimates from Philadelphia Water Department for conducting LCR compliance monitoring and field testing of alternative tap monitoring.
f. Assumes 300 samples annually.

APPENDIX A
STATISTICAL BACKGROUND INFORMATION

When developing a statistically valid sampling program, estimates of the number of samples required and the frequency of collection should be based on an understanding or estimation of the variability of the parameter to be measured, along with the desired significance and accuracy required of the data. This appendix discusses the statistical basis used in developing the alternative tap monitoring protocol.

ACCURACY AND CONFIDENCE LEVELS

Statisticians use two parameters, accuracy and confidence level, when defining the reliability of data from a sample population. Both accuracy level and confidence level typically increase as the sample population size increases. The financial cost of gathering data, however, is typically a limiting factor in sample population size. It is, therefore, necessary to determine an adequate sample population that is both large enough to be statistically reliable and small enough to be financially feasible.

The accuracy can be expressed as a band around the mean, within which the true mean is expected to fall. This band is normally thought of as a percentage from the mean. Typical sample accuracy ranges from 1% to 20%. For example, data from a sample population that is 10% accurate has a band around it with an upper bound of the mean plus 10% and a lower bound of the mean minus 10%. A sample population designed for 10% accuracy with a sample mean of 0.02 mg/L implies that the true mean lies within the range of 0.018 mg/L and 0.022 mg/L. Since accuracy defines a band around the mean, then a narrower band indicates that a smaller difference can be detected. Relative to lead and copper monitoring, as levels decline, the desired accuracy level around the mean will need to be tighter in order to be confident that a difference can be detected.

Accuracy must be coupled with confidence in the sample population. The term "confidence level" is used to express this. Confidence levels are expressed in terms of the percent of time the conclusions are expected to be true. Usual confidence levels range from 60% to 99%. For example, a 95% confidence level for a 50% accuracy level means that we expect data to produce values within 50% of the true mean 95% of the time.

DATA VARIABILITY AND SAMPLE SIZE

Variability of the parameter of interest among the targeted population is another factor that affects the necessary sample population size. The greater the variability, the larger the sample population size needs to be for statistically reliable results. As a starting point, an estimate of the sample size is given by the following formula (McClave and Benson 1982):

$$n = (zs/Bx)^2 \quad \text{(A-1)}$$

where:

n = number of samples

z = critical region of the normal distribution, which depends on the significance level of the test (i.e., z= 1.96 for a 95 percent significance level and z= 1.645 for a 90 percent significance level)

s = standard deviation of the sample

B = the bound or accuracy of the estimated mean level (i.e., if B = 0.10, the estimate of the population mean would fall within plus or minus 10 percent of the sample average)

x = the average of the sample

This formula is based on the central limit theorem (Miller and Freund 1985), which states that for sufficiently large sample sizes, the distribution of the sample mean is approximately normally distributed regardless of the underlying distribution of the population. This theory states that the distribution of the difference between two sample means is at least approximately normally distributed. Therefore, when evaluating probabilities associated with the values of a specific test statistic, the procedures would be the same as if the population itself were normally distributed. The definition of what constitutes a 'large' sample size varies. Statistical texts have stated that sample sizes greater than 30 would be considered large enough to apply the central limit theory and assume normality.

Based on the above formula, estimates of sample size should be based on an understanding of the inherent variability of lead or copper levels measured at the tap and the desired significance and accuracy of the estimated mean level. This will provide the number of samples necessary for statistically significant results. From a statistical viewpoint, increasing the number of samples that are collected will increase the level of confidence that the results are an accurate predictor of the actual water quality condition to be sampled.

The coefficient of variation is a measure of the variability of a data set, and is defined as:

Coefficient of Variation = Standard Deviation/Mean

To determine the number of samples that need to be taken and the frequency of collection, estimates of the variability in lead and copper levels should be made. These estimates can be based on historical data from the utility or a similar system or pilot data. On-going evaluations of lead and copper data sets can be used as they become available to further refine the variability estimates.

Tables A.1 through A.3 list sample sizes for a range of coefficients of variation and accuracy and confidence levels, based on this sample estimation formula. The frequency of sample collection may be reduced as the monitoring program progresses if the variability of the data is lowered. However, maintaining a conservative estimate of the required sample size throughout the study period may be most appropriate. Once the data has been generated from a specific monitoring protocol, it is desirable to determine the confidence and accuracy of the sample. Using the same equation for estimating sample size, but solving for (z/B), the actual confidence and accuracy of the data generated can be calculated.

Table A.1

Sample sizes for various coefficients of variation and accuracy levels – 95% confidence

Accuracy level, %	Coefficient of Variation (standard deviation/mean)										
	0.1	0.2	0.3	0.4	0.5	0.6	0.7	0.8	1	1.5	2
5	15	61	138	246	384	553	753	983	1537	3457	6147
10	4	15	35	61	96	138	188	246	384	864	1537
15	2	7	15	27	43	61	84	109	171	384	683
20	1	4	9	15	24	35	47	61	96	216	384
25	1	2	6	10	15	22	30	39	61	138	246
30	0	2	4	7	11	15	21	27	43	96	171
40	0	1	2	4	6	9	12	15	24	54	96
50	0	1	1	2	4	6	8	10	15	35	61

Table A.2

Sample sizes for various coefficients of variation and accuracy levels – 90% confidence

Accuracy level, %	Coefficient of Variation (standard deviation/mean)										
	0.1	0.2	0.3	0.4	0.5	0.6	0.7	0.8	1	1.5	2
5	11	43	97	173	271	390	530	693	1082	2435	4330
10	3	11	24	43	68	97	133	173	271	609	1082
15	1	5	11	19	30	43	59	77	120	271	481
20	1	3	6	11	17	24	33	43	68	152	271
25	0	2	4	7	11	16	21	28	43	97	173
30	0	1	3	5	8	11	15	19	30	68	120
40	0	1	2	3	4	6	8	11	17	38	68
50	0	0	1	2	3	4	5	7	11	24	43

Table A.3

Sample sizes for various coefficients of variation and accuracy levels – 80% confidence

Accuracy level, %	Coefficient of Variation (standard deviation/mean) 0.1	0.2	0.3	0.4	0.5	0.6	0.7	0.8	1	1.5	2
5	7	26	59	105	164	237	322	420	657	1478	2628
10	2	7	15	26	41	59	80	105	164	370	657
15	1	3	7	12	18	26	36	47	73	164	292
20	0	2	4	7	10	15	20	26	41	92	164
25	0	1	2	4	7	9	13	17	26	59	105
30	0	1	2	3	5	7	9	12	18	41	73
40	0	0	1	2	3	4	5	7	10	23	41
50	0	0	1	1	2	2	3	4	7	15	26

APPENDIX B
THEORETICAL FUNDAMENTALS OF THE EXPERIMENTAL APPROACH

Electrochemical corrosion is the primary driving force of metal release in drinking water distribution systems. However, corrosion processes cause the formation of both soluble particles and colloidal corrosion scales of the interior surface of pipes. Unless tests are performed for a steady-state condition of the corrosion surface, corrosion current (i_{corr}) measurements do not provide a direct estimate of the rate of metal release. An additional limitation of accepted methods for measuring i_{corr} values is associated with the need to employ fairly sophisticated equipment to measure relationships between the current and its electrochemical corrosion potential (E_{corr}) through the surface and to obtain relevant i_{corr} values based on the numerical processing of these data.

This appendix presents the development of the Corrosion Potential Stagnation/Flow (CPSF) theory and outlines trends predicted by it for varying water qualities. The unique feature of the CPSF method is it does not involve the use of electrochemical equipment to generate data for evaluation of metal release and i_{corr}.

Theory of Behavior of Corrosion Potential in Stagnation Episodes in the Absence of Chlorine

The behavior of E_{corr} in the absence of chlorine species for the case of copper corrosion will be the focus of this appendix (the theory for lead-containing materials is fundamentally similar). The majority of laboratory data generated to evaluate trends in metal release associated with variations of water quality parameters (pH, alkalinity, phosphate) have been obtained for chlorine-free water. Chlorine is usually absent in water systems that distribute groundwater. In these situations, dissolved oxygen (DO) is the only oxidant. The standard electrochemical potential of the O_2/H_2O couple is 1.23 V vs. normal hydrogen electrode (n.h.e.) at pH 0 and equilibrium DO molar concentration corresponding to partial pressure for oxygen of 1 Bar.

In accordance with the general theory of corrosion (as outlined in Internal Corrosion 1996), the current of electrochemical reduction of DO to H_2O is, at any moment, coupled with the current of copper oxidation to Cu (I) and Cu (II) species. Although Cu^+ species are very important for the formation of surface cuprite (e.g., Merkel 2003, Merkel et. al. 2002, Baitz and Follner 2002, Ferguson et. al. 1996) and are likely to be formed or consumed during water stagnation, soluble copper is predominantly Cu^{2+}. In order to simplify the discussion that follows, it has been presumed that the reduction of oxygen proceeds simultaneously with the formation of Cu^{2+} species only. However, this process can involve intermediate steps, in which Cu^+ is generated and consumed.

Based on this hypothesis, we can examine the behavior of E_{corr} of the copper surface during stagnation episodes. As stated, the corrosion potential of the metal is defined at any moment of time by the steady-state condition that the sum of oxygen consumption and copper oxidation currents equals zero:

$$i_{Cu} + i_{O_2} = 0 \qquad \text{(B.1)}$$

In equation B.1, i_{Cu} is the electrochemical current of copper oxidation (associated with copper release) and i_{O_2} is the electrochemical current of oxygen reduction. The magnitudes of these currents are defined by the value of E_{corr} and intrinsic properties of the oxygen reduction and copper oxidation processes. In the absence of mass transfer limitations, this can be expressed in the form of the Butler-Volmer equation. For copper, this equation can be written as shown:

$$i_{cu} = i_{cu}^{0}\left(10^{T_{Cu}^{a}(E_{corr}-E_{cu})} - 10^{-T_{Cu}^{c}(E_{corr}-E_{Cu})}\right) \quad \text{(B.2)}$$

Equation B.2 describes the dependence of the copper oxidation current on the experimentally measured E_{corr}, the nominal equilibrium electrochemical potential of the copper couple (E_{cu}), and the electrochemical parameters of that couple including the standard exchange current i^{0}_{Cu} and anodic and cathodic Tafel coefficients for the Cu/Cu^{2+} couple (T^{a}_{Cu} and T^{c}_{Cu}, respectively). The standard electrochemical exchange currents and Tafel coefficients of the respective reactions vary widely depending on water quality parameters, surface age, presence and nature of surface scales, etc. Potentially, they can be determined experimentally using a number of techniques that employ a perturbation of E_{corr} and measure the resulting electrochemical current such as PDS and similar electrochemical methods (Reiber et. al. 1996).

The expression for the magnitude of electrochemical current of oxygen reduction is similar to that for copper:

$$i_{O_2} = i_{O_2}^{0}[O_2]\left(10^{T_{C_2}^{a}(E_{corr}-E_{O_2})} - 10^{-T_{C_2}^{c}(E_{corr}-E_{O_2})}\right) \quad \text{(B.3)}$$

In Equation B.3, $T^{a}_{C_2}$ and $T^{c}_{C_2}$ are anodic and cathodic Tafel coefficients for the O_2/H_2O couple, respectively, and $i^{0}_{O_2}$ is the standard exchange current of oxygen reduction. In contrast with Equation B.2, Equation B.3 also contains a pre-exponential term of oxygen concentration ($[O_2]$) because its activity is not necessarily one, in contrast with that for copper (0).

An idealized depiction of conditions described by Equations B.2 and B.3 and the coupling of the oxygen reduction and copper oxidation currents is shown in Figure B.1. In this figure, the logarithms of the absolute values of currents are represented as a function of the electrochemical potential (Evans plot). The point where the cathodic and anodic branches of the oxygen reduction and copper oxidation currents intersect corresponds to the actual corrosion potential and current (denoted as E_{corr} and i_{corr}, respectively) observed for that system. The value of i_{corr} determines the rate of metal oxidation and formation of both soluble species and, in the case of precipitation, surface corrosion scales.

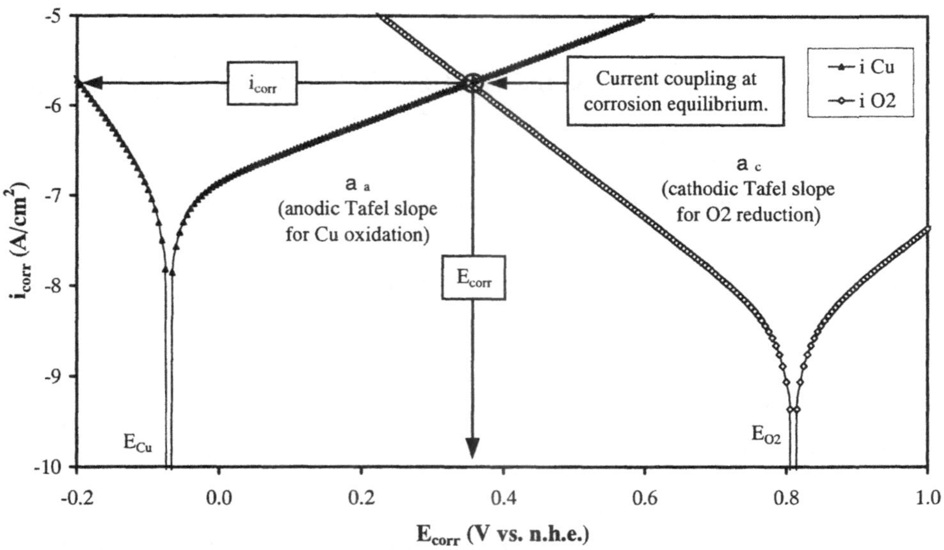

Figure B.1 Idealized depiction of the coupling of electrochemical currents of copper oxidation and oxygen reduction using an Evans plot (log($|i_{corr}|$) vs. E_{corr}) representation.

The positions of E_{cu} and E_{O_2} correspond to the thermodynamically predicted values of the standard potentials of the Cu/Cu^{2+} and O_2/H_2O couples, respectively. Tafel slopes shown on the graph close to the linear branches of the log($|i_{corr}|$) vs. E_{corr} curves correspond to the coefficients in Equations B.2 and B.3. Positions of i_{corr} and E_{corr} indicated by arrows correspond to the coupling of the oxygen reduction and copper oxidation currents.

Important to the CPSF method, the positions of E_{cu} and E_{O_2} potentials shown in Figure B.1 depend on the concentrations of these components at any given time. For example, the value of E_{O_2} is affected by the pH of the solution and the concentration of DO at the surface as shown in Equation B.4:

$$E_{O_2} = E^0_{O_2} - \Phi \cdot pH + 0.25\Phi \log[O_2] \qquad \text{(B.4)}$$

Φ in Equation B.4 is the Nernst factor, which equals 0.059 V at the 25ºC. The expression of the value of E_{Cu} is:

$$E_{Cu} = E^0_{Cu} + 0.5\Phi \log[Cu^{2+}] \qquad \text{(B.5)}$$

In Equation B.5, $[Cu^{2+}]$ is the activity of free copper rather than the total concentration of all Cu^{2+} species at the surface. As will be discussed below, the activity of free Cu^{2+} ions can be calculated using data on the formation of typical complexes of copper (II).

In a general case, the system of Equations B.1 to B.5 cannot be solved analytically. However, if the corrosion potential is far enough from the nominal electrochemical potentials of the copper and oxygen couples (which is likely) and the contributions of the currents of cathodic reduction of Cu^{2+} to Cu(0) and water oxidation to oxygen are negligible, then Equations B.1 through B.5 can be simplified and processed as shown:

$$i_{cu}^{0}10^{T_{Cu}^{a}(E_{corr}-E_{cu})} = [O_2]i_{O_2}10^{-T_{O_2}^{c}(E_{corr}-E_{O_2})} \tag{B.6}$$

After taking the logarithms of the right and left hand sides of this equation we obtain Equation B.7:

$$\log i_{cu}^{0} + T_{Cu}^{a}(E_{corr} - E_{cu}) = \log([O_2]i_{O_2}) - T_{O_2}^{c}(E_{corr} - E_{O_2}) \tag{B.7}$$

Reducing Equation B.7 to express an explicit formula for E_{corr} can be obtained as shown below:

$$E_{corr} = \frac{1}{T_{Cu} + T_{O_2}}\left(\log\left(\frac{[O_2]i_{O_2}}{i_{cu}^{0}}\right) + T_{Cu}E_{cu} + T_{O_2}E_{O_2}\right) \tag{B.8}$$

T_{Cu} and T_{O_2} in Equation B.8 denote the Tafel coefficients for the copper oxidation and oxygen reduction currents, respectively (indexes a and c will be henceforth omitted).

Upon substitution of the expressions for E_{Cu} and E_{O_2} (Equations B.4 and B.5 above) into Equation B.8, we can find that:

$$E_{corr} = \frac{1}{T_{Cu} + T_{O_2}}\left(\log\left(\frac{[O_2]i_{O_2}}{i_{cu}^{0}}\right) + T_{Cu}\left(E_{Cu}^{0} + 0.5\Phi\log[Cu^{2+}]\right) + T_{O_2}\left(E_{O_2}^{0} - \Phi pH + 0.25\Phi\log[O_2]\right)\right) \tag{B.9}$$

This expression relates the value of E_{corr} with the concentrations of DO and free copper at the surface at any moment of time. Equation B.9 can be rewritten to separate the contributions of the fundamental parameters of the corroding system that do not change during stagnation and flow, and concentrations of species that change with time (e.g., free copper, total oxygen). This rearrangement is shown in Equation B.10:

$$E_{corr} = \frac{1}{T_{Cu} + T_{O_2}}\left(\log\left(\frac{i_{O_2}}{i_{cu}^{0}}\right) + T_{Cu}E_{Cu}^{0} + T_{O_2}E_{O_2}^{0} - T_{O_2}\Phi \cdot pH\right) + \\ + \frac{0.25\Phi}{T_{Cu} + T_{O_2}}\left(2T_{Cu}\log[Cu^{2+}] + \left(T_{O_2} + \frac{4}{\Phi}\right)\log[O_2]\right) \tag{B.10}$$

So far, the discussion has been centered on the use of the activity (or concentration) of the free copper instead of total copper (Cu_{total}). The free copper constitutes only a fraction of Cu_{total}, and its contribution is affected by pH, concentration of inorganic carbonate and organic ligands. Although the formation of complexes with organic ligands (notably, natural organic matter, NOM) may be of importance in some surface waters (Rehring and Edwards 1996, Broo et. al. 1999, Korshin et. al. 2000, Boulay and Edwards. 2001), the influence of pH and total concentrations of carbonate is deemed to be more universal and have been considered to be the primary factors affecting free copper.

Thus, total copper released upon stagnation is assumed to be predominantly comprised of carbonate and hydroxo complexes of Cu^{2+} (Pankow 1991, Schecher and McAvoy 1998, Benjamin 2002). The formation of sulfate and chloride complexes at levels of these anions typical for drinking water qualities is largely unimportant. Therefore, the concentrations of total soluble copper is the sum of the concentrations of mono-, di-, tri- and tetrahydroxo complexes of copper and its carbonate complexes ($[Cu(CO_3)]^0$, ($[Cu(CO_3)_2]^{2-}$, ($[Cu(HCO_3)]^+$) (Schecher and McAvoy 1998, Schecher 2001):

$$Cu_{total} = [Cu^{2+}] + [Cu(OH)^+] + [Cu(OH)_2{}^0] + [Cu(OH)_3{}^-] + [Cu(OH)_4{}^{2-}] \\ + [Cu(CO_3)^0] + [Cu(CO_3)_2{}^{2-}] + [Cu(HCO_3)^+] \tag{B.11}$$

Using known formation constants for all these complexes (Schecher 2001) and equations that relate their concentrations with that of free copper (Benjamin 2002), the expression for total copper can be written as shown below:

$$Cu_{total} = [Cu^{2+}]\left(1 + \frac{{}^*\beta_1}{H} + \frac{{}^*\beta_2}{H^2} + \frac{{}^*\beta_3}{H^3} + \frac{{}^*\beta_4}{H^4} + \beta_{CuCO_3}[CO_3^{2-}] + \beta_{Cu(CO_3)_2}[CO_3^{2-}]^2 + \beta_{CuHCO_3}H[CO_3^{2-}]\right)$$

(B.12)

In this expression, H and $[CO_3^{2-}]$ correspond to the activities of proton and carbonate ions, respectively. Formation constants ${}^*\beta_1$, ${}^*\beta_2$, ${}^*\beta_3$, and ${}^*\beta_4$ are for mono-, di-, tri- and tetrahydroxo complexes of copper, respectively. Formation constants β_{CuCO_3}, $\beta_{Cu(CO_3)_2}$ and β_{CuHCO_3} define the respective carbonate complexes of Cu^{2+}. The activity of the H^+ ion can always be calculated as:

$$H = 10^{-pH} \tag{B.13}$$

Using speciation formulas for the concentration of the carbonate ion in a closed system (Benjamin 2002), the expression that correlates the total and free copper is obtained and can be written as shown:

$$[Cu^{2+}] = \frac{Cu_{total}}{\left(1 + \frac{{}^*\beta_1}{H} + \frac{{}^*\beta_2}{H^2} + \frac{{}^*\beta_3}{H^3} + \frac{{}^*\beta_4}{H^4} + \frac{C_T K_1 K_2 (\beta_{CuCO_3} + \beta_{CuHCO_3} H)}{H^2 + K_1 H + K_1 K_2} + \frac{\beta_{Cu(CO_3)_2}(C_T K_1 K_2)^2}{(H^2 + K_1 H + K_1 K_2)^2}\right)}$$

(B.14)

K_1 and K_2 are the first and second deprotonation constants of carbonic acid. An extended form of Equation B.10 that reflects the effects of varying water quality parameters on the correlation between E_{corr} and total copper, can be obtained by substituting Equation B.14 into Equation B.10:

$$E_{corr} = \frac{1}{T_{Cu} + T_{O_2}} \left(log\left(\frac{i_{O_2}}{i^0_{Cu}}\right) + T_{Cu} \cdot E^0_{Cu} + T_{O_2} \cdot E^0_{O_2} - T_{O_2} \cdot \Phi \cdot pH \right) +$$

$$+ \frac{0.25 \cdot \Phi}{T_{Cu} + T_{O_2}} \left(2 \cdot T_{Cu} \cdot log[Cu_{total}] + \left(T_{O_2} + \frac{4}{\Phi} \right) log[O_2] \right) -$$

$$- \frac{0.5 \cdot \Phi \cdot T_{Cu}}{T_{Cu} + T_{O_2}} log\left(1 + \frac{{}^*\beta_1}{H} + \frac{{}^*\beta_2}{H^2} + \frac{{}^*\beta_3}{H^3} + \frac{{}^*\beta_4}{H^4} + \frac{C_T K_1 K_2 (\beta_{CuCO_3} + \beta_{CuHCO_3} H)}{H^2 + K_1 H + K_1 K_2} + \frac{(C_T K_1 K_2)^2 \beta_{Cu(CO_3)_2}}{(H^2 + K_1 H + K_1 K_2)^2} \right)$$

(B.15)

As written above, Equation B.15 is a function of two time-dependent variables, namely DO and total dissolved copper (Cu_{total}); both measured at the surface of the corroding metal rather than in the bulk solution. However, this expression can be simplified because in a closed system the concentrations of DO and total copper do not change independently. In a steady state condition, each mole of consumed oxygen causes two moles of Cu^{2+} species to be released. If the initial concentration of DO prior to the start of a stagnation episode is $C^0_{O_2}$ and the concentration of dissolved copper at the beginning of stagnation is virtually zero, then an expression of E_{corr} as a function of total copper can be obtained:

$$E_{corr} = \frac{1}{T_{Cu} + T_{O_2}} \left(\log\left(\frac{i_{O_2}}{i^0_{cu}}\right) + T_{Cu} E^0_{Cu} + T_{O_2} E^0_{O_2} - T_{O_2} \Phi \cdot pH \right) -$$

$$- \frac{0.5 \Phi T_{Cu}}{T_{Cu} + T_{O_2}} \log\left(1 + \frac{{}^*\beta_1}{H} + \frac{{}^*\beta_2}{H^2} + \frac{{}^*\beta_3}{H^3} + \frac{{}^*\beta_4}{H^4} + \frac{C_T K_1 K_2 (\beta_{CuCO_3} + \beta_{CuHCO_3} H)}{H^2 + K_1 H + K_1 K_2} + \frac{\beta_{Cu(CO_3)_2} (C_T K_1 K_2)^2}{(H^2 + K_1 H + K_1 K_2)^2} \right)$$

$$+ \frac{0.25 \Phi}{T_{Cu} + T_{O_2}} \left(2 T_{Cu} \log[Cu_{total}] + \left(T_{O_2} + \frac{4}{\Phi} \right) \log[C^0_{O_2} - 0.5 Cu_{total}] \right)$$

(B.16)

Equation B.16 constitutes the "master" expression for the CPSF system that relates changes in the corrosion potential of copper during stagnation with the release of copper. Equation B.16 correlates the values of E_{corr} measured at any moment of time with the total concentration of copper at the surface at that moment.

Equation B.16 does not address the kinetics of metal release, which has to be ascertained using a different set of equations. However, it predicts how E_{corr} will change as soluble copper is both released due to electrochemical corrosion and accumulated in non-flowing water in a copper pipe. The behavior of E_{corr} can be modeled for idealized corroding copper surfaces. Typical results of such calculations are shown in Figure B.2. The value of all complex formation and deprotonation constants used in these calculations are shown in Table B.1.

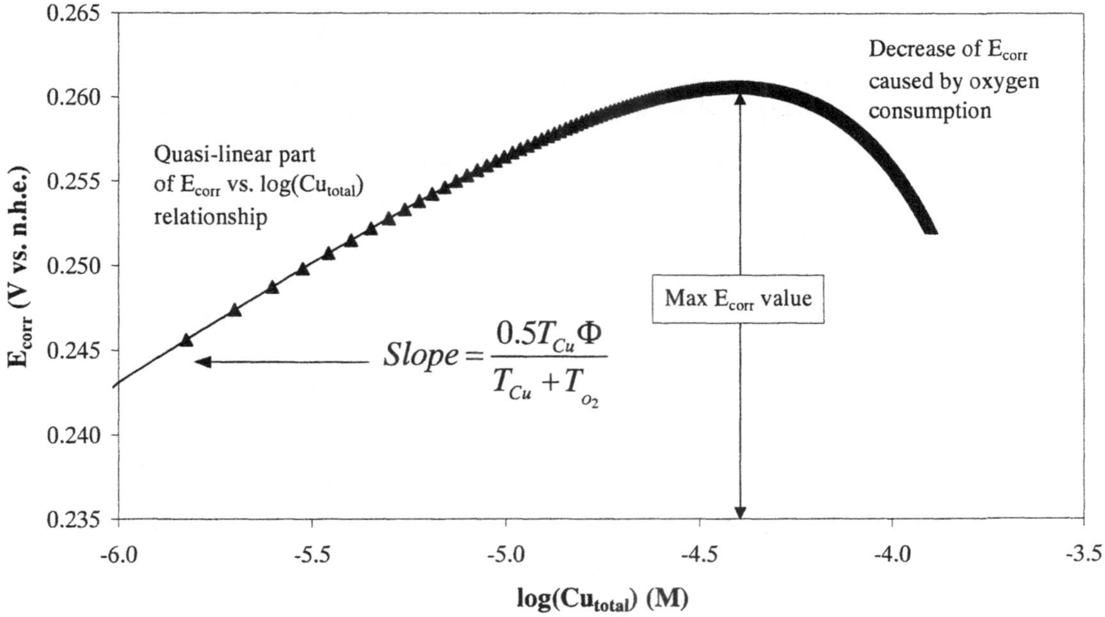

Figure B.2. Theoretical calculations of E_{corr} vs. log(Cu_{total}) relationship. Alkalinity 200 mg/L, pH7.5. Intrinsic copper oxidation and oxygen reduction currents $1{\cdot}10^{-5}$ A/cm^2 and $3{\cdot}10^{-7}$ A/cm^2, respectively. Anodic Tafel slope of copper oxidation and cathodic Tafel slope of oxygen reduction are 4.5 V^{-1} and 3.5 V^{-1}, respectively.

Table B.1
Equilibria constants used in calculations of E_{corr} vs. total copper correlations.
(Adopted from Schecher 2001)

Species	Relevant symbol in Equations B.15 and B.16	Numerical value of formation and deprotonation constants
$[Cu(OH)^{+}]$	$^{*}\beta_1$	$1.00 \cdot 10^{-8}$
$[Cu(OH)_2^0]$	$^{*}\beta_2$	$6.31 \cdot 10^{-17}$
$[Cu(OH)_3^-]$	$^{*}\beta_3$	$1.58 \cdot 10^{-27}$
$[Cu(OH)_4^{2-}]$	$^{*}\beta_4$	$1.26 \cdot 10^{-40}$
$[CuCO_3^0]$	β_{CuCO_3}	$5.89 \cdot 10^{6}$
$[Cu(CO_3)_2^{2-}]$	$\beta_{Cu(CO_3)_2}$	$1.58 \cdot 10^{10}$
$[CuHCO_3^+]$	β_{CuHCO_3}	$1.26 \cdot 10^{12}$
$[HCO_3^-]$	K_1	$5.01 \cdot 10^{-7}$
$[CO_3^{2-}]$	K_2	$5.01 \cdot 10^{-11}$

The graph has two main features. First, in the region of copper concentrations below approximately $10^{-4.5}$ M of Cu_{total} (2.0 mg/L), the relationship between $\log(Cu_{total})$ and E_{corr} is virtually linear. This situation corresponds to the initial phase of a stagnation period, during which species of copper (II) rapidly accumulate in the stagnating water, while the concentration of DO at the surface has not yet undergone significant changes. The slope of the linear relationship between E_{corr} and $\log(Cu_{total})$ is, in accord with Equation B.16, defined by a combination of the Tafel slopes of the cathodic and anodic half-reactions as shown in Equation B.17:

$$Slope = \frac{0.5T_{Cu}\Phi}{T_{Cu} + T_{O_2}} \qquad \text{(B.17)}$$

The next feature of the relationship shown in Figure B.2 is a maximum of E_{corr}. For the conditions used in the calculations being discussed, this maximum is observed when $\log(Cu_{total})$ is approximately $10^{-4.4}$ M (2.5 mg/L). At higher concentrations of Cu_{total}, E_{corr} decreases rather than increases.

This non-monotonic behavior is explained by a combination of effects of copper release and DO consumption that occur simultaneously at the surface. At the beginning of stagnation, the concentration of DO is largely stable but the concentration of Cu_{total} and the values of E_{corr} increase. When both DO and Cu_{total} concentrations change simultaneously and relatively rapidly, E_{corr} tends to reach a maximum. Finally, when the concentration of copper has largely stabilized but DO concentration changes rapidly due to nearly complete oxygen consumption at the metal surface, E_{corr} values start to decrease as shown.

The level of DO consumption and Cu_{total} release at which the value of E_{corr} reaches a maximum can be determined analytically using Equation B.16. In the calculation, it is presumed that the maximum E_{corr} will be observed at an unknown concentration of copper (x) and corresponding to a change in the oxygen concentration ($C^0_{O_2}$-0.5x):

$$E_{corr} = \frac{1}{T_{Cu}+T_{O_2}}\left(\log\left(\frac{i_{O_2}}{i^0_{cu}}\right)+T_{Cu}E^0_{Cu}+T_{O_2}E^0_{O_2}-T_{O_2}\Phi\cdot pH\right)-$$
$$-\frac{0.5\Phi T_{Cu}}{T_{Cu}+T_{O_2}}\log\left(1+\frac{{}^*\beta_1}{H}+\frac{{}^*\beta_2}{H^2}+\frac{{}^*\beta_3}{H^3}+\frac{{}^*\beta_4}{H^4}+\frac{C_TK_1K_2(\beta_{CuCO_3}+\beta_{CuHCO_3}H)}{H^2+K_1H+K_1K_2}+\frac{\beta_{Cu(CO_3)_2}(C_TK_1K_2)^2}{(H^2+K_1H+K_1K_2)^2}\right)$$
$$+\frac{0.25\Phi}{T_{Cu}+T_{O_2}}\left(2T_{Cu}\log[x]+\left(T_{O_2}+\frac{4}{\Phi}\right)\log\left[C^0_{O_2}-0.5x\right]\right)$$

(B.18)

The value of x can be found when the derivative of E_{corr} vs. Cu_{total} (denoted as x in Equation B.18) is zero:

$$\frac{dE_{corr}}{dx} = \frac{d\left(2T_{Cu}\log x+\left(T_{O_2}+\frac{4}{\Phi}\right)\log\left(C^0_{O_2}-0.5x\right)\right)}{dx}=0 \qquad \text{(B.19)}$$

The solution of Equation B.19 yields the following values of the Cu_{total} concentration corresponding to the E_{corr} maximum and the value of the latter:

$$[Cu_{total}]_{max} = \frac{4T_{Cu}C^0_{O_2}}{T_{O_2}+\frac{4}{\Phi}+2T_{Cu}} \qquad \text{(B.20)}$$

$$E^{max}_{corr} = \frac{1}{T_{Cu}+T_{O_2}}\left(\log\left(\frac{i_{O_2}}{i^0_{cu}}\right)+T_{Cu}E^0_{Cu}+T_{O_2}E^0_{O_2}-T_{O_2}\Phi\cdot pH\right)+$$
$$+-\frac{0.5\Phi T_{Cu}}{T_{Cu}+T_{O_2}}\log\left(1+\frac{{}^*\beta_1}{H}+\frac{{}^*\beta_2}{H^2}+\frac{{}^*\beta_3}{H^3}+\frac{{}^*\beta_4}{H^4}+\frac{C_TK_1K_2(\beta_{CuCO_3}+\beta_{CuHCO_3}H)}{H^2+K_1H+K_1K_2}+\frac{\beta_{Cu(CO_3)_2}(C_TK_1K_2)^2}{(H^2+K_1H+K_1K_2)^2}\right)$$
$$-\frac{0.25\Phi}{T_{Cu}+T_{O_2}}\left(2T_{Cu}\log\left(\frac{4T_{Cu}C^0_{O_2}}{T_{O_2}+\frac{4}{\Phi}+2T_{Cu}}\right)+\left(T_{O_2}+\frac{4}{\Phi}\right)\log\left(\frac{\left(T_{O_2}+\frac{4}{\Phi}\right)C^0_{O_2}}{T_{O_2}+\frac{4}{\Phi}+2T_{Cu}}\right)\right)$$

(B.21)

These calculations show that the linearity between log(Cu_{total}) and E_{corr} is observed as long as less than 5% of the initial DO concentration is consumed by copper oxidation (for the initial DO concentration close to 8 mg/L, which is typical for air-saturated surface water qualities). The maximum of E_{corr} is likely to be observed at the moment when approximately 10% of the initial DO has been consumed in corrosion processes.

To summarize, the main result of the CPSF theory is that in the initial phase of stagnation, the logarithm of total copper concentration at the corroding surface is linearly correlated with E_{corr}. E_{corr} can be measured experimentally with virtually any desired precision and time resolution using very inexpensive and easy to use equipment, such as a digital voltmeter. By measuring E_{corr} changes during the episodes of stagnation, it is possible to monitor in situ the accumulation of copper in the water. This theory also predicts that the behavior of E_{corr} during stagnation should be largely similar of that of a copper (II) ion-selective electrode (ISE) in the same water.

E_{corr} values are also affected by the intrinsic electrochemical parameters of the system. These parameters include the values of standard copper oxidation and oxygen reduction currents and relevant anodic and cathodic Tafel coefficients. Electrochemical parameters determine the absolute value of E_{corr} but do not necessarily affect its correlation with $\log(Cu_{total})$ in stagnation episodes.

However, the theory presented above has significant limitations. First, the CPSF theory of E_{corr} behavior for copper does not take into consideration the formation of Cu(I) species, which is likely to take place and affect the kinetics of copper release and E_{corr} values. Second, all theoretical analysis presumed that all metal oxidized by dissolved oxygen is released to form soluble copper. The formation or dissolution of solid phases was not considered in the above theory. The dissolution of the solid copper-containing phases (e.g., malachite) will also contribute to soluble copper but this process was not accounted for. Mass transfer limitations have not been explicitly accounted for in the expressions for the values of currents of the copper oxidation and oxygen reduction currents. Mass transfer limitations are likely to arise during stagnation episodes. The presented theory presumes that the concentration of oxygen and copper is the same at a corroding surface and in a bulk solution. However, they may be different, and non-uniform distribution of DO and dissolved copper is an important factor that must be addressed.

CPSF Theory with Elements of Corrections for Diffusion Limitations

The mass transfer limitations associated with diffusion are likely to play a major role in metal release. The treatment of model CPSF data presumed that all relevant species are uniformly distributed inside the corrosion cell. However, this is not the case for most electrochemical systems without forced convection (Newman 1972). The following discussion introduces a fairly simple correction to reflect a non-uniform distribution of oxidants and reduced species at the corroding surface.

In the simplest case, a correction can be done presuming that the layer of solution adjacent to the surface (diffusion layer) contains all released metal. The concentration of DO also changes within the same layer but remains stable outside of it. In the case of stationary diffusion, the relationship between both copper and DO concentrations and the distance from the electrode surface can be described by the error function (Cussler 1984). Below, the model is simplified and it is presumed that the distribution of DO and copper species in the diffusion layer is uniform, but the thickness of this hypothetically uniform layer changes with time. In accord with the theory of diffusion (Cussler 1984), the thickness of the diffusion layer δ_{diff} can be estimated using Equation B.22:

$$\delta_{diff} = \sqrt{Dt} \tag{B.22}$$

In this equation, D is the diffusion coefficient of the species of interest, while t is the time from the start of diffusion (that is, stagnation time in the case of CPSF measurements). The diffusion coefficients of DO and copper at the normal temperature are $2.1 \cdot 10^{-5}$ cm^2/s and $7.14 \cdot 10^{-6}$ cm^2/s, respectively.

Based on the presumption of the uniformity of the species distribution in the diffusion layer, its growth with time and release of metal associated with the oxidation of copper by DO, ordinary differential equations that describe the process can be developed. For example, the differential equation for the total copper and DO concentration in the uniform diffusion layer can be written as follows:

$$\frac{dCu_{diff}}{dt} = \frac{1}{\sqrt{D_{Cu}t}}\left(\frac{i_{corr}(i)}{2F}\right) \tag{B.23}$$

$$\frac{dO_{2\_diff}}{dt} = -\frac{1}{\sqrt{D_{O_2}t}}\left(\frac{i_{corr}(i)}{4F}\right) \tag{B.24}$$

where D_{O_2} and D_{Cu} are diffusion coefficients of oxygen and copper (respectively), Cu_{diff} and $O_{2\ diff}$ are the concentrations of these species in the diffusion layer, $i_{corr}(t)$ is the rate of corrosion at any moment of time during a stagnation episode, and F is the Faraday constant. It is the concentration of copper and DO in the diffusion layer (rather than the concentration in the bulk solution) that affect the values of E_{corr} and i_{corr}. Therefore, Cu_{diff} and $O_{2\ diff}$ values have to be used in calculations of E_{corr} and i_{corr} using the equations developed earlier.

Metal release measurements do not normally allow quantifying metal concentrations in the diffusion layer. Instead, only average concentrations in a homogenized sample (for example, in a one-liter sample withdrawn according to stipulations of the Lead and Copper Rule) can be experimentally determined. In this case, the predicted average concentration of the total copper for a rectangular cell with a known thickness (denoted as l_{cell} in Equation B.25) can be calculated presuming the applicability of Equation B.22:

$$Cu_{total} = \frac{\sqrt{Dt}}{l_{cell}} Cu_{diff} \tag{B.25}$$

Using the concept formally expressed in Equations B.22 through B.25, calculations were carried out for the kinetics of metal release and a comparison was made of the correlation between the concentration of total soluble copper and changes of E_{corr} during stagnation episodes. These model calculations were carried out using the values of standard copper oxidation and oxygen reduction currents $1 \cdot 10^{-6}$ A/cm^2 and $3 \cdot 10^{-7}$ A/cm^2, respectively and Tafel anodic and cathodic coefficients 4.5 V^{-1} and 3.0 V^{-1}, respectively. The thickness of the corrosion cell was 1 cm, pH 7.5, and alkalinity was 20 mg/L.

Results of the model calculations for DO and total copper concentrations in the diffusion layer are shown in Figure B.3, while the behavior of E_{corr} and i_{corr} over time is demonstrated in Figure B.4. The relationship between both the averaged concentrations of total copper in the corrosion cell and those in the diffusion layer versus E_{corr} is shown in Figure B.5. Data presented in Figures B.4 and B.5 were calculated applying the hypothesis of uniformity of the diffusion layer.

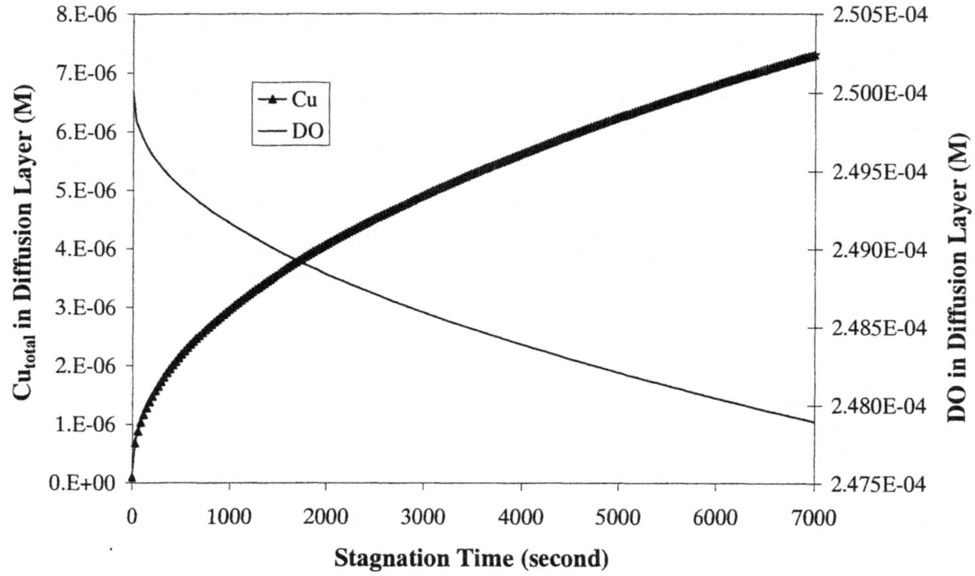

Figure B.3 Accumulation of copper and consumption of DO in the hypothetically uniform diffusion layer that develops during stagnation episodes.

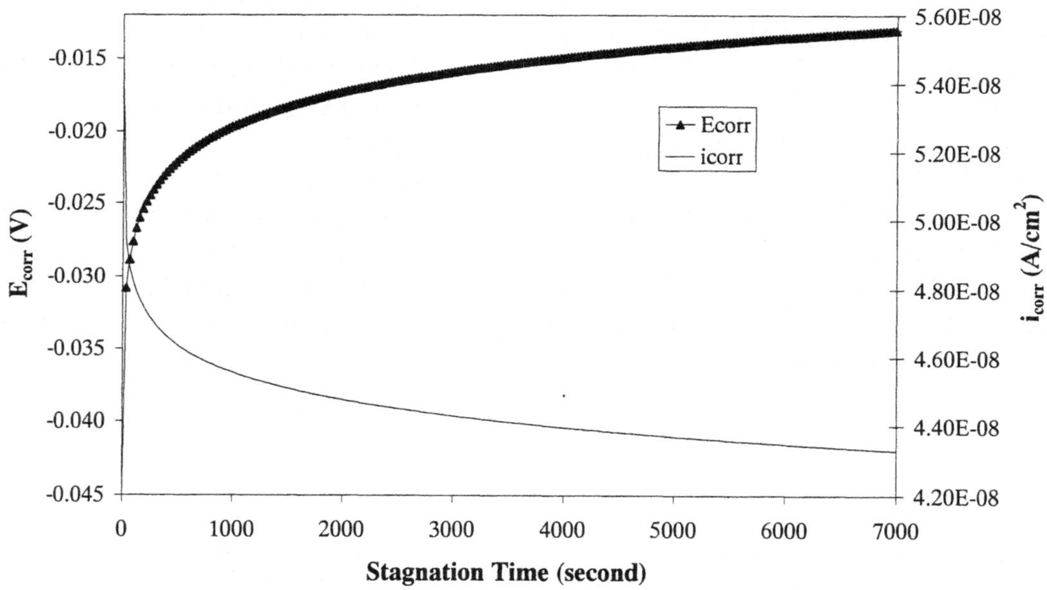

Figure B.4 Results of numerical modeling of the behavior of E_{corr} and i_{corr} during stagnation.

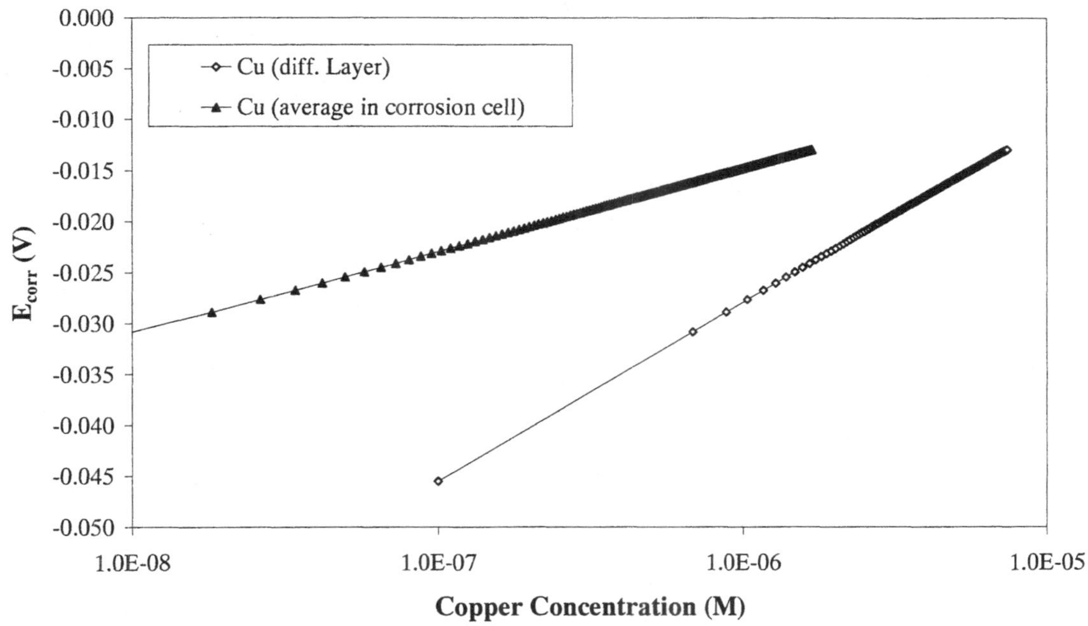

Figure B.5 Results of numerical modeling of correlations between E_{corr} and copper concentrations in the diffusion layer and in experimentally processed homogenized samples.

For the conditions used in the model calculations, copper rapidly accumulates in the diffusion layer, while the DO concentration is largely stable. Consistent with the predictions of the general CPSF theory, E_{corr} undergoes a noticeable change during stagnation, especially during the first 20 minutes of stagnation. The change of i_{corr} during the same period of time is also observed.

Examination of the correlation between the concentrations of copper in the uniform diffusion layer and that averaged for the whole volume of water inside the CPSF cell (Figure B.5) shows that despite pronounced differences in the absolute values of Cu_{total} and in the diffusion layer, the correlations between their logarithms and E_{corr} are almost perfectly linear. This observation is applicable for the phase of corrosion in which the consumption of oxygen in the diffusion layer does not exceed approximately 5% of its initial concentration, as was determined in the development of the general theory of the CPSF method.

Despite the simplicity of the employed hypothesis concerning the properties of the diffusion layer, the model allows estimating the nature of trends caused by the existence of a non-uniform distribution of soluble metal in the CPSF cell. The relevant data of numerical modeling indicate that the development of a uniform diffusion layer is not expected to affect the linearity of correlation between E_{corr} and $\log(Cu_{total})$ values, although it may affect its slope and position.

Evaluation of Precipitation Effects on the Behavior of Metal Release and Corrosion Potential in the CPSF Cell

Following the introduction of a hypothesis concerning the development of a diffusion layer during stagnation in the CPSF cell, it is relevant to examine the possibility of formation of solid phases in that diffusion layer. This project does not address the effects of dissolution of copper-containing solid phases that might have already been formed on the corroding surface. Although this process might be of importance in the kinetics of metal release (Merkel 2003, Merkel et. al. 2002, Lytle and Schock 2000), the formation of over-saturated solutions at the vicinity of the surface and growth of corrosion scales is more typical for surfaces that have not yet achieved a steady state.

Theoretical analysis of the predominance of solid phases in drinking water as well as experimental observations indicate that in the absence of phosphate and precipitation of its minerals, the basic copper carbonate, malachite ($Cu_2(OH)_2CO_3$) is the prevalent solid formed on corroding copper surfaces (Merkel et. al. 2002, Edwards et. al. 1996a, Edwards et. al. 1996b).

The objective of the following exercise was to evaluate the magnitude of effects of this process on the behavior of E_{corr} and its relationship with experimentally measured metal concentrations. The equation developed to estimate the concentration of copper in the hypothetically uniform diffusion layer was presented earlier (Equations B.22 through B.25). To introduce the kinetic term that corresponds to the precipitation of malachite, Equation B.23 can be modified as shown below:

$$\frac{dCu_{diff}}{dt} = \frac{1}{\sqrt{D_{Cu}t}}\left(\frac{i_{corr}(i)}{2F}\right) - k_p\left(Cu_{diff} - Cu_p\right) \qquad \text{(B.26)}$$

In this equation, Cu_p is the threshold concentration that has to be reached to initiate the precipitation of malachite. This concentration can be determined based on the solubility product of malachite ($10^{-33.4}$) at any pH and total concentration of carbonate. The rate of formation of malachite in over-saturated solutions is denoted as k_p (quantified in s^{-1} units). It will be varied in model calculations to mimic different scenarios of malachite precipitation.

Equation B.26 presumes that once the threshold concentration of malachite precipitation in the diffusion layer is reached, precipitation of malachite will occur. The model is rather simple and more sophisticated solid phase formation kinetics can be alternatively used. However, for the purposes of exploring the effects of solid phase formation in the diffusion layer, this model (Equation B.26) was determined to be adequate.

Model calculations to be discussed below were carried out for conditions similar to those employed in the discussion of effects of a uniform diffusion layer. Namely, the standard current of copper oxidation and oxygen reduction were $1\cdot10^{-5}$ A/cm^2 and $3\cdot10^{-6}$ A/cm^2, respectively. The Tafel slopes were 4.5 V^{-1} and 3.0 V^{-1} for copper and oxygen, respectively. The initial oxygen concentration was $2.5\cdot10^{-4}$ M (8 mg/L), pH and alkalinity were 7.5 mg/L and 20 mg/L, respectively. The thickness of the CPSF cell was 1 cm. The rate of malachite precipitation varied from 0.00 s^{-1} to 0.1 s^{-1}.

The kinetics of the formation of malachite, total soluble copper and free copper (all in the diffusion layer) and average copper concentration measured experimentally (that is, presuming complete mixing of the solution in the diffusion layer and the rest of the sample) for malachite precipitation rates 0.01 s^{-1} and 0.05 s^{-1} are presented in Figure B.6 and Figure B.7, respectively. For kp=0.01 s^{-1} a significant level of malachite formation was observed at a stagnation time of greater than approximately 1000 seconds, while for kp=0.05 s^{-1} this takes place at a stagnation time greater than 200 seconds. This time range is typical for the onset of solid phase formation upon stagnation in copper pipe in drinking water qualities.

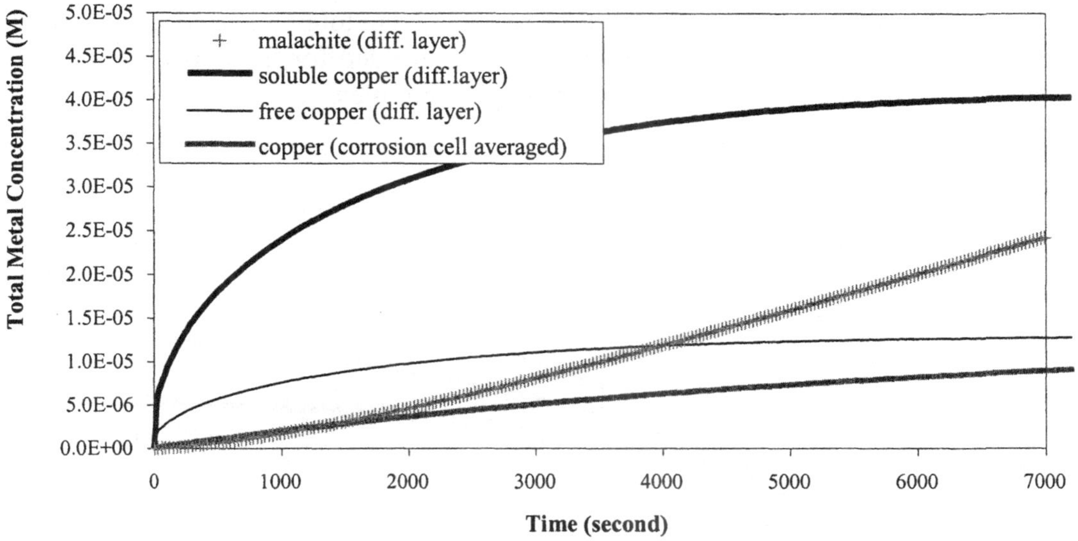

Figure B.6 Results of modeling of copper release and formation of malachite in the uniform diffusion layer. k_p=0.01 s^{-1}.

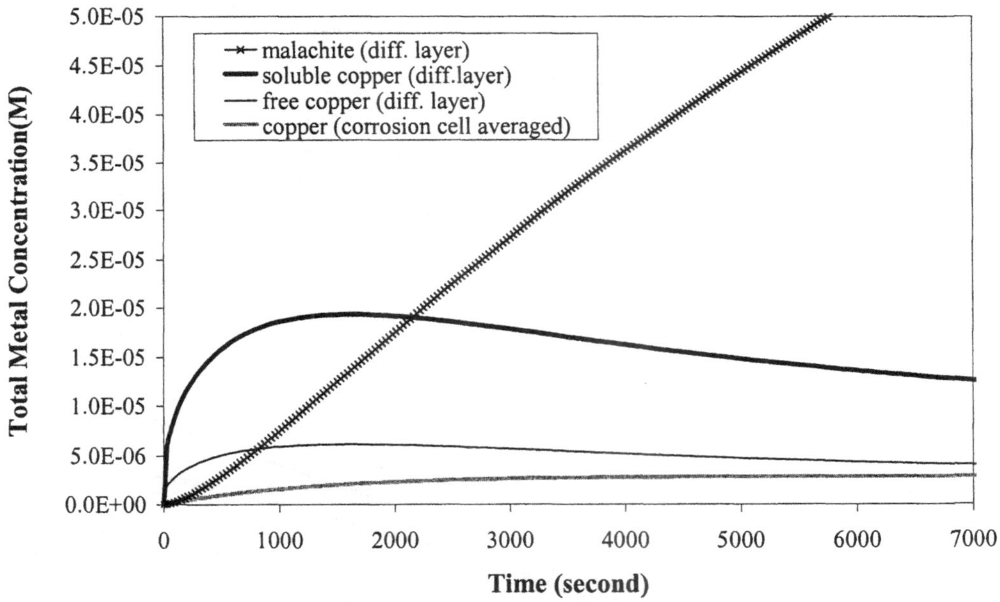

Figure B.7 Results of modeling of copper release and formation of malachite in the uniform diffusion layer. k_p=0.05 s^{-1}.

Figure B.8 and Figure B.9 compare the kinetics of accumulation of total soluble copper in the diffusion layer and in the CPSF cell, respectively, at varying malachite precipitation rates. For formal precipitation rates greater than 0.05 s^{-1}, the concentration of copper in the diffusion layer at some point reaches a maximum after which it can rapidly decrease. The effect of precipitation does not seem to dramatically affect the kinetics of E_{corr} changes in the CPSF cell (Figure B.10). This result needs to be confirmed and explored in more detail. However, the correlation between the values of E_{corr} and total soluble metal released inside the CPSF cell indicates that the formation of malachite can affect this relationship (Figure B.11), especially for the presumed rates of malachite precipitation greater than 0.02 s^{-1}. At slower rates of the precipitation of the solid phase, the correlation between the total released copper and E_{corr} remains largely unaffected. For these conditions, the time needed for the malachite contribution to reach 10% of total oxidized copper is approximately 13 minutes.

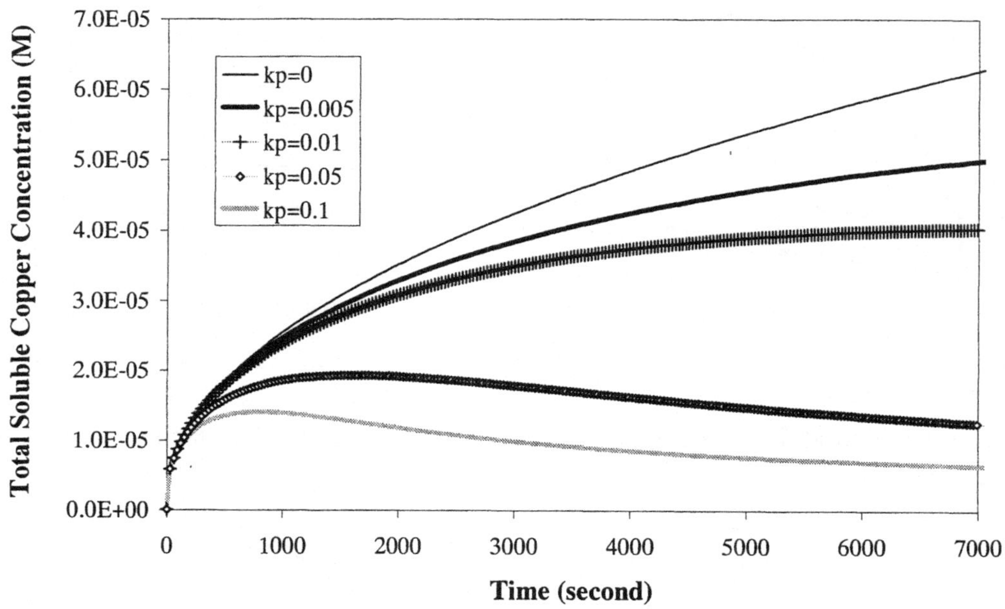

Figure B.8 Comparison of the kinetics of total soluble copper release in the uniform diffusion layer at the corroding copper surface at varying kinetic constants of malachite precipitation.

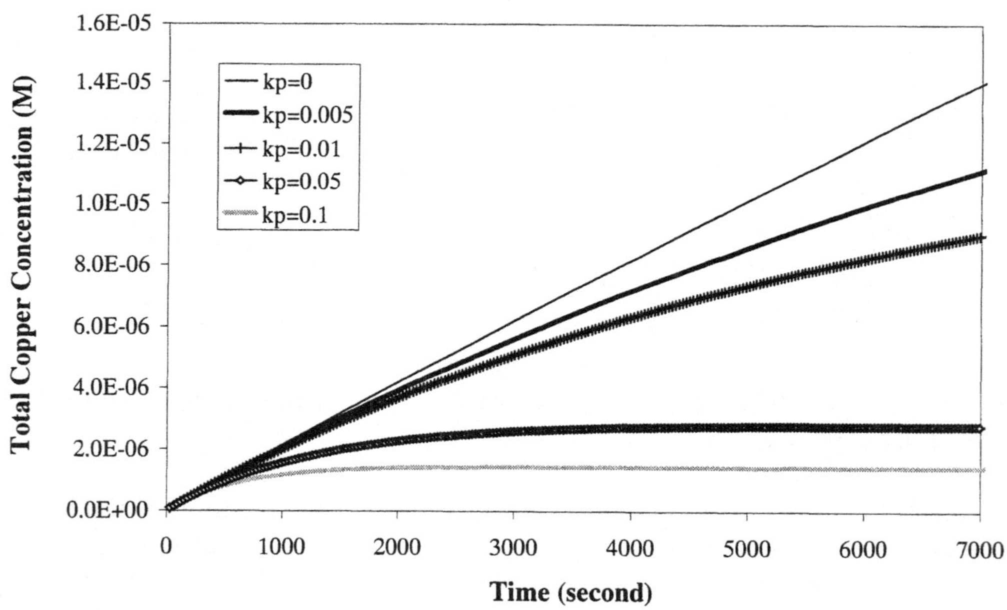

Figure B.9 Comparison of the kinetics of total copper release in the CPSF cell at varying kinetic constants of malachite precipitation.

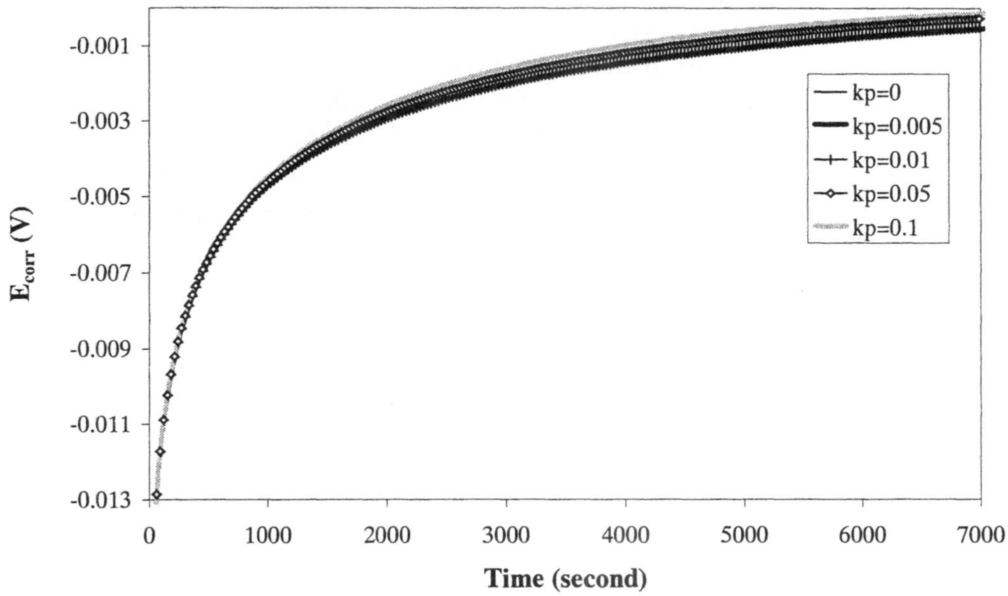

Figure B.10 Comparison of the dynamics of changes of E_{corr} at varying kinetic constants of malachite precipitation.

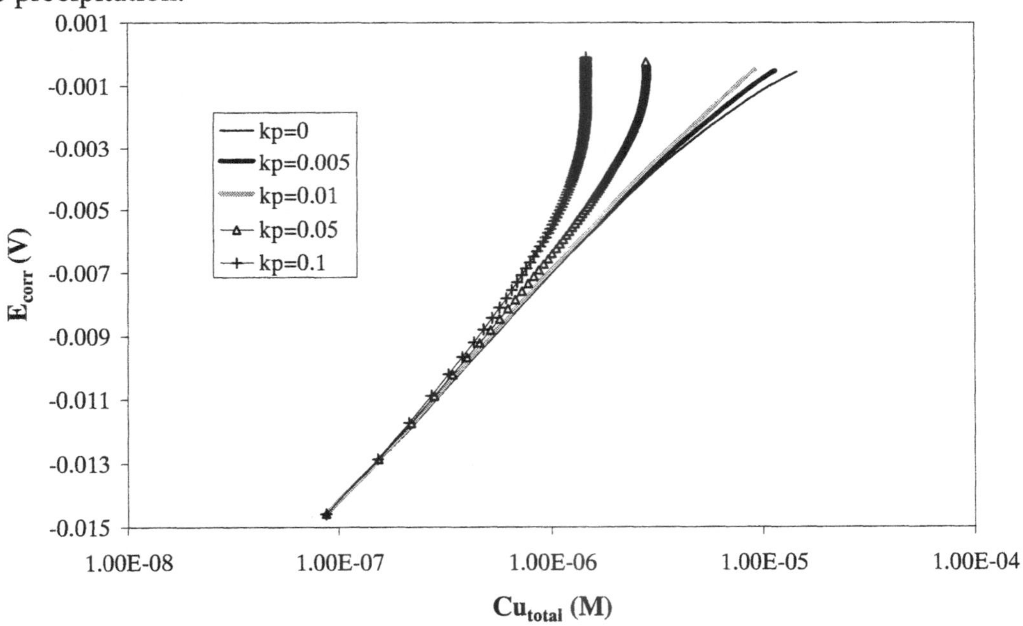

Figure B.11 Comparison of the relationship between E_{corr} and release of total soluble copper at varying kinetic constants of malachite precipitation.

SUMMARY

The theoretical analysis development of CPSF cell kinetics yielded several key results. First, in the absence of precipitation of solid phases, a strong linear correlation between the logarithm of total soluble copper and corresponding E_{corr} values was expected. Because E_{corr} can be easily measured, the concentration of copper in CPSF cell can therefore be determined provided that the cell has been calibrated to determine the actual $\log(Cu_{total})$ vs. E_{corr} correlation. Second, the formation of solid phases can present a problem for predictions of the total copper based on E_{corr} measurements because it may cause deviations of linearity of the $\log(Cu_{total})$ vs. E_{corr} correlation. The magnitude of this deviation depends on the solid phase precipitation rate. At present, experimental data are not detailed enough to ascertain whether this may severely limit the performance of the CPSF method. Our estimate based on the results of model calculations indicates that there might be a threshold of approximately 15 minutes, after which formation of non-soluble copper has to be accounted for. However, these limitations do not contradict the general applicability of the predicted trends.

APPENDIX C
ALTERNATIVE TAP MONITORING FIELD STUDIES

Four participating utilities conducted field study evaluations that implemented the draft monitoring protocol for alternative tap sampling; the City of Hastings, NE; Oak Lodge Water District, Clackamas, OR; Philadelphia Water Department, Philadelphia, PA; and Detroit Water and Sewerage Department, Detroit, MI. The objective of these field studies was to generate data to assess the variability and sensitivity of the alternative lead and copper tap monitoring protocol. The following sections contain descriptions of the alternative tap monitoring programs implemented by each utility, summaries of the data collected from these systems, and comparisons between the data collected from this monitoring and available regulatory tap monitoring data.

CITY OF HASTINGS, NE

Background

The City of Hastings, NE provides water service to approximately 23,000 people from an untreated groundwater supply. Twenty wells pump water directly into the main distribution system. Typical pH is 7.5 and alkalinity levels for the water from these wells is 200 to 300 mg/L as $CaCO_3$. The majority of Hastings monitoring sites contain lead service lines. In-home tap monitoring completed for the Lead and Copper Rule (LCR) resulted in 90th percentile lead levels well below the action level of 15 µg/L, however copper levels have exceeded the 1.3 mg/L copper action level in all sampling rounds, as shown in Table C.1.

Table C.1
LCR regulatory monitoring results – Hastings, NE

Year	No. of Samples	Mean	Median	Min.	Max.	90th Percentile
			Lead Levels, µg/L			
1993	188	1.7	1.0	0.0	15.0	3.0
1997	57	1.9	1.0	0.0	16.0	5.0
2000	29	2.1	1.4	1.0	13.0	3.0
			Copper Levels, mg/L			
1993	188	0.67	0.45	0.0	3.9	1.6
1997	57	0.73	0.47	0.0	2.2	1.8
2000	29	0.71	0.46	0.02	2.1	2.1

Alternative Tap Monitoring Program

Hastings conducted an alternative tap monitoring program for standing lead and copper levels at ten sites in their service area in March and April of 2002. The focus of the monitoring program was to evaluate copper levels; therefore all of the selected sites contained copper plumbing with lead solder. None of the sites selected for the alternative monitoring program had been sampled for regulatory monitoring. The age of the in-home plumbing was estimated by the date of installation of the service line to the house. Eight of the sites had services installed over 20 years ago, and two sites had services installed less than 20 years ago (Table C.2). One-liter standing samples were collected after a 6 to 8 hour standing time and analyzed for both copper and lead levels. Samples were collected weekly for a six week period (i.e. six rounds of sampling). All analyses were completed by a certified laboratory using standard methods specified in the LCR.

Table C.2
Sampling sites for Hastings, NE

Site Number	Address	Service Size (in.)	Service Type	Length (ft)	Installation Date	Main Size (in.)
1	1126 N. Washington	3/4	Cu	50	1960	6
2	1118 N. Washington	3/4	Cu	52	1955	6
3	614 Ringland Rd.	3/4	Cu	78.33	1944	4
4	1102 E. 9th	1	Cu	243.5	1968	6
5	509 Garfield	3/4	Cu	56	1979	4
6	221 Kent	3/4	Cu	-	1997	6
7	915 Webster	3/4	Cu	85	1952	6
8	1308 S Wabash	1	Cu	50	1986	10
9	2201 West 9th	3/4	Cu	50	1963	6
10	947 S. Colorado	3/4	Cu	48	1965	6

Results and Discussion

Lead and copper levels measured from all data collected during the alternative monitoring program are listed in Table C.3. Lead levels measured during the alternative monitoring program averaged 21.8 μg/L and copper levels averaged 1.1 mg/L. Lead level variability was lower than copper level variability as measured by the coefficient of variation (CV) which is the ratio of the standard deviation over the mean. Tables C.4 and C.5 list summary statistics for lead and copper for each site sampled, and for each sampling round.

Table C.3
All lead and copper data - Hastings, NE

	Number of samples	Mean	Median	90th percentile	Standard deviation	CV
Lead, μg/L	60	21.8	21.2	26.9	6.1	0.3
Copper, mg/L	60	1.1	1.1	1.9	0.65	0.6

Table C.4
Lead and copper levels from each site during sampling – Hastings, NE

Parameter	Number of sites	Mean	Median	90th percentile	Standard deviation	CV
Lead, μg/L	10	16.0 – 30.5	19.0 – 29.6	23.5 – 49.0	2.8 – 10.9	0.1 - 0.4
Copper, mg/L	10	0.28 – 2.44	0.29 – 2.38	0.33 – 29.7	0.05 - 0.58	0.06 - 0.90

Table C.5
Lead and copper levels for each round of sampling – Hastings, NE

Parameter	Number of samples per site	Mean	Median	90th percentile	Standard deviatio n	CV
Lead, µg/L	6	15.9 – 25.0	14.9 – 24.0	21.0 – 38.0	3.3 – 9.2	0.2 - 0.4
Copper, mg/L	6	1.0 – 1.2	0.9 –1.1	1.9 – 2.3	0.6 - 0.8	0.59 - 0.61

Figures C.1 through C.4 show the lead and copper levels measured from each of the sites and each of the sample rounds for the alternative monitoring program. Site No. 8 exhibited the highest lead and copper levels during the monitoring period. The interior piping in this home had been replaced just five months before monitoring occurred, and presumably the faucet was replaced as well. This site exhibited average lead levels greater than 30 µg/L and average copper levels greater than 2.2 mg/L, both of which were significantly higher than the lead and copper results from the other sites selected for this alternative monitoring program.

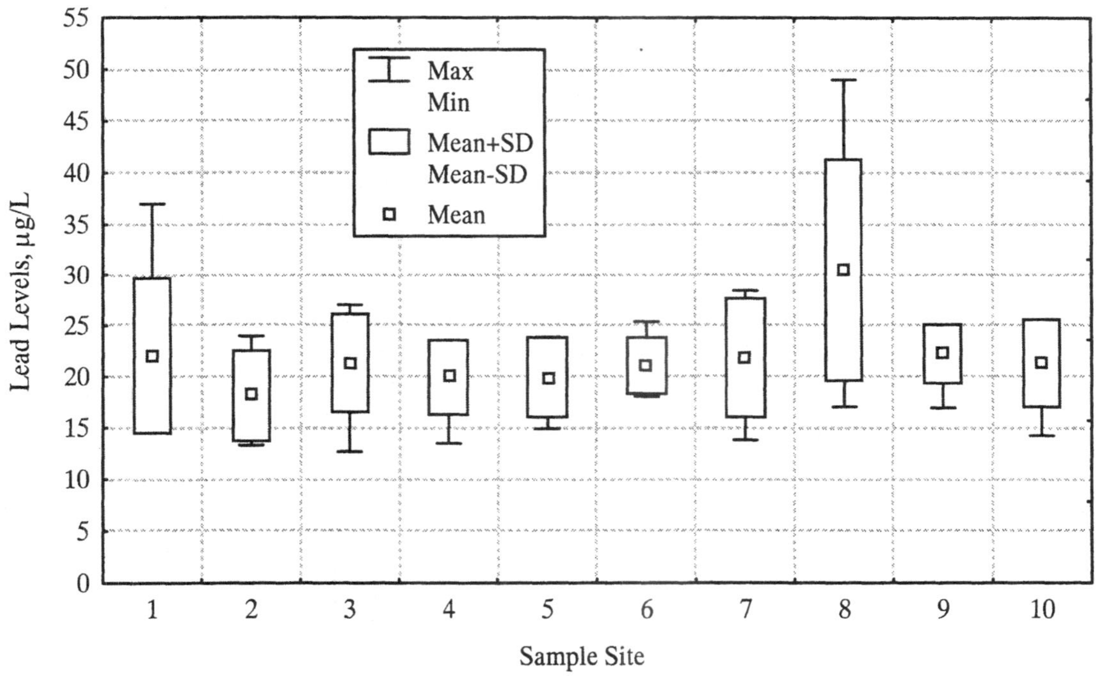

Figure C.1 Lead levels by site for Hastings tap monitoring program

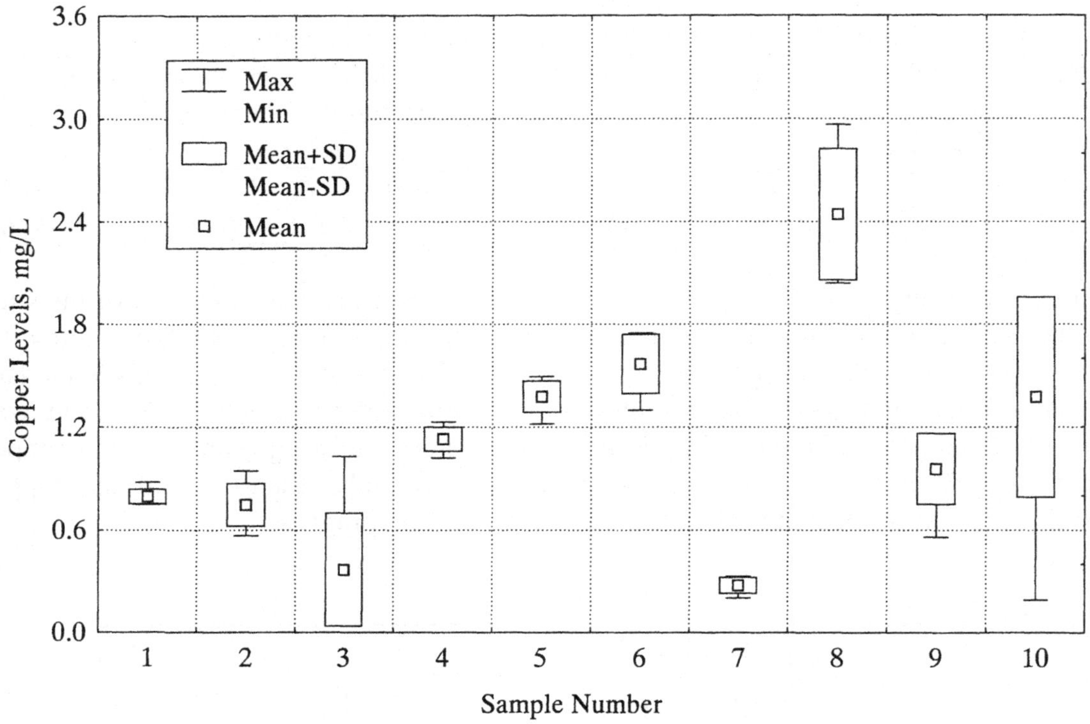

Figure C.2 Copper levels by site for Hastings tap monitoring program

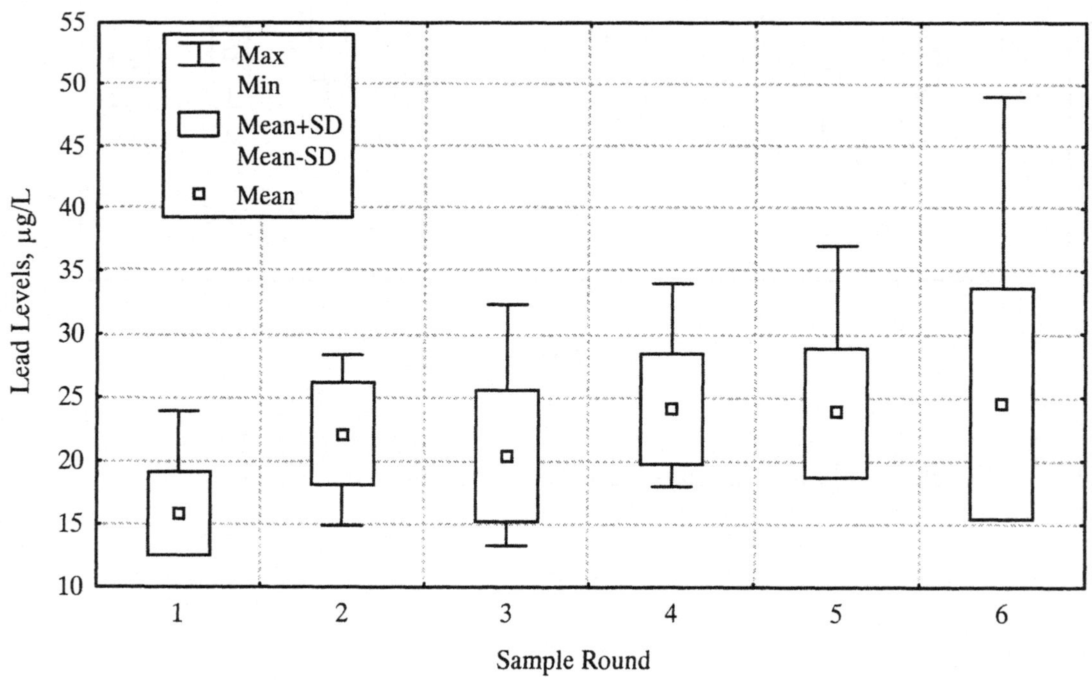

Figure C.3 Lead levels by sample round for Hastings monitoring program

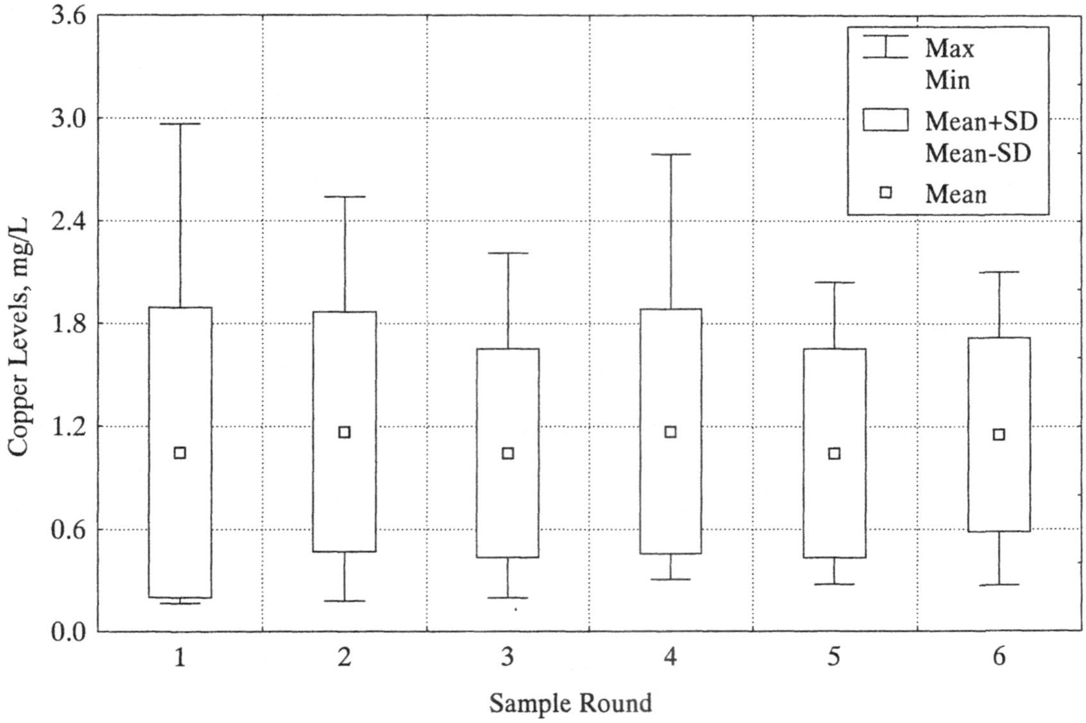

Figure C.4 Copper levels by sample round for Hastings monitoring program

None of the sites used for regulatory monitoring were used for the alternative monitoring program. Both lead and copper levels measured during the alternative monitoring program were higher than regulatory monitoring results (Table C.6), with lead levels an order of magnitude higher. Regulatory monitoring data for lead and copper levels from only sites with lead services lines are listed in Table C.7, along with results from the alternative monitoring program. Table C.8 contains regulatory and alternative monitoring data from only the sites that contained lead soldered copper plumbing. Lead levels from the alternative monitoring program were an order of magnitude higher than lead levels from both lead service line sites and lead soldered copper sites. Copper levels from alternative monitoring were similar to copper levels from lead soldered copper sites but the variability was lower.

Table C.6
Comparison of LCR and alternative monitoring results for all sites - Hastings, NE

	Number of samples	Mean	Median	Min.	Max.	90th percentile	Standard deviation	CV
				Lead, μg/L				
LCR - 1993	188	1.7	1	0	15.0	3.0	2.1	1.23
LCR - 1997	57	1.9	1	0	16.0	5.0	2.9	1.48
LCR - 2000	29	2.1	1.4	1	13.5	3.0	2.3	1.13
Alt. Mon.	60	21.8	21.2	12.7	49.0	26.9	6.1	0.28
				Copper, mg/L				
LCR - 1993	188	0.67	0.45	0.00	3.93	1.63	0.65	0.97
LCR - 1997	57	0.73	0.47	0.00	2.22	1.82	0.70	0.96
LCR- 2000	29	0.71	0.46	0.02	2.11	2.06	0.71	1.00
Alt. Mon	60	1.10	1.06	0.17	2.97	1.89	0.65	0.59

Table C.7
Regulatory monitoring results for lead service line sites and alternative monitoring program – Hastings, NE

Monitoring Program	Number of samples	Mean	Median	Min.	Max.	90th Percentile	Standard deviation	CV
				Lead Levels, μg/L				
LCR - 1993	169	1.8	1.0	0.0	15.0	4.0	2.2	1.2
LCR - 1997	37	1.7	1.0	0.0	16.0	4.0	2.8	1.6
LCR - 2000	28	2.1	1.4	1.0	13.5	3.0	2.4	1.1
Alt. Mon.	60	21.8	21.2	12.7	49.0	26.9	6.1	0.3
				Copper Levels, mg/L				
LCR - 1993	169	0.58	0.37	0.0	2.37	1.47	0.56	0.97
LCR - 1997	37	0.40	0.19	0.0	1.85	1.17	0.51	1.3
LCR - 2000	28	0.68	0.43	0.02	2.11	2.06	0.71	1.7
Alt. Mon.	60	1.1	1.1	0.17	3.0	1.9	0.65	0.6

Table C.8
Regulatory monitoring results for copper service line sites– Hastings, NE

Year	Number of samples	Mean	Median	Min.	Max.	90th percentile	Standard deviation	CV
				Lead Levels, μg/L				
1993	18	1.5	1.0	0.0	4.0	3.0	1.1	0.7
1997	20	2.3	1.0	0.0	10.0	8.0	6.1	1.3
2000	1	2.4	--	--	--	--	-	-
Alt. Mon.	60	21.8	21.2	12.7	49	26.9	2.9	0.3
				Copper Levels, mg/L				
1993	19	1.53	1.47	0.62	3.93	2.42	0.79	0.5
1997	20	1.34	1.38	0.14	2.22	2.15	0.59	0.4
2000	1	1.55	-	-	-	-	-	-
Alt. Mon.	60	1.1	1.1	0.17	3.0	1.9	0.65	0.6

Some variations between the results of the LCR and alternative monitoring may be due to changes in the LCR monitoring procedures in Nebraska. The original LCR sampling protocol required the entire household plumbing to be stagnant for 6 to 8 hours. The samples were taken by the homeowner in the early morning to obtain the first draw sample. The point of collection was typically a kitchen or bathroom faucet. In 2001 the Nebraska Health Department authorized a revised LCR protocol that allows for normal household use of water at all non-sampling taps. On the morning of scheduled sampling, utility staff contact residents and tag a faucet to be sampled so no water will be used. After 6 hours a utility staff collects a 2-liter water sample. During alternative tap monitoring, the entire household plumbing remained stagnant for 6 to 8 hours. This difference in procedures could have attributed to the variation between the results from the two sampling protocols.

Finally, an evaluation of the impact of service line age on copper levels was conducted. Table C.9 lists the summary statistics for sites where the service line age was less than or equal to 20 years old, and for sites where the service line age was greater than 20 years old. As shown in Figure C.5, copper levels decreased with increasing service line age.

Table C.9
Age comparison for alternative monitoring results – Hastings, NE

Service line age	Parameter	Number of samples	Mean	Median	90th percentile	Standard deviation	CV
<=20 Years	Pb, μg/L	12	25.7	22.9	34.0	9.1	0.4
	Cu, mg/L	12	2.0	1.9	2.8	0.54	0.3
>20 years	Pb, μg/L	48	20.8	21.0	25.4	4.7	0.2
	Cu, mg/L	48	0.9	0.9	1.5	0.46	0.5

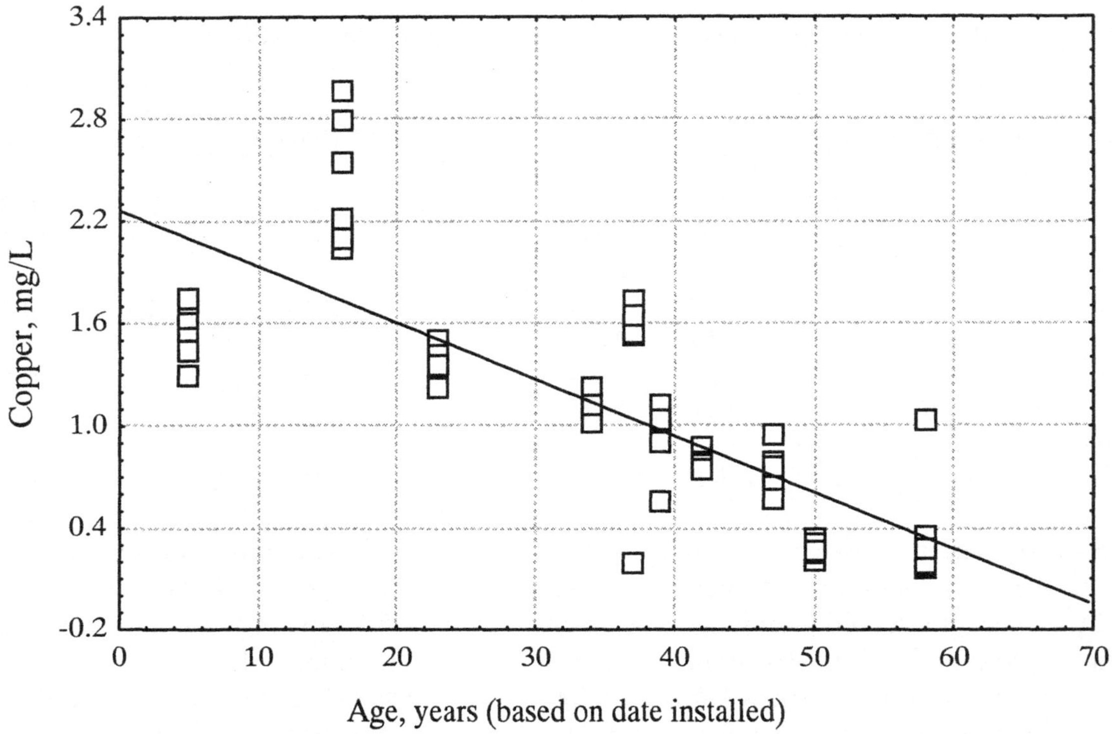

Figure C.5 Copper levels and age of plumbing from Hastings monitoring program

Overall comparisons showed that copper levels were similar and lead levels much higher in the alternative monitoring program when compared to regulatory monitoring. Some variations may be attributed to the changes to the LCR monitoring procedures which occurred in 2001. In the alternative monitoring program, all taps in a home remained stagnant, whereas in the LCR monitoring program, only the sampling tap was required to remain stagnant. More comparable data for lead levels may have occurred if the sites selected for alternative monitoring had been a subset of the sites used for regulatory monitoring. The majority of regulatory sites contained lead services whereas the alternative monitoring sites were lead soldered copper sites since the primary focus of Hasting's alternative monitoring program was to evaluate elevated copper levels, therefore lead soldered copper sites were targeted. The much higher lead levels when compared to regulatory monitoring could be due to lead leaching from faucets or higher leaching of lead from lead solder at the alternative monitoring sites when compared to regulatory sites containing lead soldered copper. The age evaluation indicates that copper levels continue to decrease with increasing age, and that newer copper piping will be particularly problematic with respect to elevated copper levels. There was essentially no difference in lead levels based on age of the service piping.

Based on the monitoring results obtained at Hastings, the alternative monitoring program exhibited copper levels similar to those measured during regulatory monitoring, particularly for sites with similar piping types (i.e. lead soldered copper piping). The variability measured in copper levels from alternative monitoring was similar or slightly higher than variability measured in copper levels from regulatory monitoring.

OAK LODGE WATER DISTRICT, OR

Background

The Oak Lodge Water District (OLWD) receives water from the North Clackamas County Water Commission (NCCWC) which provides potable water to a service population of approximately 30,000 customers northwest of Portland, Oregon. The water system, maintained by the OLWD, was optimized for corrosion control under the Lead and Copper Rule (LCR) during the period from 1994 to 1999. A new slow sand filtration treatment facility was put into service in 1999. The pH and alkalinity of water entering the distribution system are typically 7.6-7.9, and 20-35 mg/L as $CaCO_3$, respectively. This treatment facility provides 95% of the water used by OLWD.

With the installation of the new treatment facility, the OLWD was required to conduct monitoring under the LCR to demonstrate that corrosion control was still optimized. This monitoring took place in the spring of 2000, and results indicated that the lead action level was exceeded. The NCCWC installed a caustic feed system in September 2000 to raise the pH of the water to 8.0 and reduce the solubility of lead in home plumbing. Lead and copper levels were below their respective action levels during monitoring conducted in Fall 2000, but lead levels exceeded the action level in the spring of 2001. Table C.10 presents the 90th percentile lead and copper levels from the OLWD from 1992 through 2001.

Table C.10
90th percentile lead and copper levels, Oak Lodge Water District

Year	Number of homes	Lead, µg/L	Copper, mg/L
1992	62	6.0	0.42
1994	32	10.0	0.32
1995	31	4.0	0.32
1996	31	4.0	0.39
1999	31	6.0	0.16
Spring 2000	61	46.0	0.18
Fall 2000	60	14.0	0.09
Spring 2001	60	31.0	0.05

Alternative Tap Monitoring Program

OLWD collected standing lead and copper samples from 10 sites in their distribution system during the Fall of 2001. Ten rounds of samples were collected over a five week period, with samples collected approximately every 72 hours. Nine of the sample sites were Tier 1 sites with copper piping and lead solder built between 1982 and 1984. One site was a Tier 3 site with copper piping and lead solder built in 1974. All ten of the sites selected for alternative monitoring were also included in the regulatory sampling pool. A total of 100 samples were analyzed for lead and copper levels.

Results and Discussion

Lead and copper data collected during the alternative monitoring program are listed in Table C.11. Lead levels measured during the alternative monitoring program averaged 17.4 µg/L and copper levels were very low (average of 0.09 mg/L). Tables C.12 and C.13 and Figures C.6 through C.9 present summary statistics for lead and copper for each site sampled, and for each sampling round. Site No. 8 exhibited the highest lead levels during the monitoring period.

Table C.11
All data lead and copper – OLWD

	Number of samples	Mean	Median	90th Percentile	Standard Deviation	CV
Lead, µg/L	100	17.4	9.0	38.5	19.0	1.1
Copper, mg/L	100	0.09	0.09	0.17	0.05	0.6

Table C.12
Lead and copper levels from each site - OLWD

	Number of sites	Mean	Median	90th Percentile	Standard Deviation	CV
Lead, µg/L	10	0.9 – 54.8	1.0 – 48.5	2.0 – 86.5	0.74 – 26.2	0.2 – 1.7
Copper, mg/L	10	0.03 - 0.18	0.04 - 0.19	0.05 - 0.20	0.01 - 0.04	0.0 - 0.5

Table C.13
Lead and copper levels for each sample round - OLWD

	Number of samples per site	Mean	Median	90th Percentile	Standard Deviation	CV
Lead, µg/L	10	9.0 – 33.4	7.0 – 27.5	26.0 – 91.0	9.5 – 33.7	0.8 – 1.2
Copper, mg/L	10	0.07 - 0.10	0.07 - 0.11	0.13 - 0.19	0.05 - 0.07	0.5 - 0.7

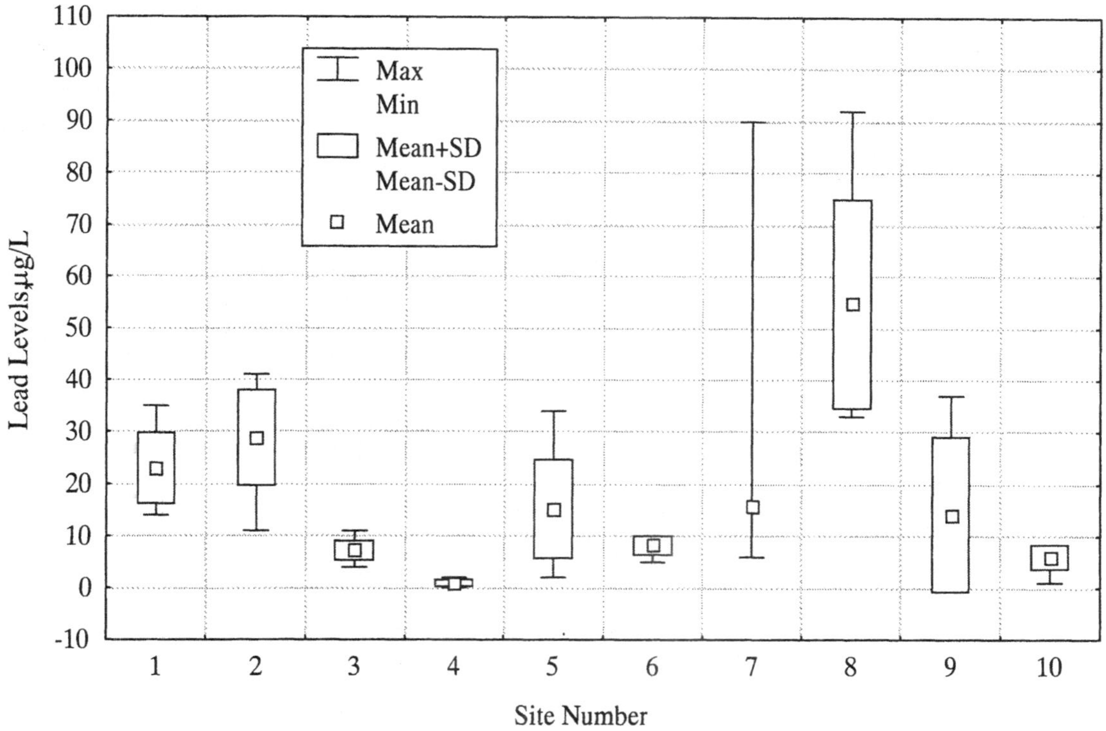

Figure C.6 Lead levels by site for Oak Lodge Water District

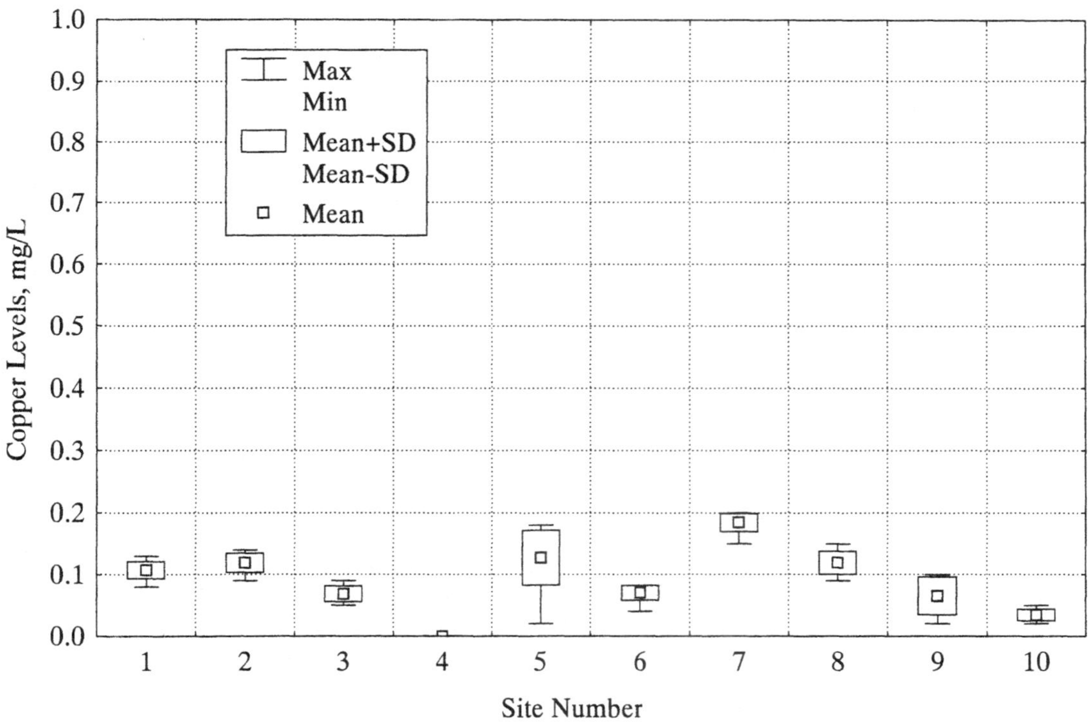

Figure C.7 Copper levels by site for Oak Lodge Water District

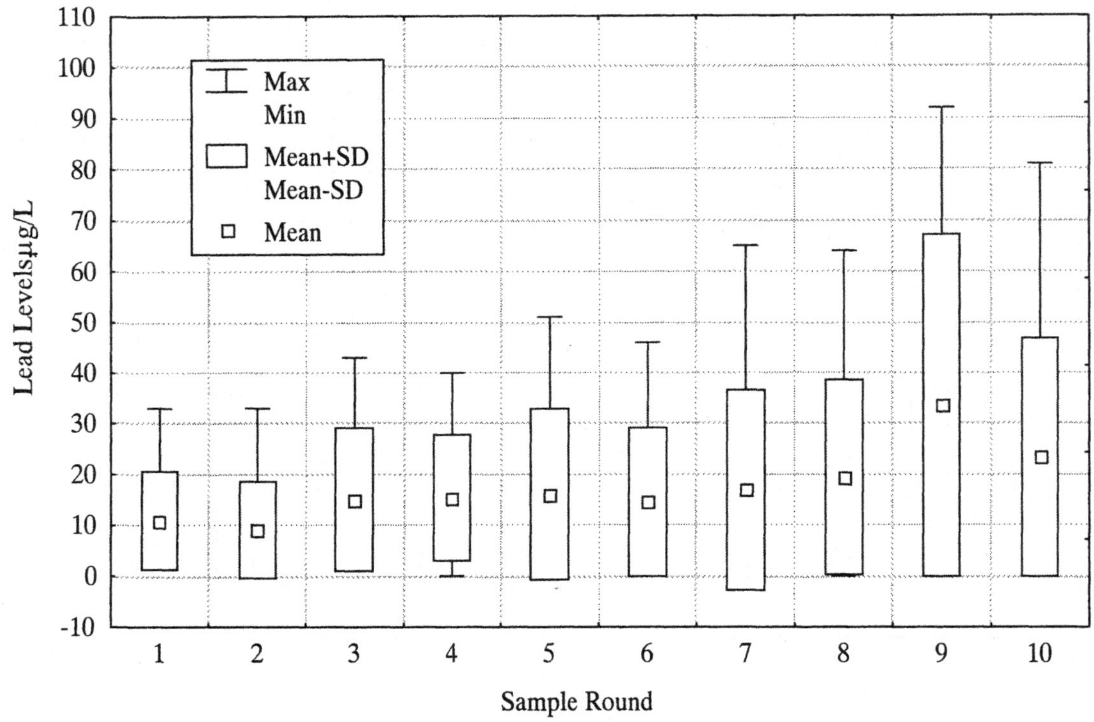

Figure C.8 Lead levels by round for Oak Lodge Water District

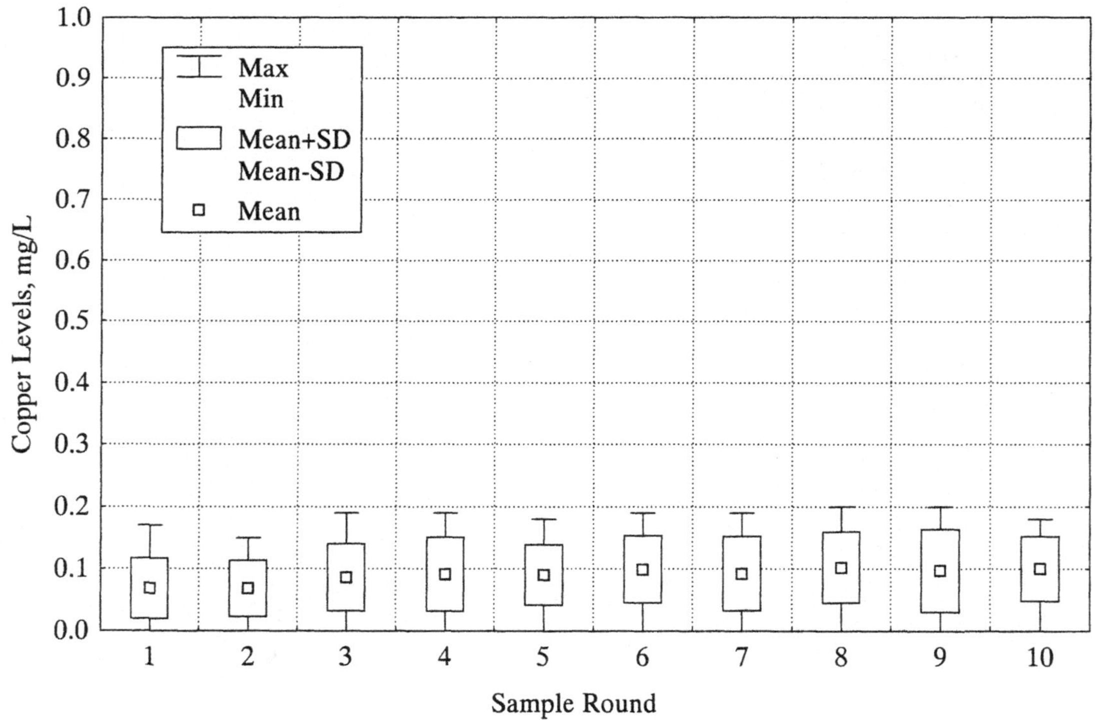

Figure C.9 Copper levels by round for Oak Lodge Water District

A comparison of summary statistics for lead and copper levels from alternative monitoring and regulatory results is presented in Table C.14. Lead and copper levels measured during the alternative monitoring program were similar to levels measured during the last three regulatory monitoring rounds from Spring 2000 through Spring 2001 (Figures C.10 and C.11) and variability was also similar.

Table C.14
Comparisons of summary statistics for lead levels from alternative tap monitoring field studies and all LCR data - OLWD

Lead, µg/L	Number of samples	Mean	Median	Min - Max	90th percentile	Standard deviation	CV
1992	62	3.4	2	2 – 17	6	2.5	0.7
1994	32	3.6	2	1 – 16	10	3.7	1.0
1995	31	2.5	1	1 – 17	4	3.4	1.4
1996	31	2.1	1	1 – 11	4	2.0	0.9
1999	31	3.2	2	0 – 8	6	2.4	0.8
Spring 2000	61	13.8	5	0 – 100	46	20.5	1.5
Fall 2000	62	6.9	3	0 – 134	14	17.3	2.5
Spring 2001	60	12.5	7.5	0 – 59	33	14.4	1.2
2001 Field Study	100	17.3	10	0 - 92	38.5	19	1.1

Copper, mg/L	Number of samples	Mean	Median	Min - Max	90th percentile	Standard deviation	CV
1992	62	0.23	0.2	0.01 -0.55	0.42	0.13	0.6
1994	32	0.19	0.16	0.04 - 0.56	0.32	0.14	0.7
1995	32	0.18	0.16	0.02 - 0.55	0.32	0.12	0.7
1996	30	0.2	0.16	0.02 - 0.72	0.41	0.17	0.9
1999	31	0.08	0.07	0.02 - 0.24	0.16	0.06	0.8
Spring 2000	61	0.09	0.08	0.0 - 0.42	0.18	0.07	0.8
Fall 2000	62	0.04	0.04	0.0 - 0.14	0.09	0.03	0.8
Spring 2001	60	0.03	0.03	0.0 – 0.10	0.05	0.02	0.7
2001 Field Study	100	0.09	0.09	0.0 - 0.20	0.17	0.05	0.6

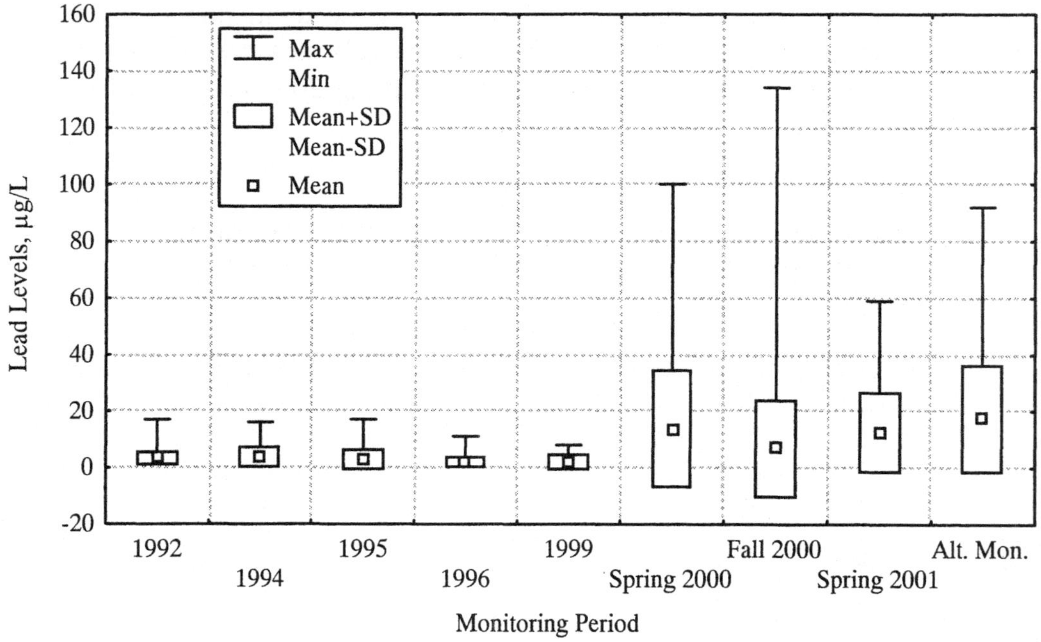

Figure C.10 Lead levels for regulatory and alternative monitoring programs – OLWD

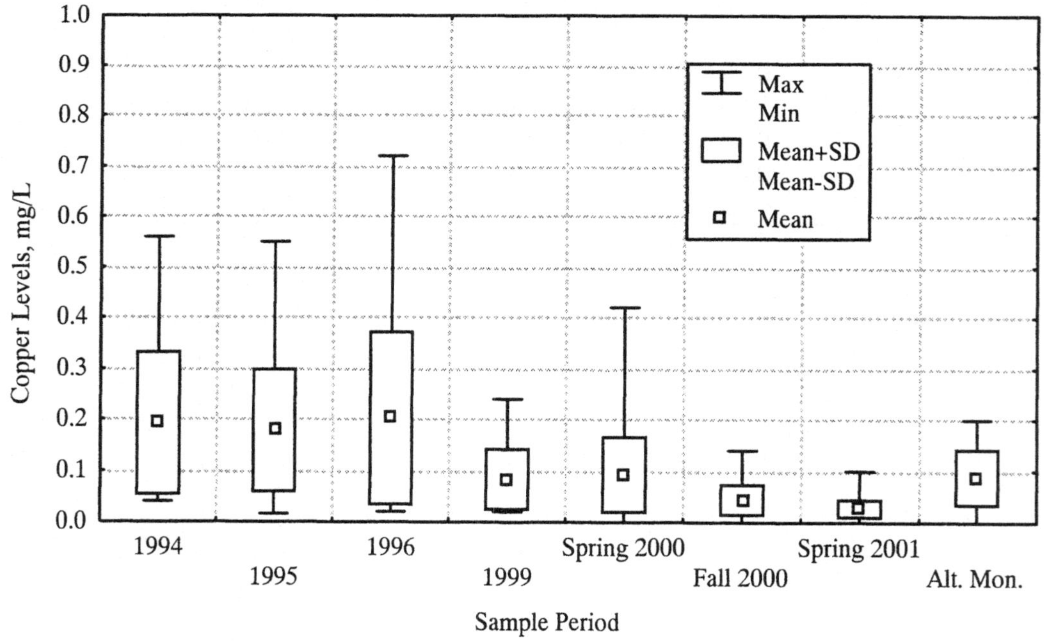

Figure C.11 Copper levels for regulatory and alternative monitoring programs - OLWD

Since all ten of the sites used for the alternative monitoring program were also used for regulatory monitoring, a comparison of lead and copper levels measured from those sites was also completed. Table C.15 lists the summary statistics of this comparison. Again, the lead and copper levels measured during the field study were similar to regulatory monitoring levels from these same sites from Spring 2000 through Spring 2001.

Table C.15
Comparisons of summary statistics for lead levels from sites sampled in both alternative tap monitoring field studies and for the LCR - OLWD

Lead, μg/L	Number of samples	Mean	Median	Min-Max	90th percentile	Standard deviation	CV
1992	9	3.6	3.0	2 – 7	7.0	1.8	0.5
1994	4	3.8	2.0	1 – 10	10.0	4.3	1.1
1995	4	1.0	1.0	1.0	1.0	0.0	0.0
1996	4	1.0	1.0	1.0	1.0	0.0	0.0
1999	3	2.3	1.0	0 – 6	6.0	3.2	1.4
Spring 2000	10	28.9	21.5	0 – 100	75.0	30.0	1.0
Fall 2000	10	11.7	10.0	5 – 28	23.0	7.4	0.6
Spring 2001	9	26.3	24.0	5 – 59	59.0	14.7	0.6
2001 Field Study	100	17.4	9.0	0 - 92	38.5	19.0	1.1

Copper, mg/L	Number of samples	Mean	Median	Min-Max	90th percentile	Standard deviation	CV
1992	9	0.26	0.19	0.01 – 0.55	0.55	0.18	0.7
1994	5	0.21	0.16	0.06 - 0.53	0.53	0.19	0.9
1995	5	0.13	0.13	0.07 - 0.21	0.21	0.06	0.4
1996	5	0.25	0.16	0.03 - 0.70	0.70	0.26	1.0
1999	3	0.10	0.08	0.03 - 0.19	0.19	0.08	0.8
Spring 2000	9	0.11	0.09	0.02 - 0.24	0.24	0.08	0.7
Fall 2000	10	0.05	0.05	0.02 - 0.10	0.10	0.03	0.5
Spring 2001	9	0.04	0.02	0.0 – 0.10	0.10	0.03	0.9
2001 Field Study	100	0.09	0.09	0.0 - 0.20	0.16	0.05	0.6

Summary

Overall comparisons showed that both lead and copper levels were similar in the alternative monitoring program when compared to regulatory monitoring for periods with similar water quality characteristics (i.e. Spring 2000 through Spring 2001 when the slow sand filtration was implemented). Comparisons to the entire set of regulatory monitoring data as well as regulatory results from the same sites yielded similar results. Alternative tap monitoring data did not exhibit less variable results than regulatory monitoring.

PHILADELPHIA WATER DEPARMENT

Background

The City of Philadelphia Water Department (PWD) serves a population of approximately 1.7 million in Philadelphia and parts of Bucks County, PA from two surface water sources, the Delaware and Schuylkill Rivers. These waters are treated at one of three conventional surface water treatment plants (WTPs): Baxter, Queen Lane, or Belmont. The treatment processes at these plants are similar, and include raw water sedimentation, coagulation, flocculation and clarification, followed by rapid sand or dual media filtration. Historically, the quality of the treated water provided by PWD's three treatment plants varied, with the Belmont and Queen Lane WTP's producing a finished water with a lower pH and higher alkalinity, conductivity, TDS and phosphate levels than water from the Baxter WTP.

PWD exceeded the LCR action level for lead in their first round of monitoring in 1992, but have met the action level in all other sampling rounds (Table C.16). PWD conducted desktop analyses of corrosion control strategies and pipe-rack and jar testing, and determined that the optimal corrosion control strategy for all three of their plants would be to provide a unified corrosion control strategy using orthophosphates at all three plants. The following finished water pH and orthophosphate levels have been determined to be optimal for the PWD system:

- pH = 7.3 to 7.8
- orthophosphate = 0.6 - 0.8 mg/L (as PO_4)

Table C.16
90th percentile lead and copper levels, Philadelphia Water Department

Year	Samples Required	Homes Sampled	90th Percentile Lead Levels (μg/L)	90th Percentile Copper Levels (mg/L)
Jan-Jun 1992	100	162	21	0.9
Jul-Dec 1992	100	143	15	0.8
Jan-Jun 1997	100	118	14	0.4
Jul-Dec 1997	100	108	11	0.4
Jun-Sep 1998	50	79	10	0.3
Jun-Sep 1999	50	59	9	0.3

Alternative Tap Monitoring Program

The alternative tap-monitoring program for the PWD was conducted in the winter of 2001. Twenty-five single-family residential sites were selected and all sites contained copper piping. The majority of locations contained piping that was over 20 years old (Table C.17). A total of ten rounds of sampling were conducted over a ten-week period. In the first and last sample rounds, a split sample (125-mL followed by a one-liter sample) was collected to evaluate the faucet contribution to standing lead and copper levels. For the remainder of the sample rounds, a one-liter first draw sample was collected. The stagnation time was recorded for all samples. Stagnation times were greater than 5.5 hours at all sites, and less than or equal to 9.5 hours at most locations. There were some longer stagnation times recorded however, some as high as 46 hours and these samples were excluded from the analyses.

Table C.17
Approximate age of residential monitoring sites and number of samples collected for the Philadelphia Water Department

Age Category	Number of Sites	Samples Collected	
		Sample Volume	Number of Samples
≤ 10 years	6 [a]	1000 mL first draw	33
		125 mL first draw	8
		1000 mL second draw	8
> 10 and ≤ 20 years	5	1000 mL first draw	31
		125 mL first draw	9
		1000 mL second draw	9
> 20 years	14	1000 mL first draw	122
		125 mL first draw	33
		1000 mL second draw	32

(a) Two sites had mixed plumbing, i.e. part of the residence had been re-plumbed in the previous 10 years. One site contained a new faucet.

Results and Discussion

Lead and copper levels measured from all data collected during the alternative monitoring program are listed in Table C.18. The average lead level for first draw 1000 mL samples was 2.2 µg/L. Copper levels were extremely low in all samples volumes and lead levels exhibited higher variability than copper levels.

Table C.18
Lead and copper level data from alternative monitoring – PWD

		Sample Type	N	Mean	Median	90th Percentile	Standard Deviation	CV
Lead Levels (µg/L)		1000 mL first draw	186	2.2	1.0	3.0	3.4	1.6
	Split Sample	125 mL first draw	67	3.5	1.0	5.0	8.5	2.4
	Split Sample	1000 mL second Draw	66	4.3	1.0	5.0	15.3	3.6
Copper Levels (mg/L)		1000 mL first draw	186	0.16	0.15	0.32	0.10	0.6
	Split Sample	125 mL first draw	67	0.17	0.14	0.30	0.14	0.9
	Split Sample	1000 mL second Draw	66	0.15	0.12	0.29	0.11	0.7

Results from sequential sampling conducted during the first and last sample rounds indicated that similar lead and copper levels were measured from first draw 1000 mL, and first draw 125 mL/1000 mL samples. However, lead levels were more variable in the 1000 mL portion of the split sample when compared to the 125 mL portion of the split sample (Figures C.12 and C.13). Figures C.14 and C.15 display lead and copper levels for first draw 1000 mL by piping age category.

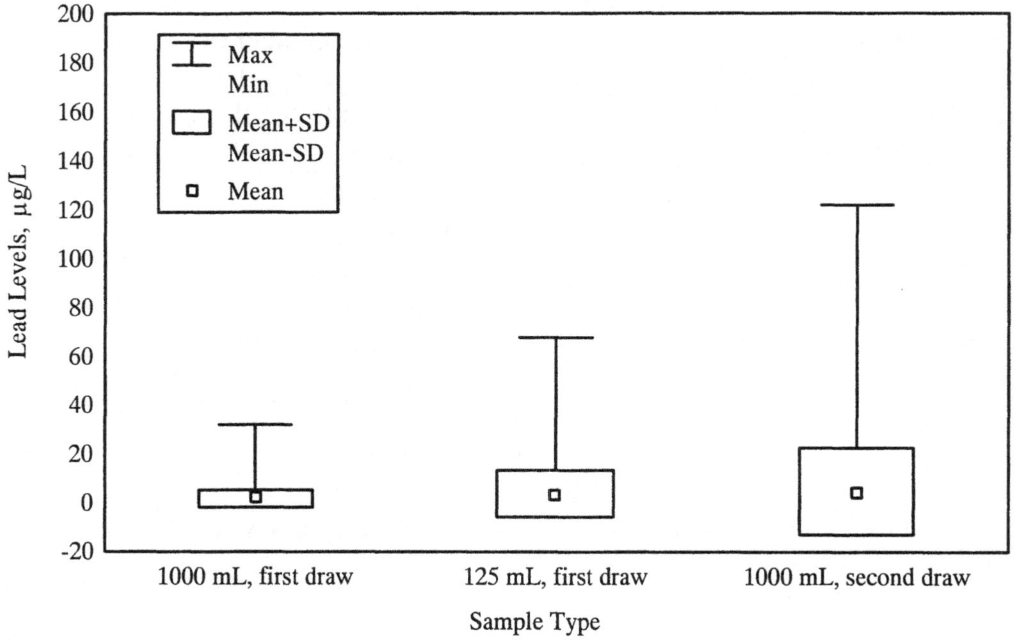

Figure C.12 Lead levels by sample type – PWD

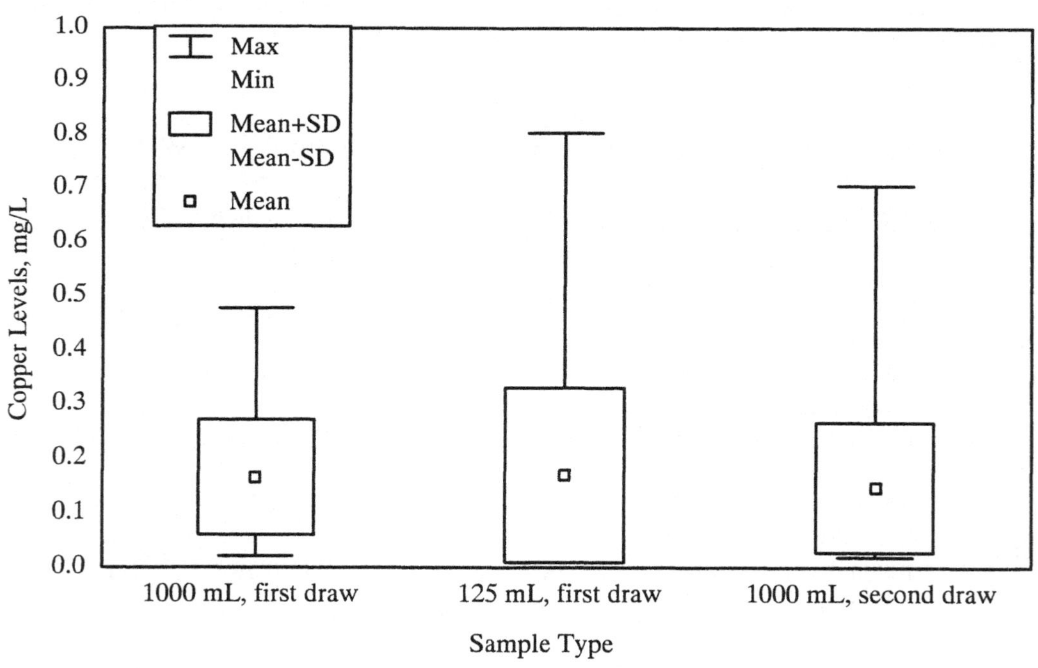

Figure C.13 Copper levels by sample type – PWD

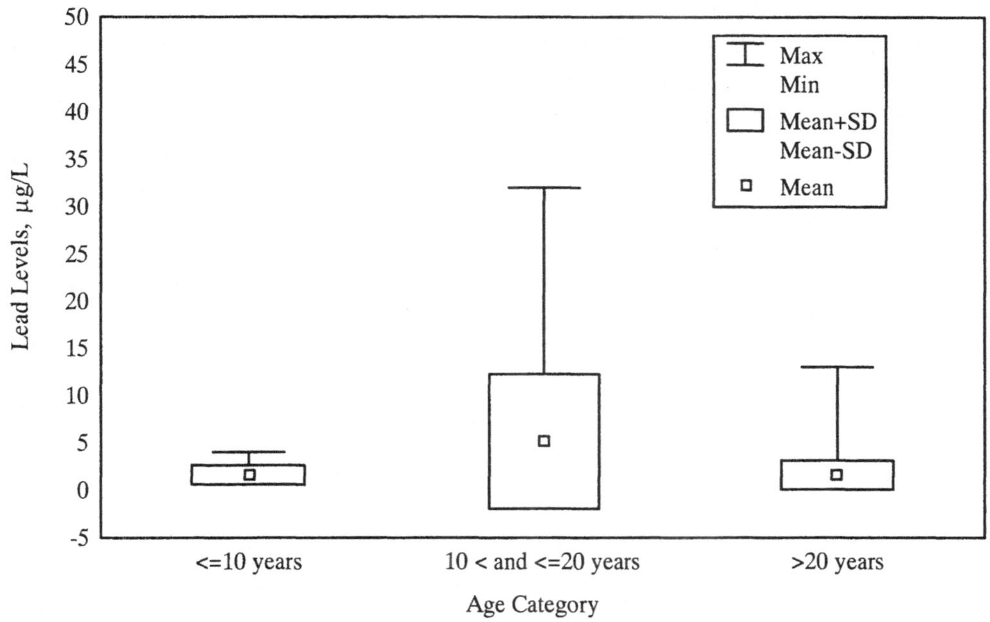

Figure C.14 Lead levels by piping age category – PWD

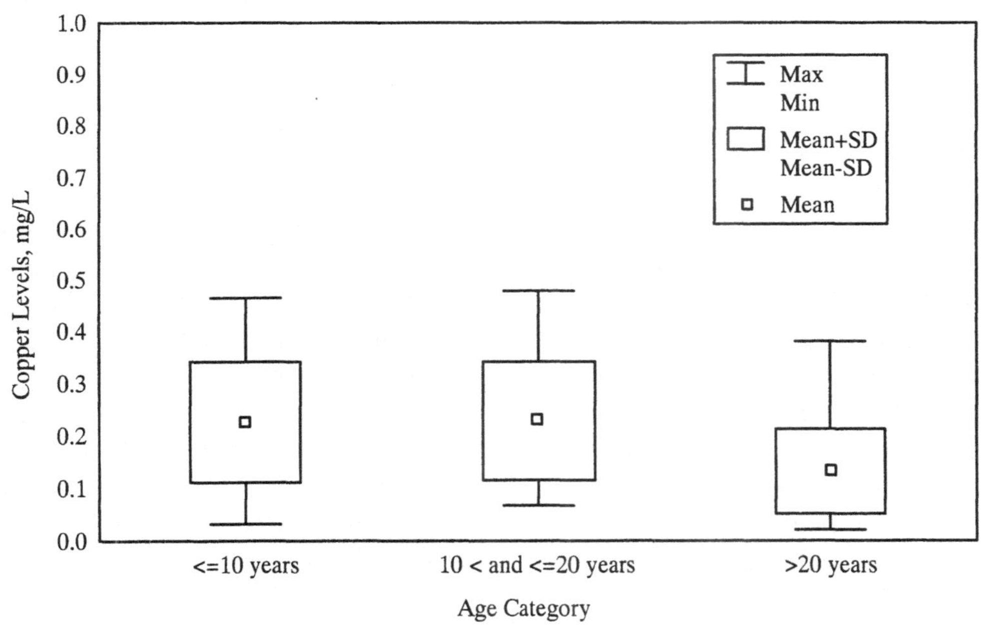

Figure C.15 Copper levels by piping age category – PWD

Similar lead and copper levels were observed between the three age categories. Lead levels were more variable in sites with piping between 10 and 20 years old. Copper levels did show some correlation with piping age, as seen in Figure C.16. A downward trend in copper levels was observed for sites with piping ages between 4 and 50 years. Site with piping that was 50 years or older exhibited somewhat elevated copper levels, indicating possible faucet or piping replacement may have occurred.

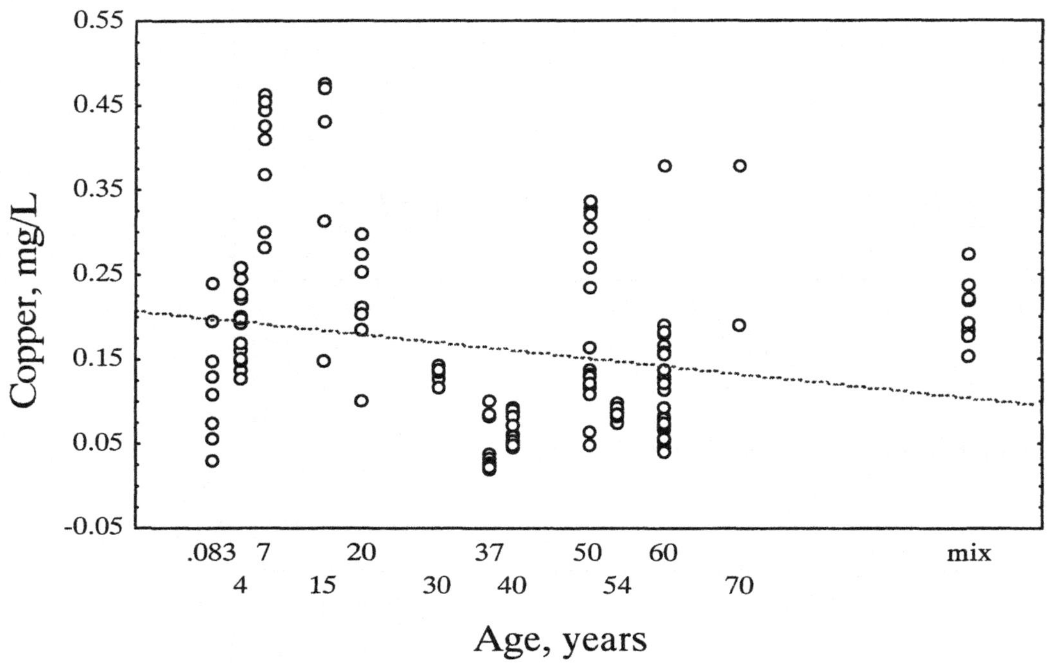

Figure C.16 Copper levels vs. age of piping for PWD

Figures C.17 through C.20 display the lead and copper levels measured from first draw 1000 mL samples for each of the sample rounds and each of the sample sites for the alternative monitoring program. Lead and copper levels were similar between sample rounds, with some sample rounds exhibiting greater variability than others. The majority of sample sites exhibited similar lead and copper levels, however there were a few sites where higher lead and copper levels were measured. Sites numbered LC-102 through LC-121 were older homes (over 10 years old) with copper plumbing and sites numbered LC-201 through LC-206 were newer homes (less than 10 years old) with new plumbing. The highest lead and copper levels were found in sample site LC-115, a fifteen year old copper piping site. Lead levels were generally low in the newer homes with a narrow range of values, but copper levels were higher and more variable.

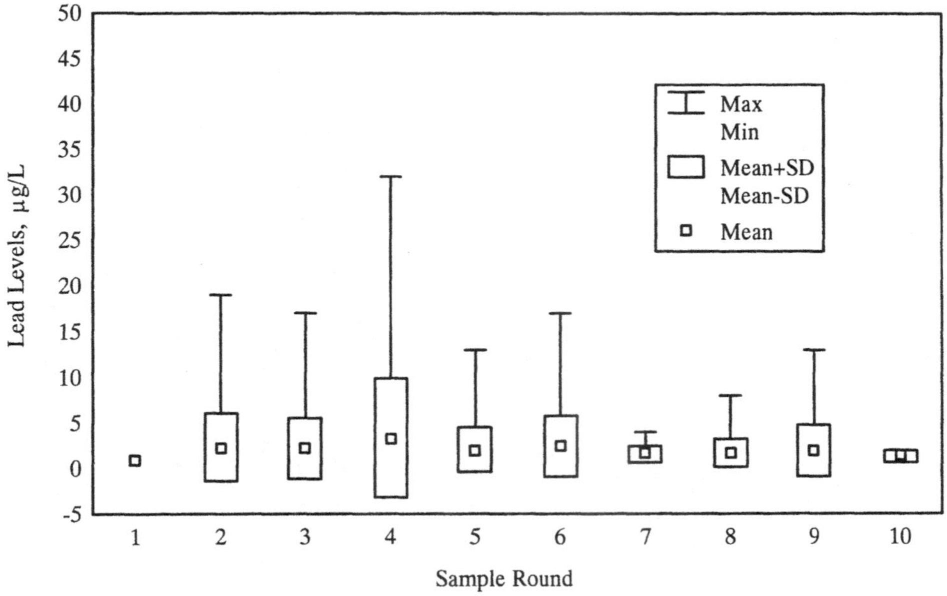

Figure C.17 Lead levels for each sampling round from PWD alternative monitoring program - 1000 mL first draw

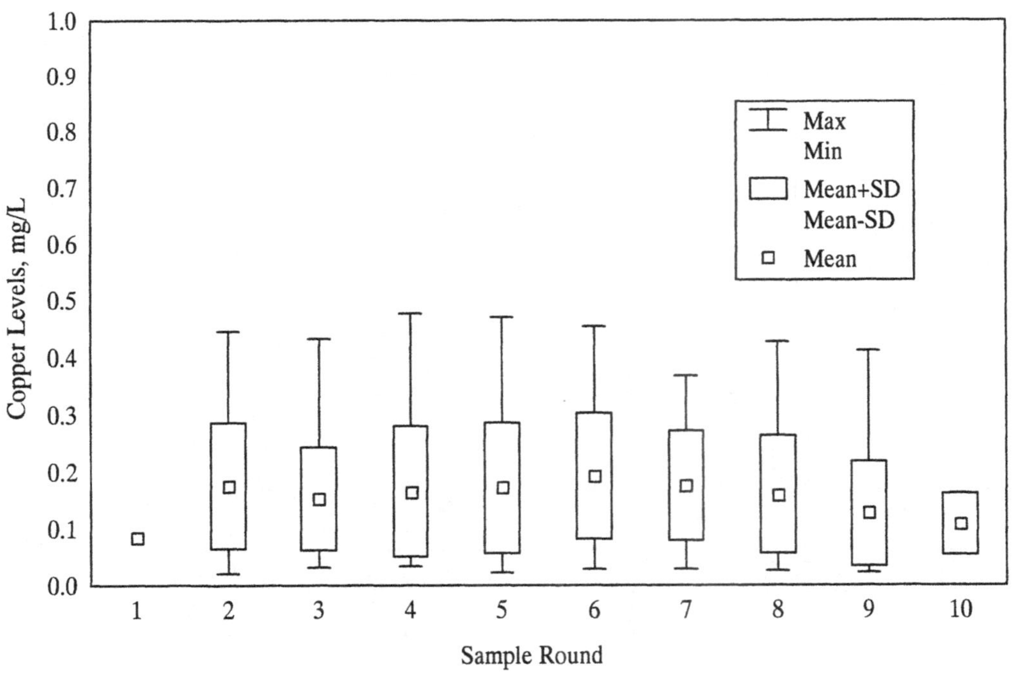

Figure C.18 Copper levels for each sampling round from PWD alternative monitoring program - 1000 mL first draw

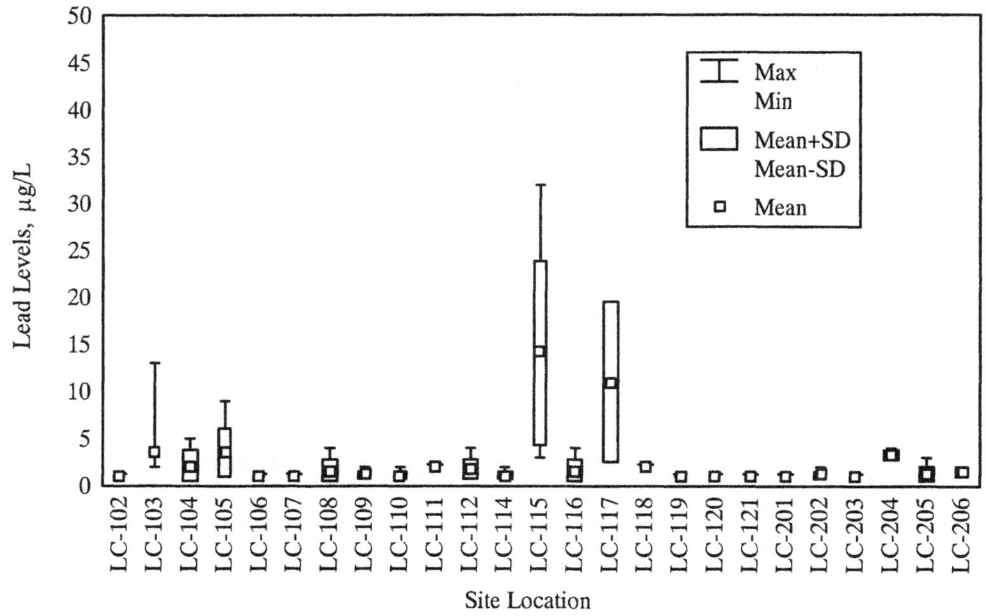

Figure C.19 Lead levels for each sampling site from PWD's alternative lead monitoring program

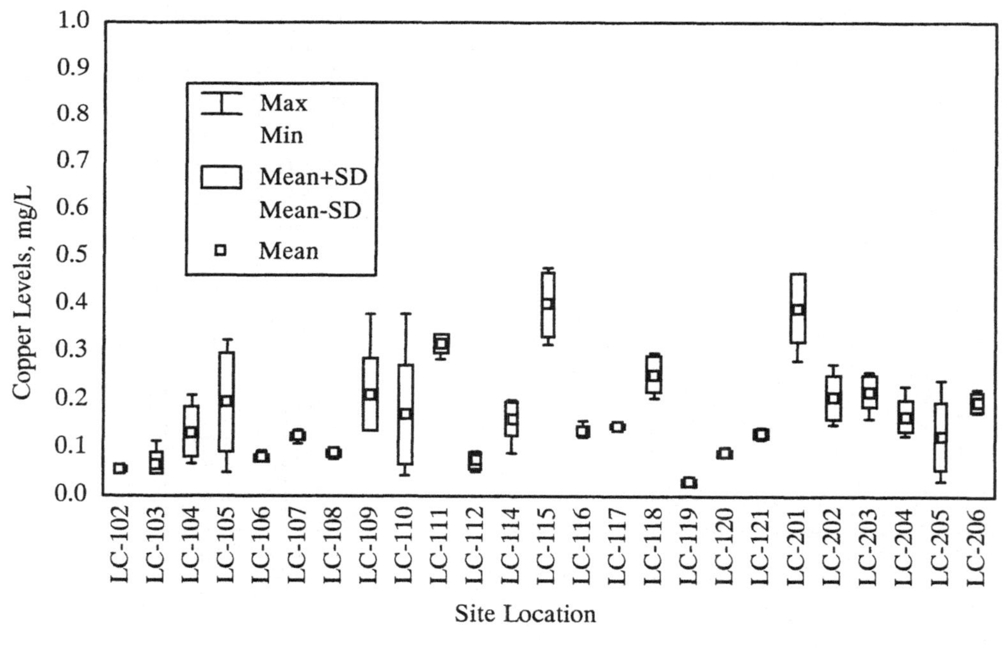

Figure C.20 Copper levels for each sampling site from PWD's alternative lead monitoring program

A comparison of summary statistics for lead and copper levels for the alternative monitoring and each of the regulatory sample rounds is presented in Table C.19. Lead levels were highly variable for all regulatory sampling periods. Lead levels from regulatory monitoring from the fall of 1997 through 2002 exhibited statistically similar results. Lead levels from the alternative tap monitoring program were much lower and exhibited a much narrower band of values (Figure C.21) when compared to lead levels measured during regulatory monitoring. The copper levels from alternative monitoring were similar to the previous three regulatory monitoring periods (1998 – 2002) and variability was slightly lower (Figure C.22).

Table C.19
Comparison of regulatory and alternative monitoring lead and copper levels – Philadelphia, PA

Monitoring program	Number of samples	Mean	Median	Min	Max	90th percentile	Standard deviation	CV
			Lead levels (μg/L)					
Jan - Jun 1992	162	14.2	6.0	1.0	205.0	22.0	28.2	2.0
Jul – Dec 1992	143	8.6	6.0	1.0	70.0	15.0	9.6	1.1
Jan - Jun 1997	118	11.0	2.0	0.0	270.0	15.0	37.5	3.4
Jul – Dec 1997	108	5.5	2.0	0.0	144.0	11.0	15.3	2.8
Jun – Sep 1998	78	6.4	2.0	1.0	95.0	11.0	14.7	2.3
Jun – Sep 1999	59	8.6	3.0	3.0	131.0	10.0	19.5	2.3
Jun – Sep 2002	62	7.9	2.0	1.0	109.0	14.0	18.5	2.4
Alt. Mon.	186	2.2	1.0	1.0	32.0	3.0	3.4	1.6
			Copper levels (mg/L)					
Jan - Jun 1992	162	0.35	0.18	0.02	1.79	0.94	0.39	1.1
Jul – Dec 1992	143	0.30	0.18	0.02	1.25	0.75	0.28	0.9
Jan - Jun 1997	118	0.18	0.13	0.03	0.88	0.39	0.16	0.8
Jul – Dec 1997	108	0.18	0.13	0.01	0.86	0.40	0.16	0.9
Jun – Sep 1998	78	0.15	0.12	0.02	0.61	0.35	0.12	0.8
Jun – Sep 1999	59	0.15	0.10	0.03	0.55	0.37	0.12	0.8
Jun – Sep 2002	43	0.12	0.09	0.01	0.32	0.25	0.08	0.7
Alt. Mon.	186	0.16	0.15	0.02	0.48	0.32	0.10	0.6

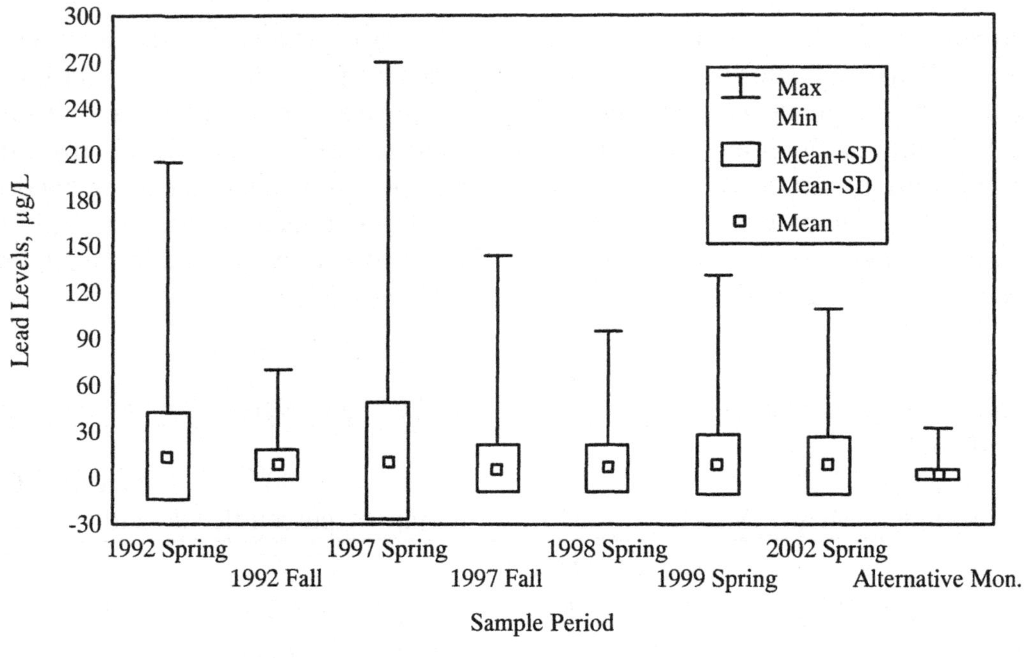

Figure C.21 Comparison of lead levels for regulatory and alternative monitoring programs – PWD

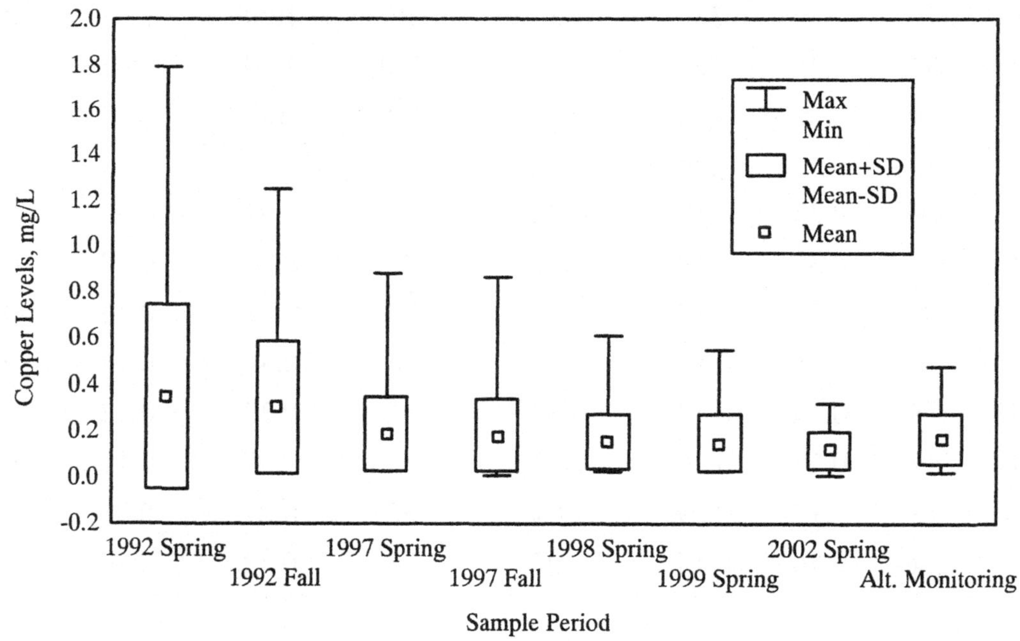

Figure C.22 Comparison of copper levels for regulatory and alternative monitoring programs – PWD

Summary

Lead and copper levels measured during the alternative monitoring program yielded very low levels of these metals, with essentially no statistical difference in metals levels based on sample volume size or age of piping. Lead levels were more variable in the second 1000 mL sample collected, which would be representative of the lead sources in the home piping rather than the faucet. Lead levels were also more variable in the sites with piping ages between 10 and 20 years.

Overall comparisons with regulatory monitoring data showed that lead levels measured during alternative monitoring were much lower and less variable. Copper levels from the alternative monitoring more closely tracked regulatory copper levels during the 1998 through 2002 time period since the PWD has been operating under 'optimized' treatment conditions for lead and copper.

DETROIT WATER AND SEWERAGE DEPARTMENT, MI

Background

The Detroit Water and Sewerage Department (DWSD) supplies water to approximately 4 million customers in 126 communities in the Detroit metropolitan area. Two water supplies, the Detroit River and Lake Huron, provide water to five water treatment plants. The WTPS use a conventional treatment process of pre-chlorination, rapid mix, flocculation, sedimentation, filtration, and chlorine disinfection. The quality of the treated waters provided by DWSD's five treatment plants are very similar, with a typical pH of 7.4 and alkalinity of 74 mg/L as $CaCO_3$.

DWSD conducted the required compliance monitoring for the Lead and Copper Rule starting in 1992. Results of this compliance monitoring, obtained from sites throughout the City of Detroit, indicated that 90th percentile lead levels exceeded the 15 µg/L action level. Detroit implemented full-scale phosphoric acid treatment at all five of their water treatment plants in the fall of 1996. A second round of monitoring was conducted during the first half of 1997, and results indicated that 27 of the 36 communities had achieved compliance. In the most recent monitoring year (2000), only 3 of 126 communities in the DWSD service area had lead levels that exceeded the 90th percentile action level (Table C.20). Table C.21 shows 90th and 50th percentile results for the City of Detroit.

Table C.20
90th percentile lead and copper levels, Detroit water and sewerage department

Year	Number of purveyor systems sampled*	90th Percentile range for lead levels (µg/L)	Number of systems exceeding action level for lead (15 µg/L)	Number of systems exceeding action level for copper (1.3 mg/L)
Round 1, 1992	97	3.0 – 70.0	36	0
Round 2, 1992	96	<2.0 – 324.0	34	0
Spring 1997	96	<2.0 – 37.0	13	0
2000	99	0.0 – 77.0	3	0

*Remaining communities conducted monitoring internally.

Table C.21
Regulatory lead and copper levels for sites in the City of Detroit

Regulatory monitoring period	Number of samples	Lead (μg/L)		Copper (mg/L)	
		90th percentile	50th percentile	90th percentile	50th percentile*
Spring 1992	89	21	7	0.10	0.04
Fall 1992	104	11.6	4	0.08	0.03
Spring 1997	118	12	2	0.04	0.01
Fall 1997	122	8	2	0.03	0.02
Fall 1998	102	8	2	0.04	0.02
Spring 1999	104	7	2	0.03	0.01
Spring 2000	119	7	2	0.04	0.01

*Median

Alternative Tap Monitoring Program

DWSD conducted monitoring at 20 residential sites in the City of Detroit system over a seven week period in March and April, 2002. A consecutive 125 mL followed by a 875 mL sample was collected during each sampling round. Average stagnation time was 7 hours, with all samples collected after the water had been standing from between 5.5 and 9.5 hours. Sixteen of the sites contained lead service lines and four of the sites had copper service lines. The type of material used in the interior plumbing was also recorded for each site. Eleven sites had copper interior piping and nine sites had galvanized plumbing. The approximate age of the interior plumbing and the faucet was recorded for many of the sites. At eleven of the sampling sites, no age was available for the interior piping and the remaining sites were split evenly between sites that had interior plumbing that was less than 10 years old, sites with interior plumbing between 10 and 20 years old, and sites with interior plumbing greater than twenty years old (three in each category). Seventeen of the sites chosen for the alternative monitoring program were also sampled during regulatory monitoring. Table C.22 lists the sampling sites, summaries of the interior, service line plumbing types, and approximate age of both the interior piping and faucet for the sampling sites. The table also indicates which sites were also sampled during regulatory monitoring.

Table C.22
Approximate age of interior plumbing and faucets at alternative monitoring sites - DWSD

Site number	Service line type	Interior plumbing type	Age of interior plumbing	Age of faucet (years)
1*	Copper	Copper	N/A	2
2*	Lead	Copper	<10	10
3*	Lead	Galvanized	N/A	7
4*	Lead	Galvanized	N/A	N/A
5*	Lead	Galvanized	N/A	N/A
6*	Lead	Copper	>20	1
7*	Lead	Copper	>10 and <=20	0.5
8*	Lead	Galvanized	N/A	N/A
9*	Lead	Galvanized/Plastic	N/A	50
10*	Lead	Galvanized	N/A	Old
11*	Lead	Copper	>10 and <=20	20
12*	Lead	Galvanized	N/A	N/A
13*	Lead	Copper	<10	4
14*	Copper	Copper	>20	15
15*	Lead	Copper	>20	2
16*	Copper	Copper	N/A	6
17	Lead	Copper	>10 and <=20	N/A
18*	Lead	Galvanized	N/A/	N/A
19	Copper	Copper	<10	10
20	Lead	Galvanized	N/A	N/A

*Site sampled during regulatory monitoring

Results and Discussion

A summary of the 125 mL and 875 mL standing lead and copper level results from the alternative monitoring program are listed in Table C.23, along with the calculated 1000 mL levels for lead and copper. Lead levels were higher and more variable in the initial 125 mL sample when compared to the 875 mL sample, but the difference was not statistically significant (where p = probability of error). Copper levels were extremely low for both samples volumes (less than 0.4 mg/L). This lead and copper data is also displayed in Figure C.23.

Table C.23
Lead and copper levels for 125 ml and 875 ml samples in Detroit

Sample volume	Number of samples	Mean	Median	Min – Max	90th percentile	Standard deviation	CV
Lead (μg/L)							
125 mL	143	6.7	3.0	3 - 221	10.0	18.6	2.8
875 mL	143	4.9	3.0	3 - 32	9.0	4.2	0.9
Calculated 1000 mL	143	5.14	3.00	3 – 38.1	8.50	4.80	0.9
Copper (mg/L)							
125 mL	143	0.06	0.03	0.02 - 0.33	0.15	0.07	1.2
875 mL	143	0.03	0.02	0.02 - 0.15	0.06	0.02	0.7
Calculated 1000 mL	143	0.03	0.02	0.02 - 0.17	0.07	0.03	0.8

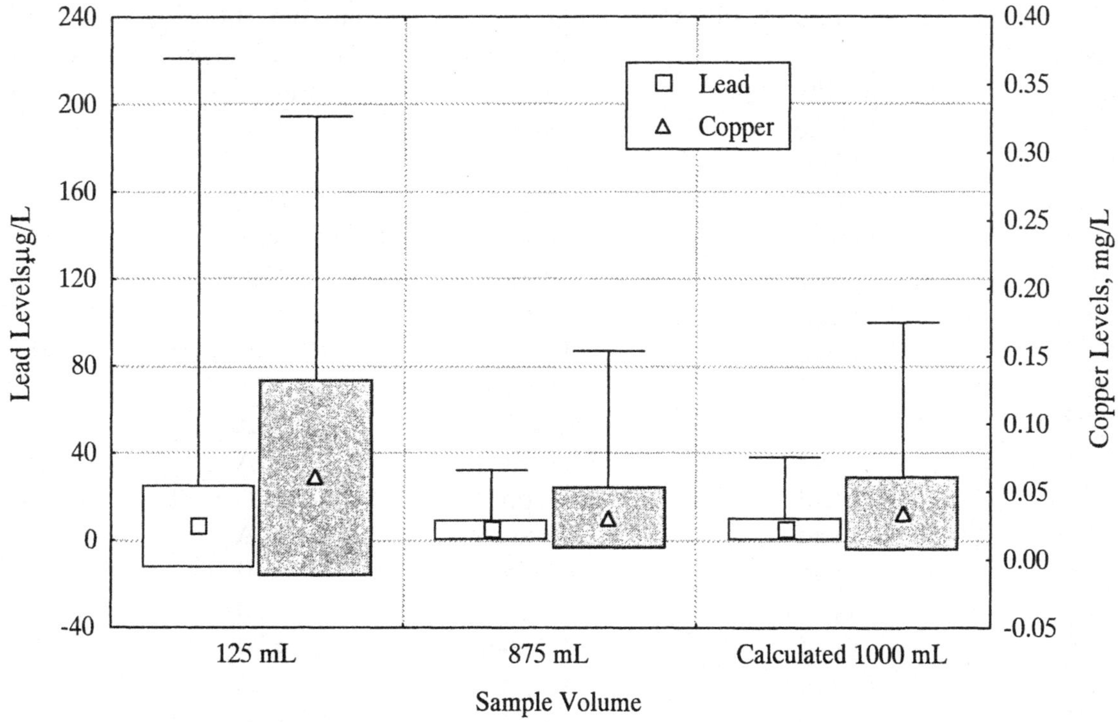

Figure C.23 Lead and copper levels for 125 ml, 875 ml and calculated 1000 ml samples – DWSD

Insufficient data was available to evaluate the effect of interior piping age on lead and copper levels, however an evaluation of faucet age was completed. Figure C.24 displays lead and copper level results for 125 mL first draw samples from the alternative monitoring program versus age of faucet at the sampling sites. Newer faucets (< 10 years) yielded higher lead and copper levels, but several higher lead levels were measured from faucets listed as 50 years old (greater than 5.0 µg/L). This could be due to incorrect faucet age categorization for these sites. Results from the second draw 875 mL samples yielded similar results, however the lead levels from faucets listed as 50 years old were all below 5 µg/L.

The majority of sample sites exhibited similar lead and copper levels, however there were a few sites where higher lead and copper levels were measured. Tables C.24 and C.25 display summary statistics for calculated 1000 mL lead and copper levels for each of the sample rounds and each of the sample sites for the alternative monitoring program. Lead and copper levels were similar between sample rounds, with some sample rounds exhibiting greater variability than others. Figures C.25 through C.28 display the data graphically.

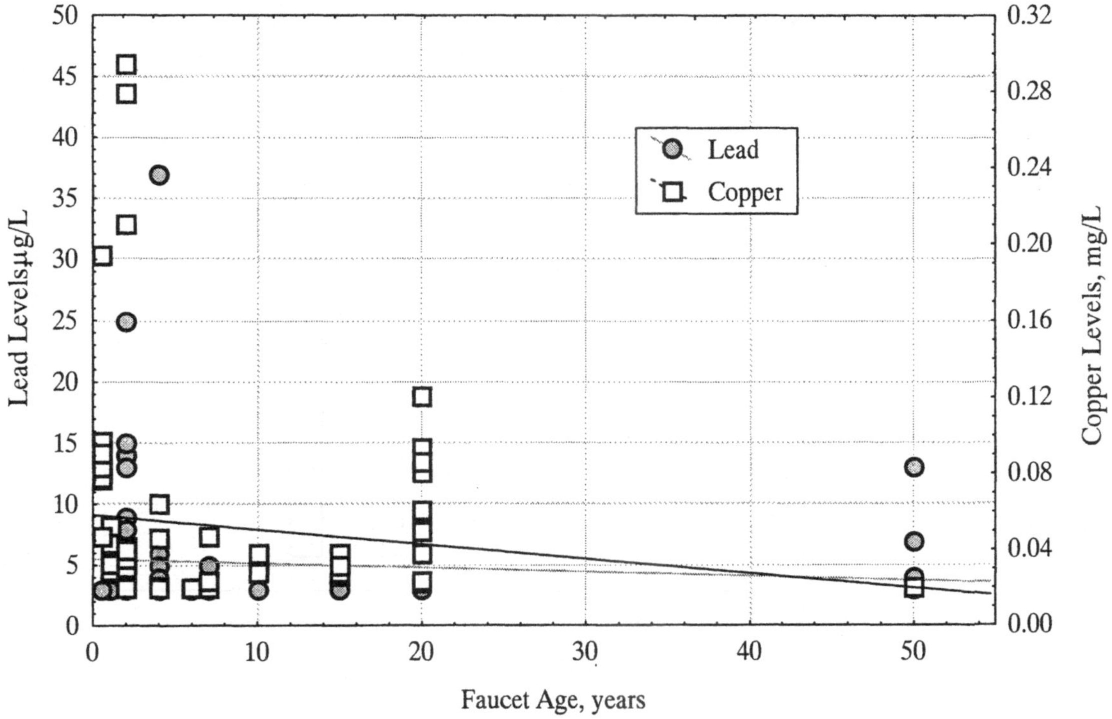

Figure C.24 DWSD lead and copper levels vs. faucet age – 125 mL samples

Table C.24
Lead and copper levels for each sample round – Detroit 1000 mL samples

	Number of samples per round	Mean	Median	90th Percentile	Standard deviation	CV
Lead, ug/L	3 – 25	1	2 – 5	0.6 – 6.5	0.4 – 1.9	0.3 - .8
Copper, mg/L	1 - 25	0.09 - 0.19	0.11 - 0.18	0.05 - 0.11	0.49 - 0.72	3 – 5.7

Table C.25
Lead and copper levels from each site - Detroit calculated 1000 mL samples

	Number of samples per site	Mean	Median	90th percentile	Standard deviation	CV
Lead (µg/L)	1- 11	3.0 – 13.6	3.0 – 14.2	3 – 22.3	0.04 – 10.5	0 – 1.2
Copper (mg/L)	1 – 11	0.02 - 0.11	0.02 – 0.11	0.02 - 0.12	0.002 – 0.03	0.05 - 0.5

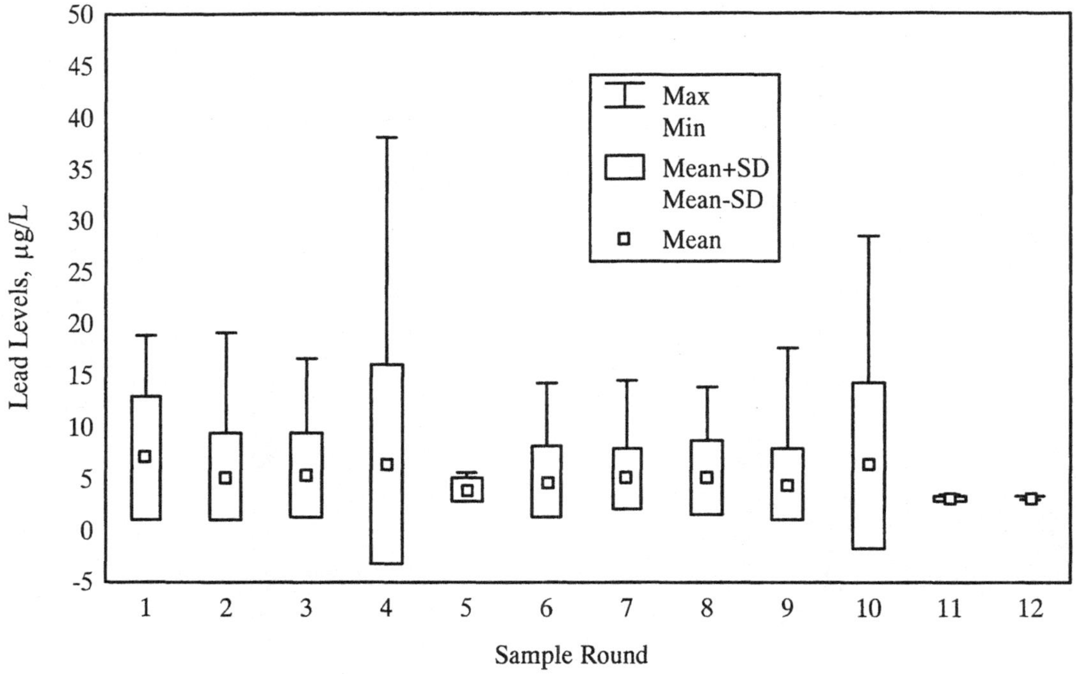

Figure C.25 Lead levels by sample round for calculated 1000 mL samples-- DWSD

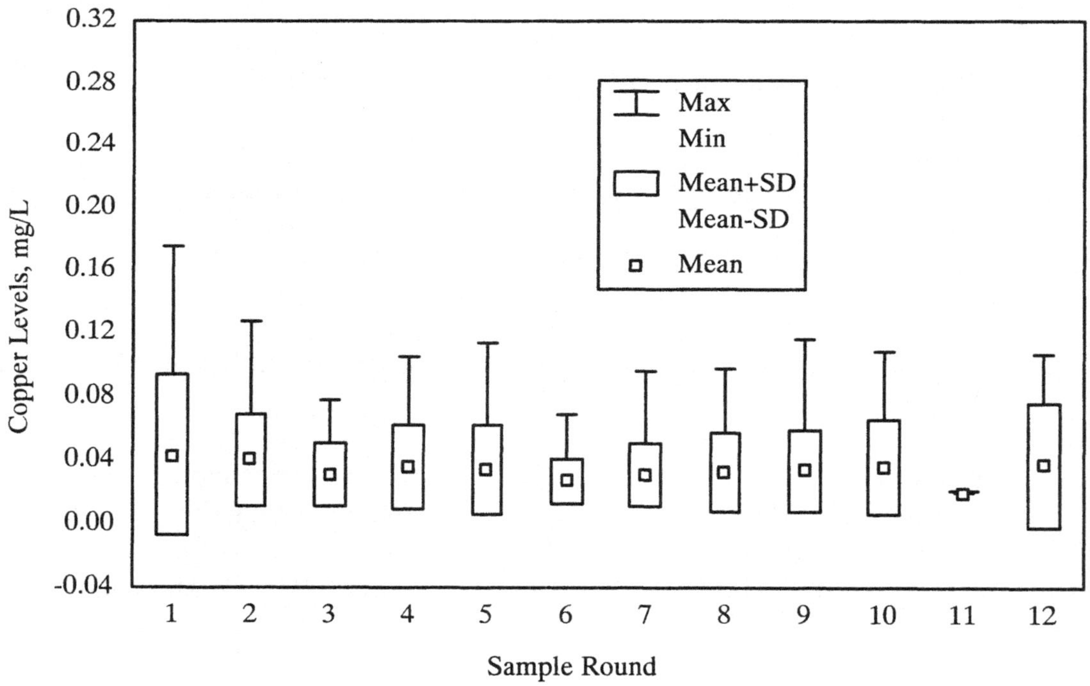

Figure C.26 Copper levels by sample round for calculated 1000 mL samples - DWSD

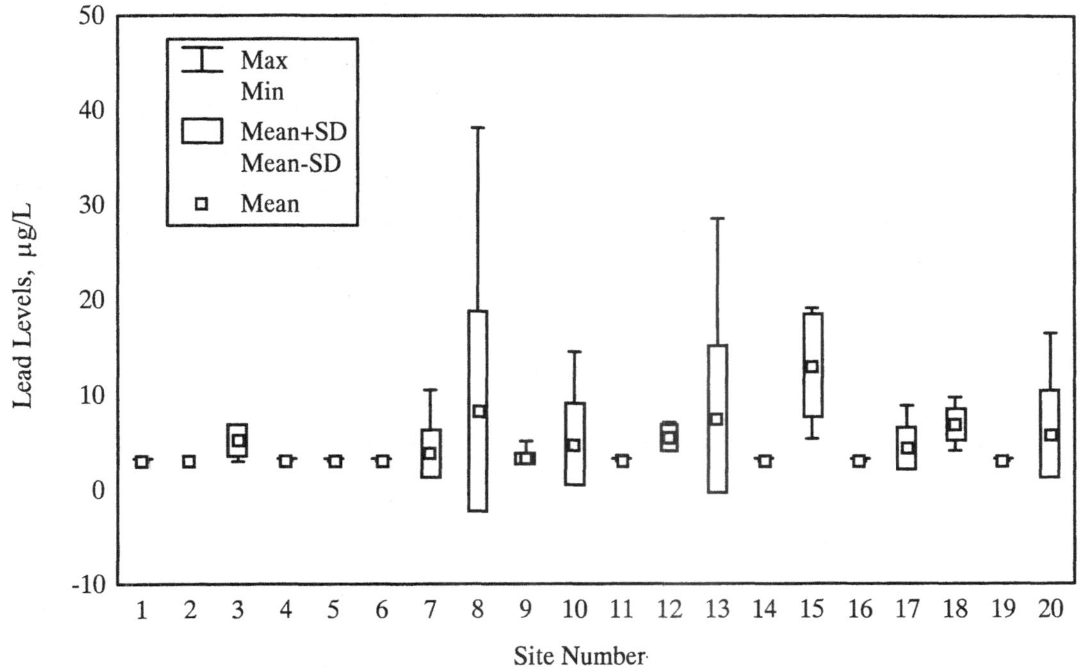

Figure C.27 Lead levels by sample site for calculated 1000 mL samples - DWSD

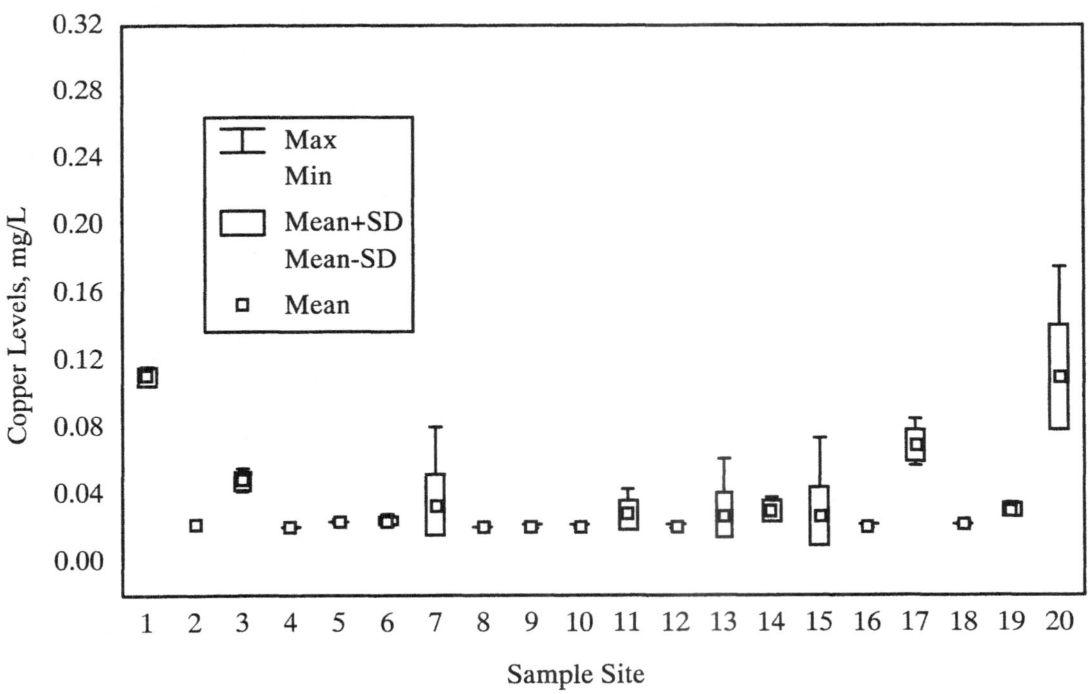

Figure C.28 Copper levels by sample site for calculated 1000 mL samples – DWSD

A comparison of summary of lead and copper levels for the City of Detroit's alternative and regulatory monitoring is presented in Table C.26 and Figures C.29 and C.30. Calculated 1000 mL lead levels from alternative monitoring were similar to regulatory monitoring results for 1997 through 1998. Lead levels from regulatory monitoring for 1999 through 2000 were lower than results obtained for the alternative monitoring program. Lead levels measured from the alternative monitoring program were less variable than regulatory monitoring results. Copper levels from alternative monitoring were similar to copper levels for the regulatory monitoring periods from 1997 through 2000 and variability was lower in the alternative monitoring data when compared to copper levels measured during regulatory monitoring.

Table C.26
Comparisons of summary statistics for lead and copper levels from alternative tap monitoring and regulatory monitoring - DWSD

Sampling program	Number of samples	Mean	Median	Min - Max	90th percentile	Standard deviation	CV
Lead (μg/L)							
Spring 1992	89	10.8	7	1 – 85	21	12.0	1.1
Fall 1992	104	8.8	4	1 – 83	116	15.0	1.7
Spring 1997	118	5.1	2	0 – 112	12	12.2	2.4
Fall 1997	122	5.8	2	2 – 246	8	22.3	3.9
Fall 1998	102	4.5	2	2 – 81	8	8.3	1.8
Spring 1999	104	3.5	2	2 – 18	7	3.3	0.9
Spring 2000	119	3.4	2	0 – 40	7	4.3	1.3
2002 Field Study*	143	5.1	3	3 - 38	8.5	4.8	0.9
Copper (mg/L)							
Spring 1992	89	0.10	0.04	0.0 - 0.69	0.10	0.13	1.4
Fall 1992	104	0.06	0.03	0.0 - 0.31	0.08	0.06	1.1
Spring 1997	79	0.04	0.01	0 .0 - 0.80	0.04	0.10	2.2
Fall 1997	122	0.04	0.02	0.0 - 0.51	0.03	0.07	2.0
Fall 1998	102	0.04	0.02	0.0 - 0.47	0.04	0.06	1.6
Spring 1999	104	0.03	0.01	0.0 - 0.28	0.03	0.05	1.7
Spring 2000	100	0.04	0.01	0.0 - 0.63	0.04	0.10	2.2
2002 Field Study*	143	0.03	0.02	0.02 - 0.17	0.03	0.03	0.8

*Calculated 1000 mL samples

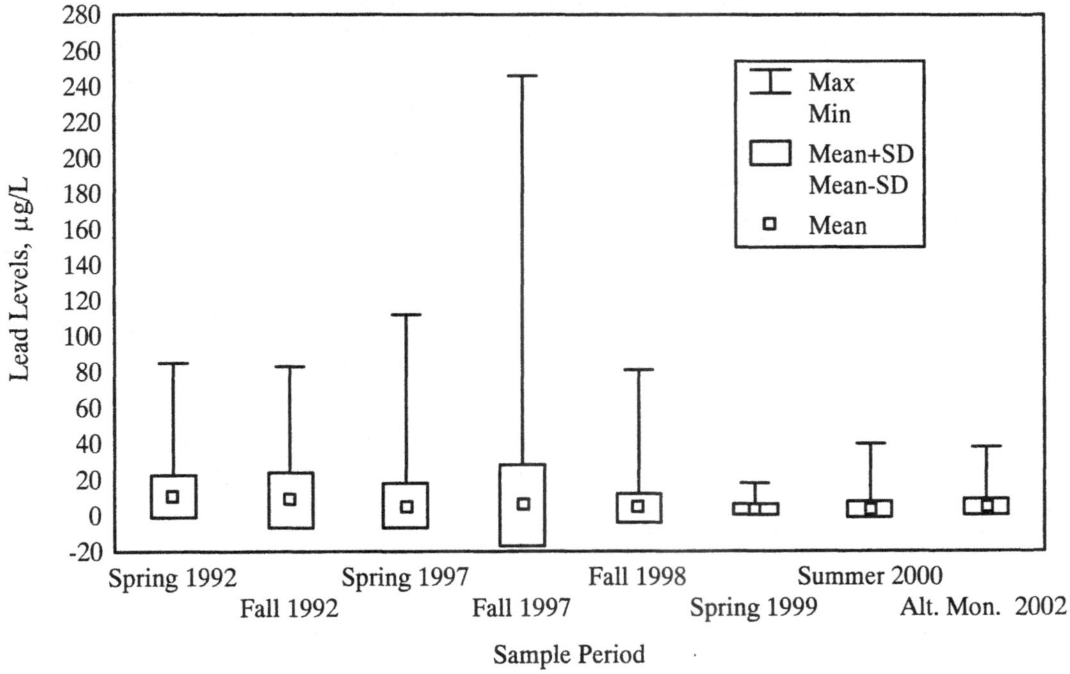

Figure C.29 LCR and field lead levels - DWSD

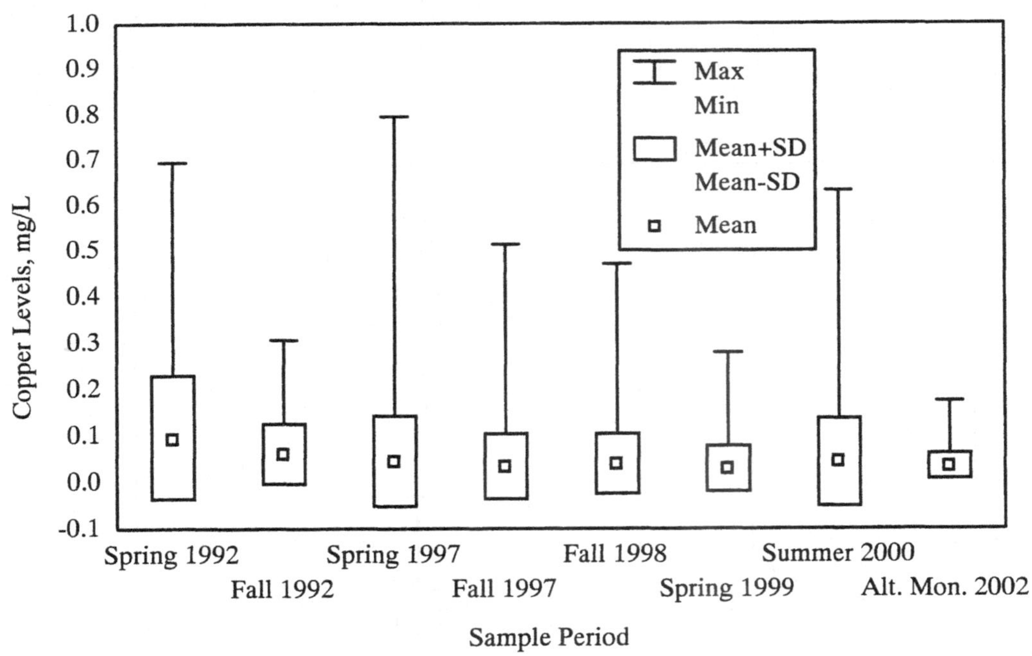

Figure C.30 LCR and field copper levels - DWSD

Lead and copper results from the seventeen sites that were sampled during the regulatory and alternative monitoring programs were also compared. Not all seventeen sites were sampled during each regulatory sampling period, except during the Spring 2000 sampling period. Table C.27 and Figure C.31 presents the summary statistics for the data collected from these sites during regulatory monitoring between 1997 and 2000, the same sites during the alternative monitoring period in 2002. Lead levels from the alternative monitoring program were similar to levels measured during regulatory monitoring and variability was also similar.

Table C.27
Comparisons of summary statistics for lead and copper levels from joint alternative tap monitoring and regulatory monitoring sites - DWSD

Sampling program	N	Mean	Median	Min - Max	90th Percentile	Standard deviation	CV
Lead (μg/L)							
Spring 1997	10	6.6	4.5	2 – 21	16.5	5.9	0.9
Fall 1997	10	4.7	4.0	2 – 9	8.5	2.5	0.5
Fall 1998	11	2.8	3.0	2 – 4	4.0	0.9	0.3
Spring 1999	15	3.4	2.0	2 – 12	7.0	2.7	0.8
Spring 2000	17	4.1	3.0	2 – 17	7.0	3.7	0.9
2002 Field Study*	121	5.2	3.0	3 - 38	8.5	5.0	1.0
Copper (mg/L)							
Spring 1997	10	0.018	0.012	0.009 - 0.04	0.035	0.011	0.6
Fall 1997	10	0.015	0.009	0.002 - 0.051	0.038	0.015	1.0
Fall 1998	11	0.028	0.010	0.002 - 0.13	0.041	0.037	1.3
Spring 1999	15	0.030	0.017	0.002 - 0.204	0.047	0.050	1.7
Spring 2000	17	0.039	0.014	0.002 - 0.352	0.080	0.083	2.1
2002 Field Study*	121	0.027	0.022	0.020 - 0.116	0.041	0.017	0.6

*Calculated 1000 mL samples

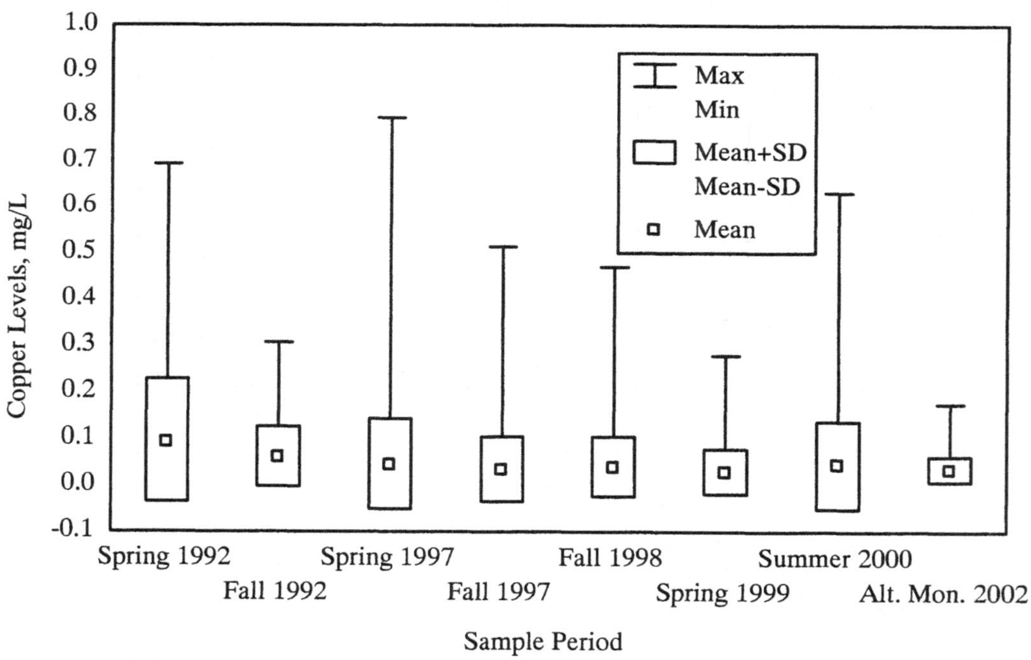

Figure C.31 LCR and field lead and copper levels from same sample sites – DWSD

APPENDIX D
PREDICTIVE TECHNOLOGY FIELD STUDIES

INTRODUCTION

Field tests that employed the Corrosion Potential-Stagnation/Flow (CPSF) approach were carried out at three sites that included the San Francisco Water Utilities' Millbrae Water Lab, the Pinellas County Kellar Water Treatment Plant in Florida, and HDR Water Research Laboratory in Bellevue, WA. CPSF Narrow Rectangular Channel (NRC) cells were installed at these sites.

The objectives of this phase of research were to identify features of the behavior of corrosion potential (E_{corr}) in drinking water containing residual chlorine, to identify trends of metal release at the end of periods of stagnation of varying duration, to explore correlations between total metal concentration and E_{corr} values, and to identify effects of water quality variations (such as pH) on metal surface aging.

HDR LABORATORY TAP WATER

Tap water was field tested in the Bellevue, Washington laboratory of HDR Inc. The tap from which samples were taken was located in the City of Bellevue. Bellevue purchases treated water from the Seattle Public Utilities (SPU) and water parameters are typically unchanged. For the tap water, pH ranged from 6.5 to 8.5, alkalinity was 16 mg/L, and the total chlorine was 1.0 mg/L. The experiments used CPSF NRC cell C-4, except as noted.

The behavior of E_{corr} of copper exposed to HDR tap water was very reproducible during stagnation episodes of varying duration (Figure D.1). This figure shows that the initial decay, or relaxation of the corrosion potential, largely coincided with a rapid accumulation of metal in the cell. This was typical for virtually all exposure times. For example, a comparison of Figure D.1 and D.2 shows that the profile of copper accumulation in the CPSF cell and E_{corr} relaxation were virtually identical from one week to another.

In both Figure D.1 and Figure D.2, it is obvious that the accumulation of copper and the relaxation of E_{corr} are strongly correlated. This correlation becomes linear between $\log(Cu_{total})$ and E_{corr}. The data that show the existence and strength of the linear correlation for HDR tap water are presented in Figure D.3.

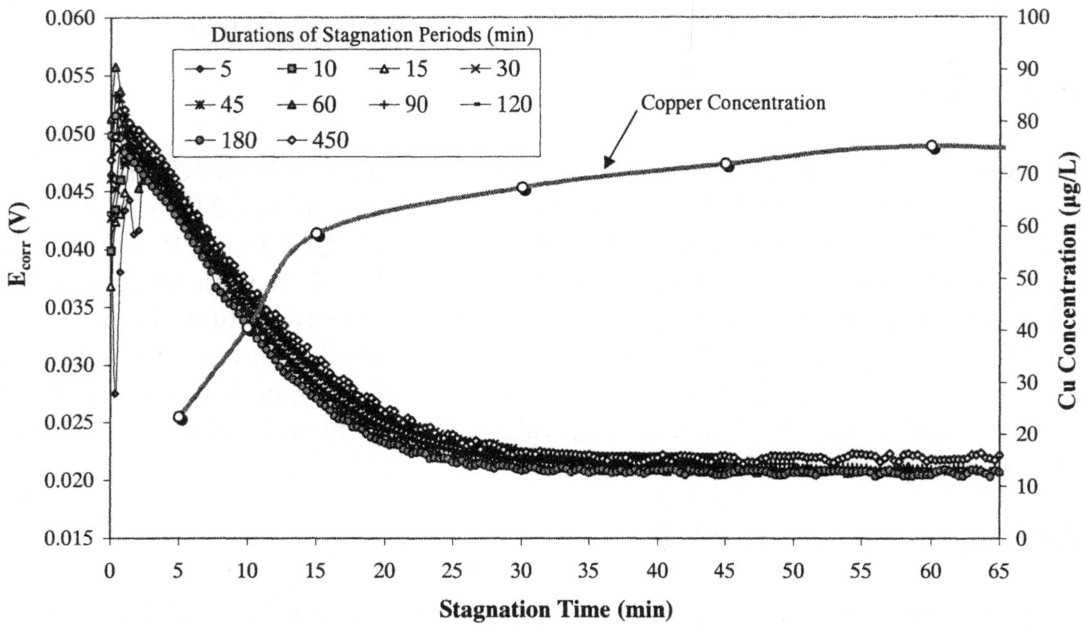

Figure D.1 E_{corr} and metal release patterns for varying stagnation periods. Exposure day 15, HDR tap water, CPSF NRC cell.

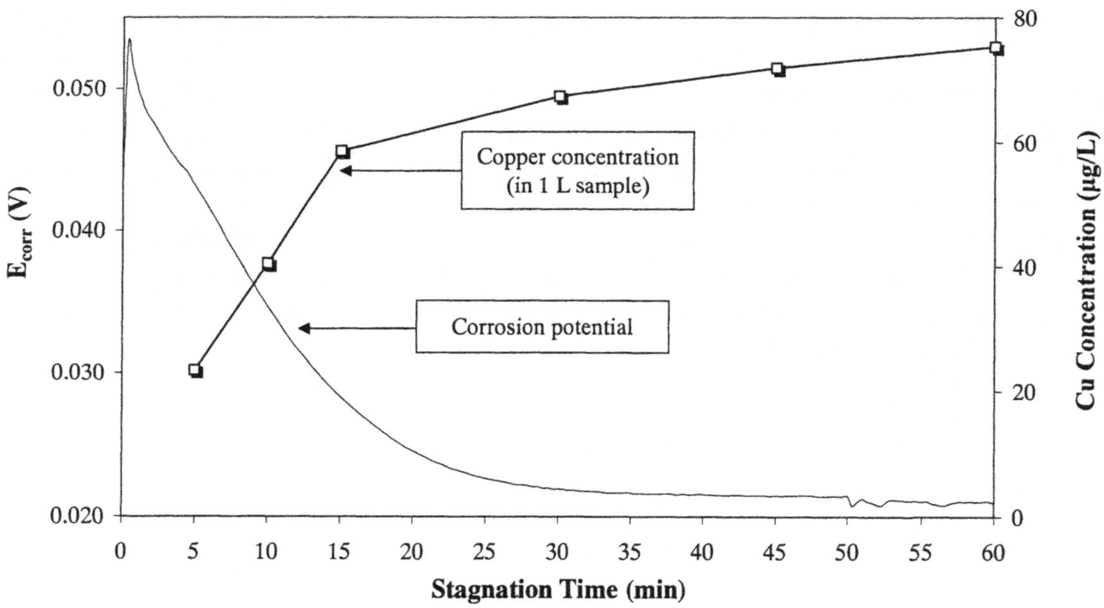

Figure D.2 E_{corr} and metal release. Day 22, HDR laboratory tap water, CPSF NRC cell.

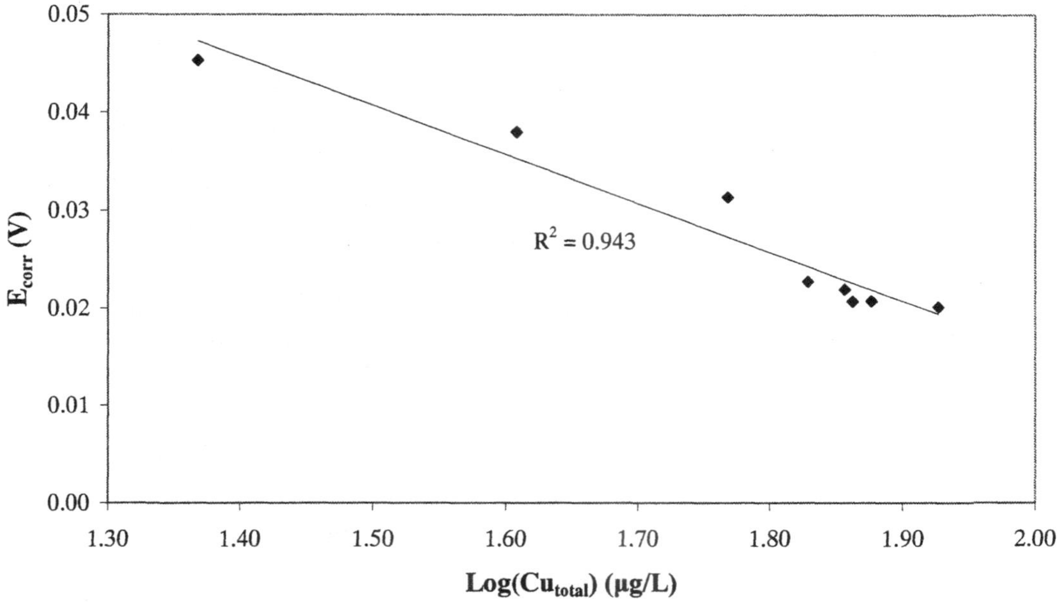

Figure D.3 Correlation between E_{corr} and total copper concentration for stagnation times varying from 5 to 60 minutes. HDR laboratory tap water, one-liter samples, CPSF NRC cell.

Although this study focused on the performance of the CPSF method for quantifying the corrosion of copper, the behavior of E_{corr} for other materials exposed to HDR laboratory tap drinking water had generally the same features as those described above (Figure D.4 and Figure D.5). The data shown in these figures were obtained using a Reiber corrosion cell, whose design and performance is different from the CPSF NRC cells developed in this study. Therefore, the kinetics of E_{corr} relaxation determined with these alternative designs of the corrosion cells can not be directly compared.

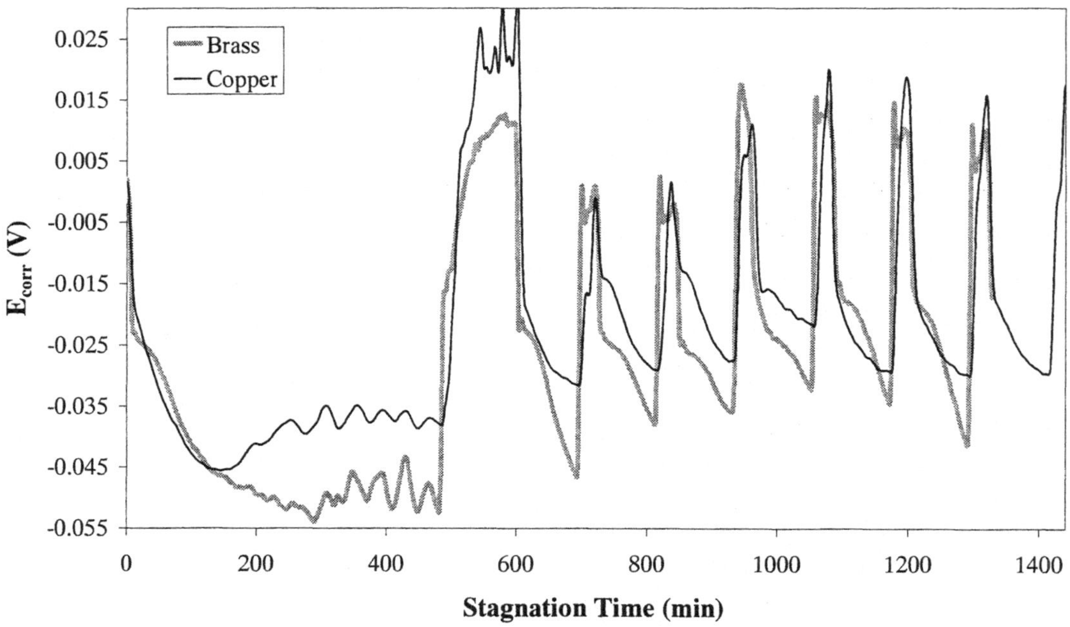

Figure D.4 Comparison of E_{corr} vs. time profiles for copper and brass. HDR laboratory tap water, Reiber corrosion cell.

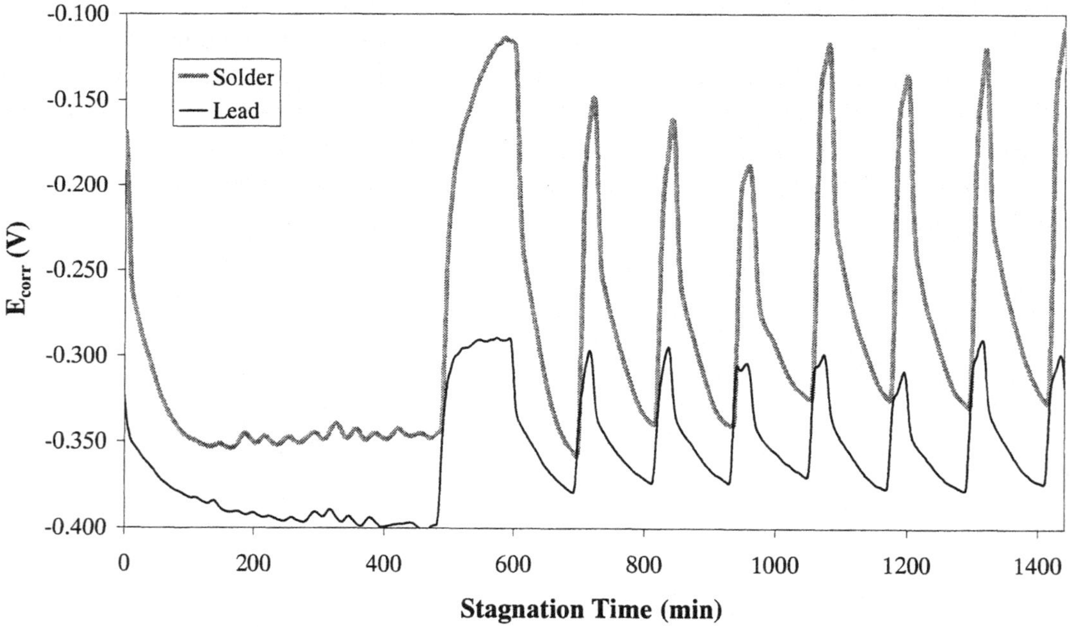

Figure D.5 Comparison of E_{corr} vs. time profiles for lead/tin solder and lead. HDR laboratory tap water, Reiber corrosion cell.

It is evident that many piping materials (copper, lead-containing brass, lead-tin solder and lead) exhibit similar trends during the relaxation of E_{corr} in stagnation episodes. The relaxation of E_{corr} is a manifestation of rapid corrosion processes that lead to chlorine consumption at the metal's surface simultaneous with the accumulation of metal in the cell. Although metal release was not measured in the Reiber cell experiments, the similarities between E_{corr} measurements obtained for a variety of metals with a Reiber cell and measurements obtained for copper with a CPSF cell suggest that the CPSF method could potentially be applied to monitor the corrosion of other metals of interest to the drinking water industry. However, CPSF's actual usability is contingent on the further development and extensive testing of that method.

In order to quantify the behavior of E_{corr} in more detail, we can introduce an alternative parameter. For chlorinated water, the time at which a minimum of E_{corr} is observed during stagnation is called the relaxation time (t_{relax}). For HDR laboratory tap water, the value of t_{relax} was in the range 25 to 40 minutes.

Although formal calculations to examine the effects of water quality and corrosion on the value of t_{relax} have not been done, preliminary data suggests that t_{relax} is affected by chlorine concentration, pH, alkalinity, and other parameters. The value of t_{relax} increased with decreasing rates of corrosion, and t_{relax} increased for lower values of i_{corr}. Because of these factors, the values of t_{relax} appear to be especially useful in identifying the influence of aging on the corrosion processes.

To identify effects of surface aging on E_{corr} and t_{relax} in CPSF experiments with HDR laboratory tap water, E_{corr} profiles were recorded at different days of the experiments with HDR laboratory tap water (Figure D.6). It can be seen that as the metal surface ages, E_{corr} values gradually increase. This is similar to theoretical results described in Chapter 5. The anodic shift of E_{corr} may be caused by a gradual decrease of the intrinsic copper oxidation current while the oxygen reduction current remains stable.

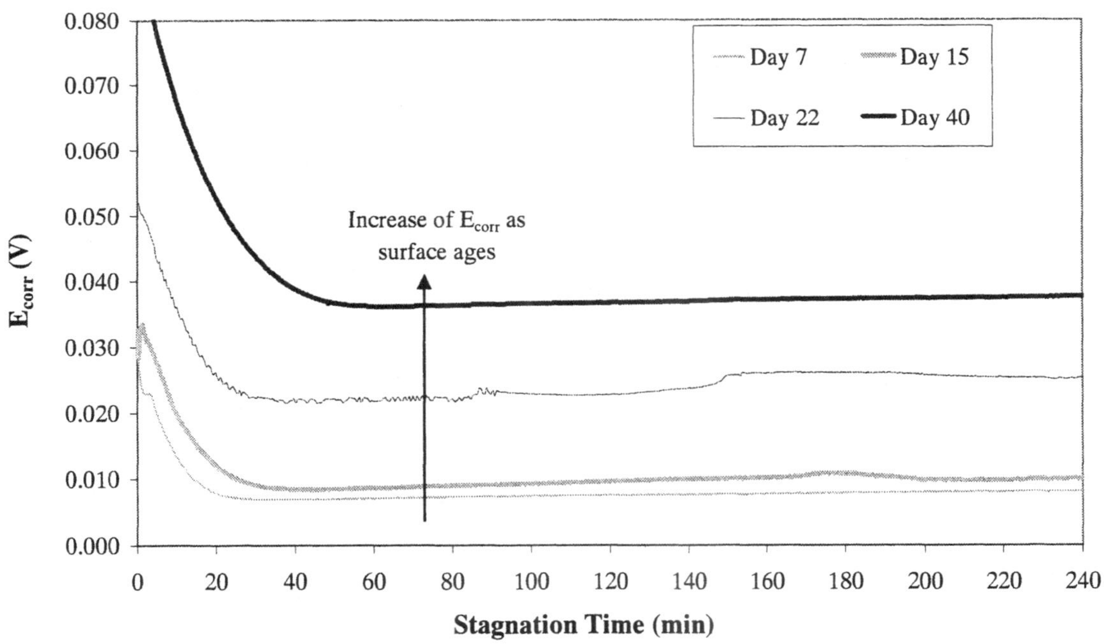

Figure D.6 Effects of surface aging on E_{corr} profiles during stagnation. HDR laboratory tap water, CPSF NRC cell.

The values of t_{relax} measured for the same series of experiments increased gradually as the surface aged (Figure D.7). This indicates a gradual decease of i_{corr} of the surface. Similar results were obtained with CPSF cell C-1, which was also installed in the HDR Water Research Laboratory in Bellevue, WA.

In the experiments with CPSF cell C-1, the pH of water was varied between 6.5 and 7.5. Measurements of the relaxation time from those experiments show that t_{relax} decreased in value following the drop in pH from 7.5 to 6.5 (day 8), indicating an increased rate of corrosion (Figure D.8). When the pH was increased from 6.5 to 7.5 (day 20), t_{relax} increased. Prolonged aging of the copper surface at that pH caused the relaxation time to increase gradually, although some variations were observed. When the pH was again reduced from 7.5 to 6.5 (day 80), the relaxation time decreased again.

These results support the significance of the measurement of t_{relax} in the presence of chlorine. The results show that t_{relax} data are useful for probing trends in the behavior of rate of corrosion associated with aging and variations of water quality parameters. However, further experiments are needed to validate this approach.

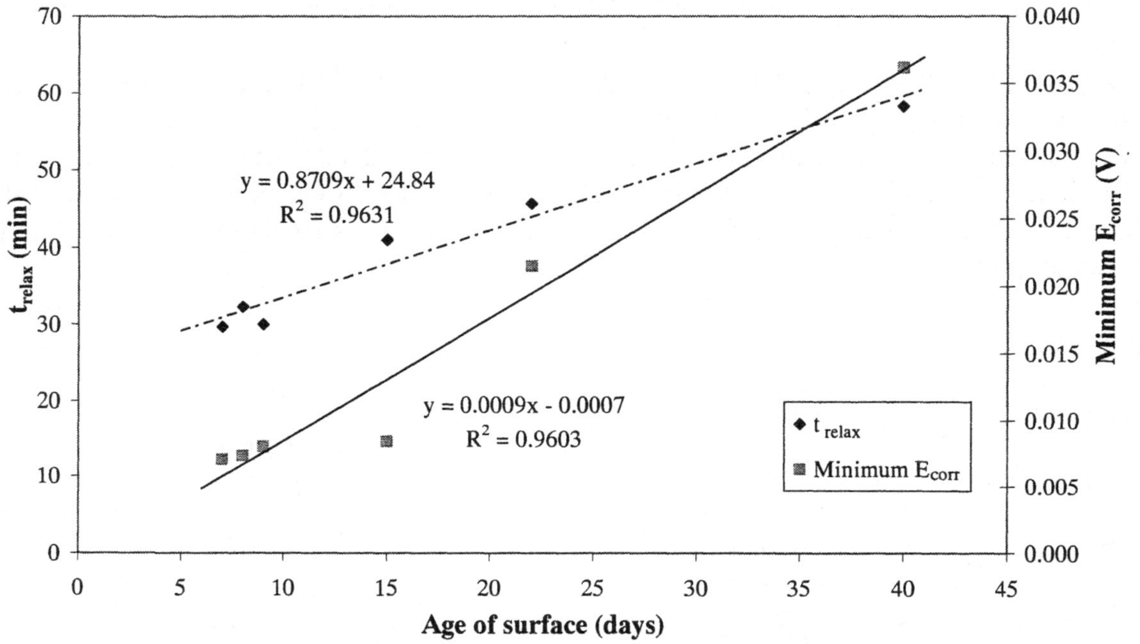

Figure D.7 Effects of surface aging on minimal values of E_{corr} and relaxation time profiles during stagnation. Seattle water, CPSF NRC cell.

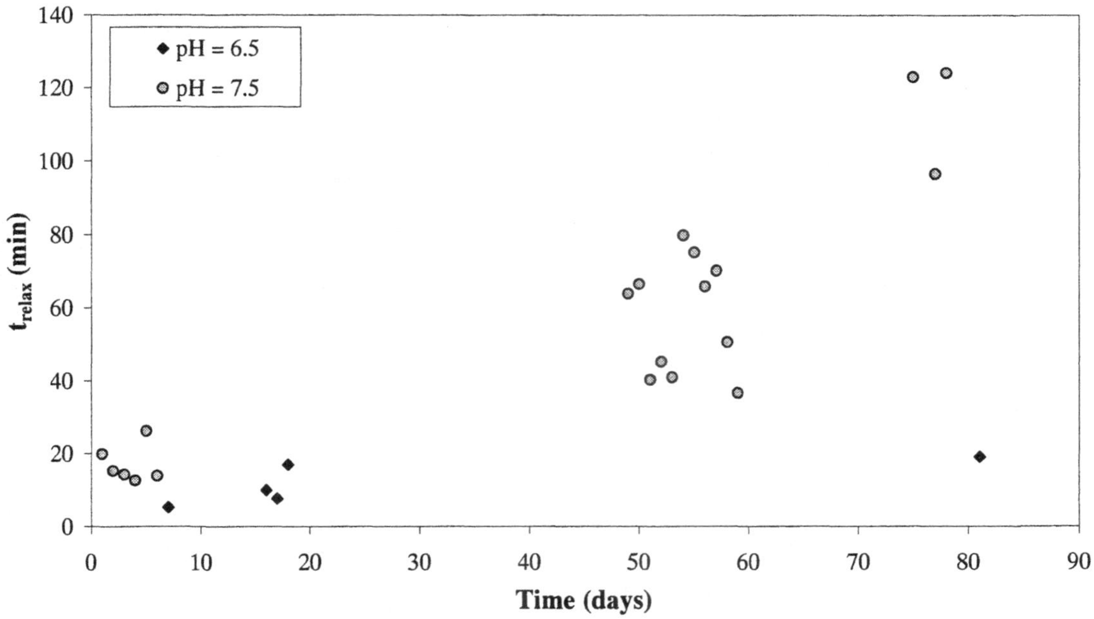

Figure D.8 Changes of the relaxation time as a function of surface aging and pH variations. HDR laboratory tap water, CPSF NRC cell.

SAN FRANCISCO TAP WATER

Tap water originating from Hetch Hetchy Reservoir, was used for field tests in San Francisco Water Utilities' Millbrae Water Lab. The average pH and chlorine concentration for the water were 8.6±0.1 and 0.9±0.1 mg/L, respectively. Both the chlorine and pH data reported by that utility exhibited some variability (Figure D.9 and Figure D.10). The experiments used CPSF NRC cell (C-4).

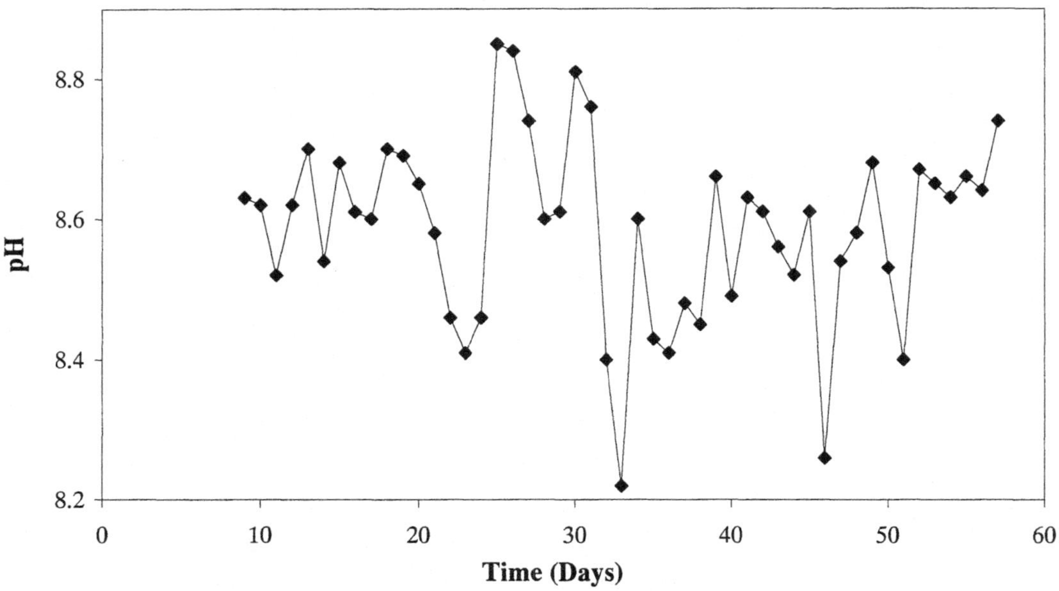

Figure D.9 Variations of pH in San Francisco tap water during field experiments.

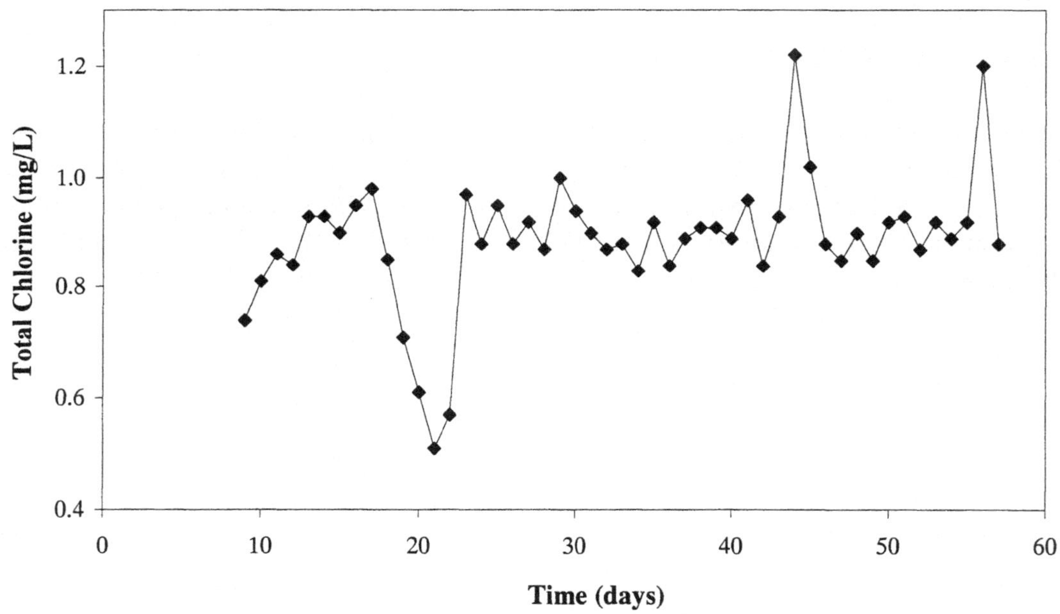

Figure D.10 Variations of total chlorine in San Francisco tap water during field experiments.

The relaxation of E_{corr} after the start of stagnation was observed in San Francisco drinking water (Figure D.11 and Figure D.12). The initial rapid decrease of E_{corr} lasted for about 60 to 90 minutes (t_{relax}) and was followed by a slow increase of E_{corr} (Figure D.11), which was also seen in other water qualities.

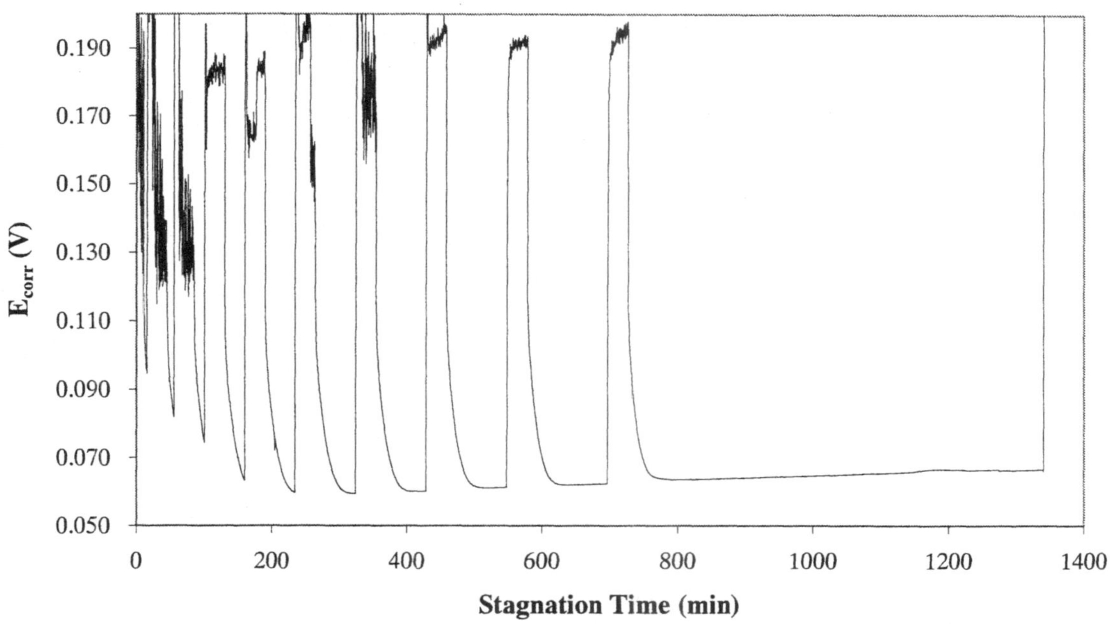

Figure D.11 E_{corr} profile obtained for varying stagnation times. Day 1, San Francisco tap water, CPSF NRC cell.

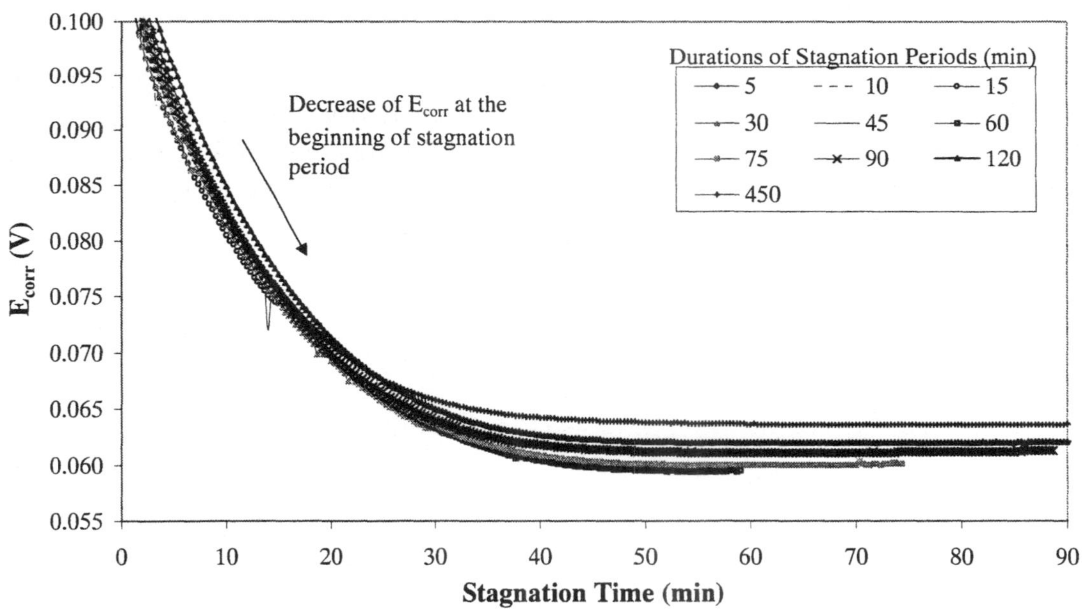

Figure D.12 Comparison of E_{corr} relaxation profiles for different stagnation periods. San Francisco tap water, CPSF NRC cell.

Overall, the CPSF data obtained for San Francisco drinking water were similar to those for HDR laboratory tap water. However, for San Francisco drinking water, there were no consistent trends associated with the aging of the surface. This is demonstrated in Figure D.13

which shows that samples withdrawn from the CPSF cell at the end of various 8-hour stagnation periods showed no obvious trend in copper concentrations as the metal surface aged.

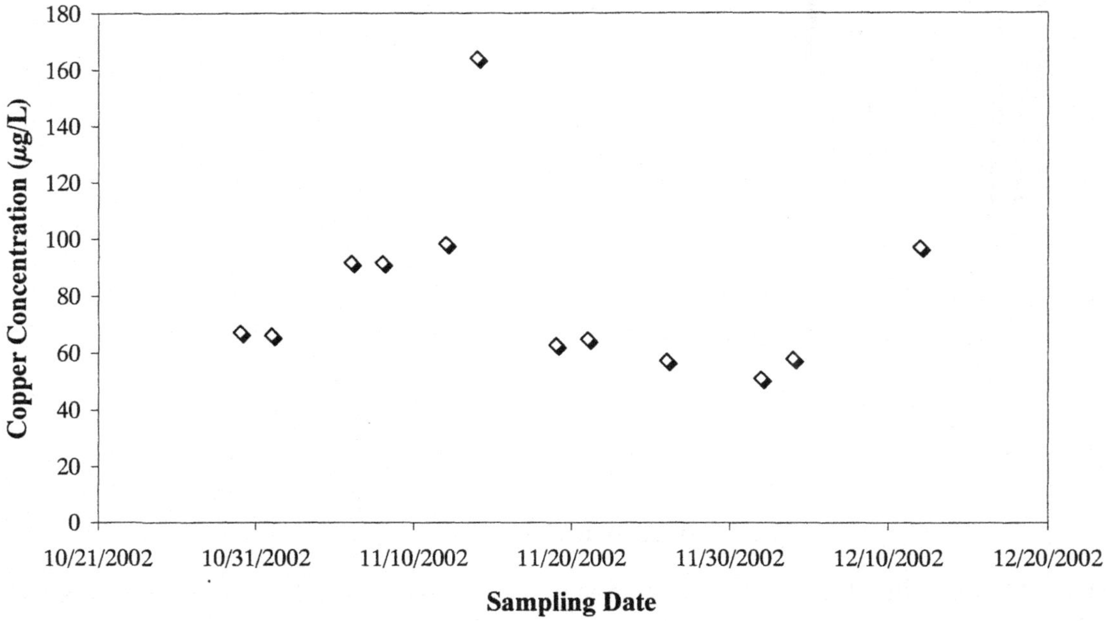

Figure D.13 Copper concentrations in samples withdrawn at the end of 8 hours stagnation. San Francisco water, one-liter sample, CPSF NRC cell.

The relaxation of E_{corr} and copper release in the CPSF cell on Day 2 are compared in Figure D.14. The data closely resemble those for HDR laboratory tap water (Figure D.1 and Figure D.2), although the copper levels in San Francisco water are considerably lower. This is in agreement with basic water quality parameters in those sources; HDR laboratory tap water has a lower alkalinity and pH compared to those of San Francisco water.

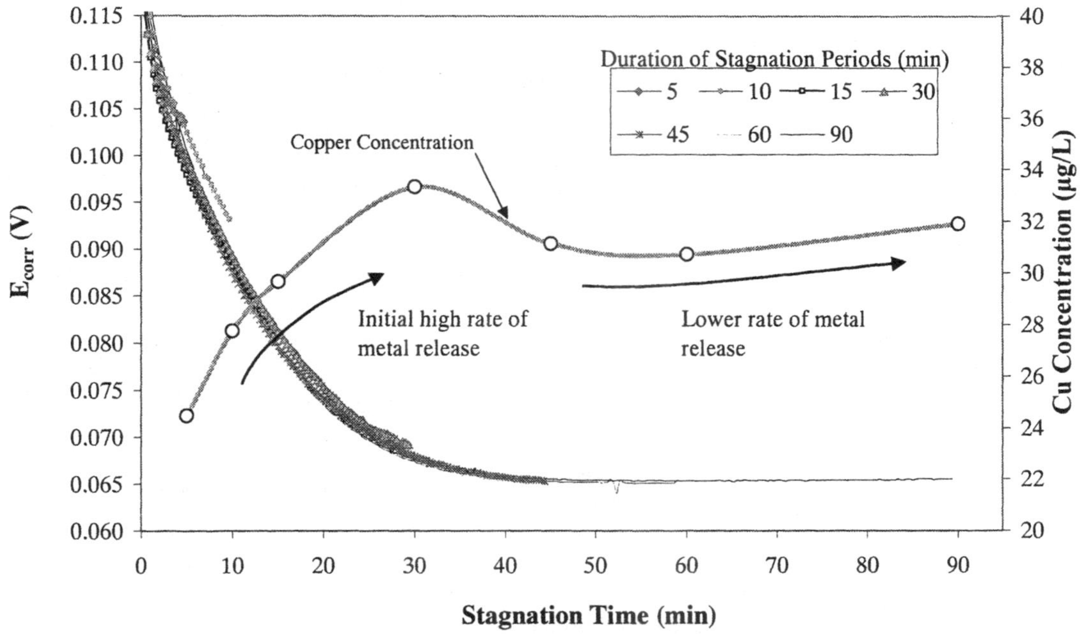

Figure D.14 E_{corr} relaxation and metal release patterns for varying stagnation periods. Day 2, San Francisco tap water, CPSF cell.

As was seen almost universally in experiments with chlorine-containing drinking waters; copper concentrations increase rapidly at the beginning of stagnation, while for longer stagnation periods, copper concentrations increase at much lower rate. In Figure D.15, the data for two successive days of measurements of E_{corr} and copper concentrations are shown. Both E_{corr} values and metal concentration are higher in Day 2 of exposure than in Day 1. Although this data expresses the overall trend, the significance of these differences should not be overestimated because the surface was still in a transient state. Day 1 and Day 2 were the only sample sets for San Francisco water which recorded metal release for varying stagnation times.

The nature of the correlation between E_{corr} and copper concentrations in the CPSF cell appears to be the same as was observed in the experiments with the synthetic water quality in the presence of chlorine and with HDR laboratory tap water (Figure D.16). However, the metal data obtained in the field experiments performed with San Francisco water were lower compared to all other water qualities. Most likely, this was caused by the transient state of the surface and possible deficiencies in the preparation of the sample for metal analysis. Nonetheless, the position of the data and the similarity of the overall trends support the conclusion that the CPSF methods can be utilized to quantify, in situ, the rate of metal release, especially for short stagnation times.

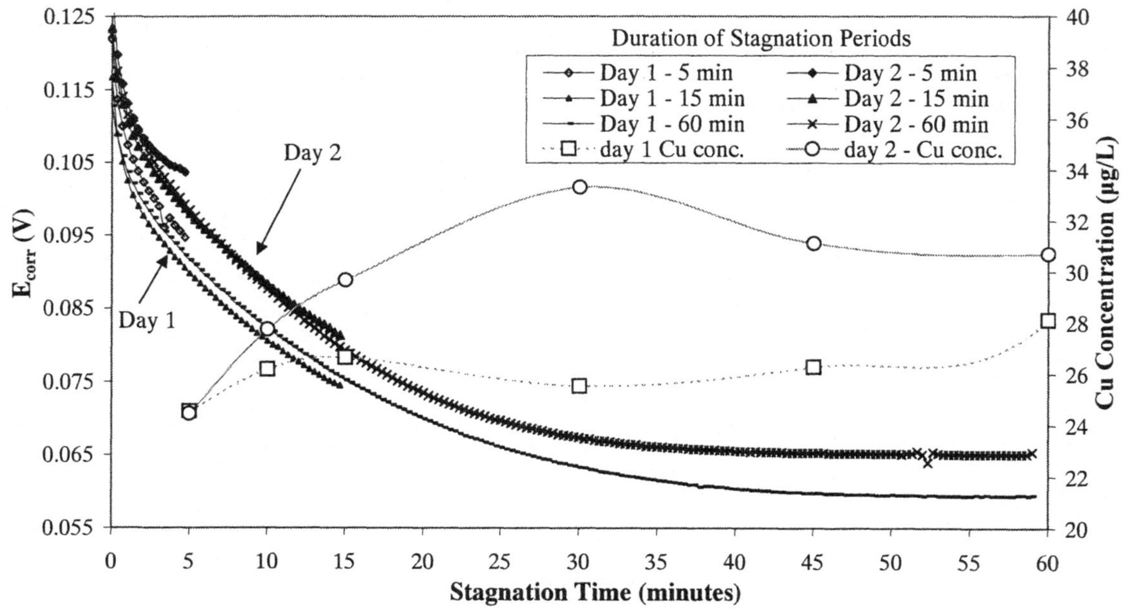

Figure D.15 Comparison of E_{corr} relaxation and metal release patterns for 2 consecutive days of exposure. San Francisco tap water, CPSF NRC cell.

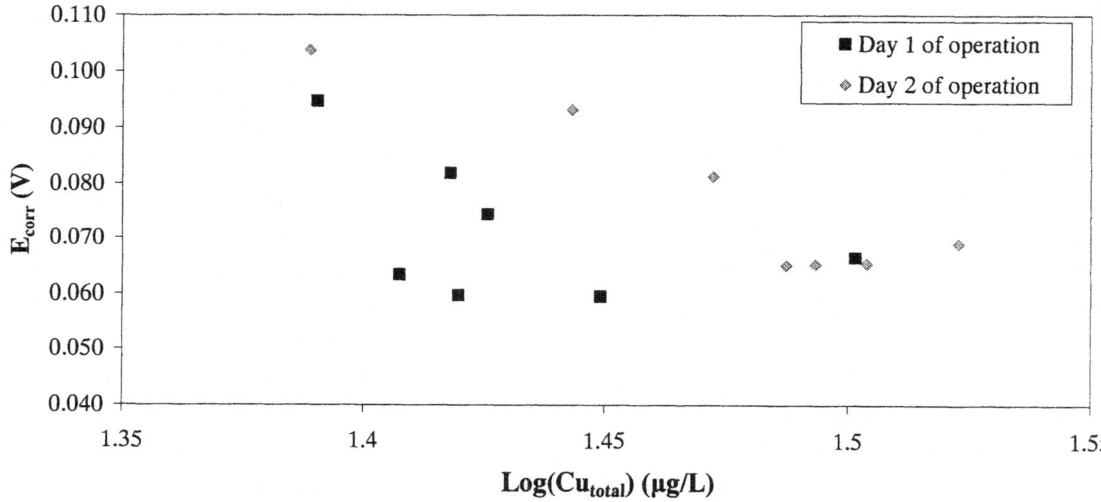

Figure D.16 Correlation between E_{corr} and copper concentration at the end of stagnation periods of varying duration. First 2 days of exposure, San Francisco tap water, CPSF NRC cell.

PINELLAS COUNTY WATER

Field experiments were performed at a water quality station at the Kellar Water Treatment Plant in Pinellas County, Florida. The concentration of the total chlorine and pH exhibited some variability throughout the period of experiments (Figure D.17 and Figure D.18). The average pH was 7.7±0.05, while the average concentration of total chlorine at the site of corrosion measurements was 3.7±0.2 mg/L. The high level of chlorine in the water was somewhat unexpected. The experiments used CPSF NRC cell C-1.

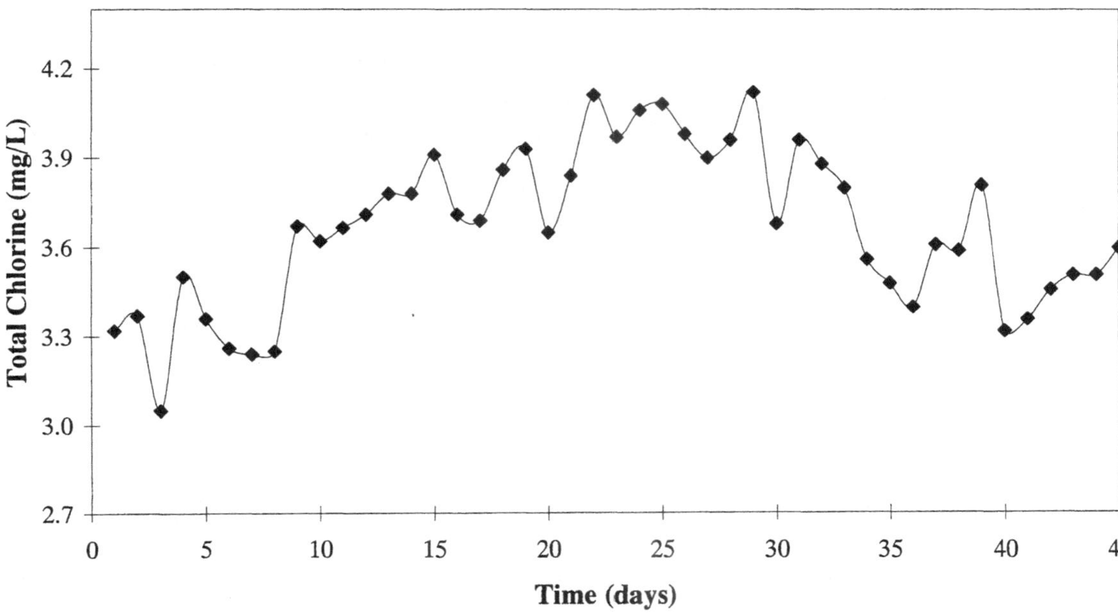

Figure D.17 Total chlorine concentration in Pinellas County tap water.

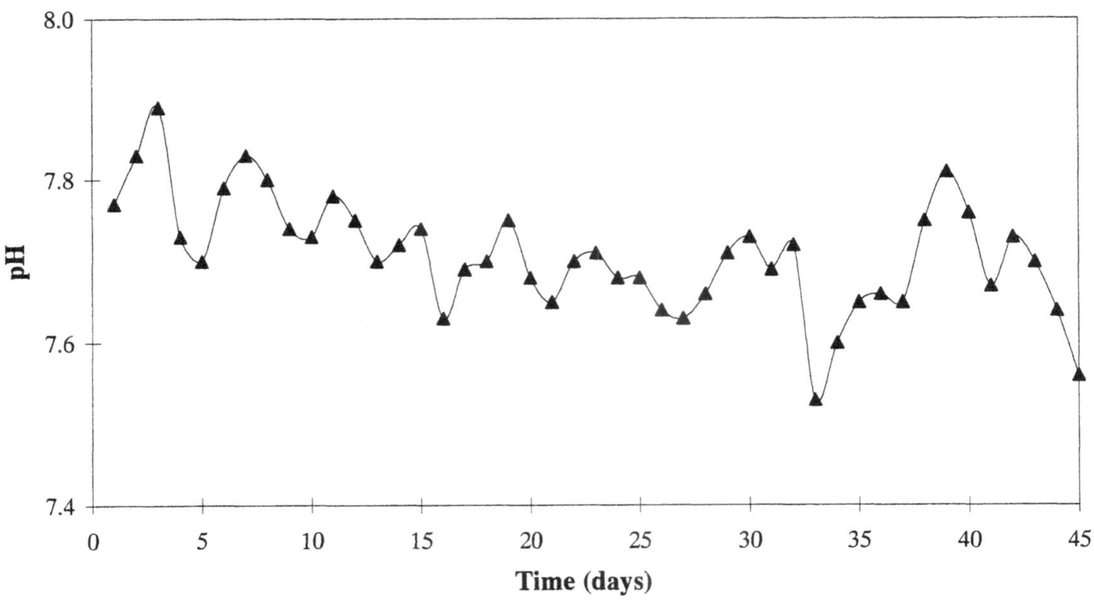

Figure D.18 Values of pH in Pinellas County tap water.

Although the majority of the chlorine was bound in the form of chloramine, the total chlorine concentration far exceeded concentrations in laboratory experiments. Consequently, no laboratory data for comparable conditions were generated during this study. Nonetheless, all major features seen in the behavior of E_{corr} during stagnation episodes in Pinellas drinking water were similar to those in other field studies. The value of E_{corr} rapidly decreased in all episodes of stagnation independent of their duration (Figure D.19). For stagnation times longer than three hours some increase of E_{corr} was seen. The increase in E_{corr} was associated with the predominance of the oxygen-driven reactions (as opposed to mainly chlorine-driven copper oxidation at shorter times).

Throughout the field experiments in Pinellas County, the aging of the surface did not appear to affect either the behavior of E_{corr} or metal release. The slight increase of E_{corr} values at stagnation times longer than 3 hours was less prominent during exposure week three than during week one (Figure D.19). However, the rapid relaxation of E_{corr} during the initial phase of stagnation was virtually identical to that throughout the experiments (Figure D.20).

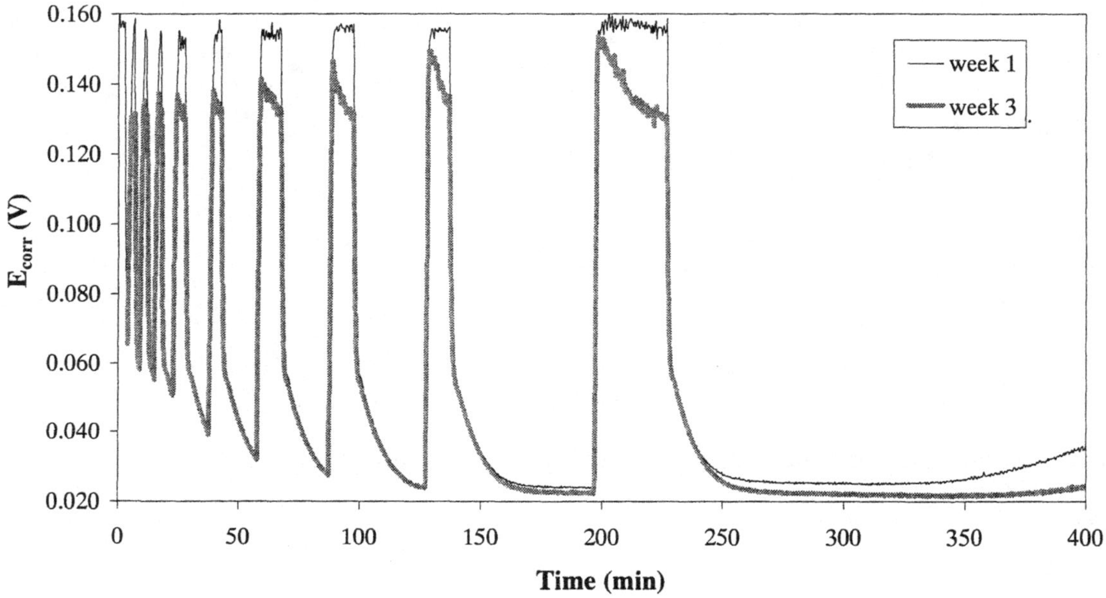

Figure D.19 E_{corr} data profiles at varying stagnation times for week one and three of exposure. Pinellas County tap water, CPSF NRC cell.

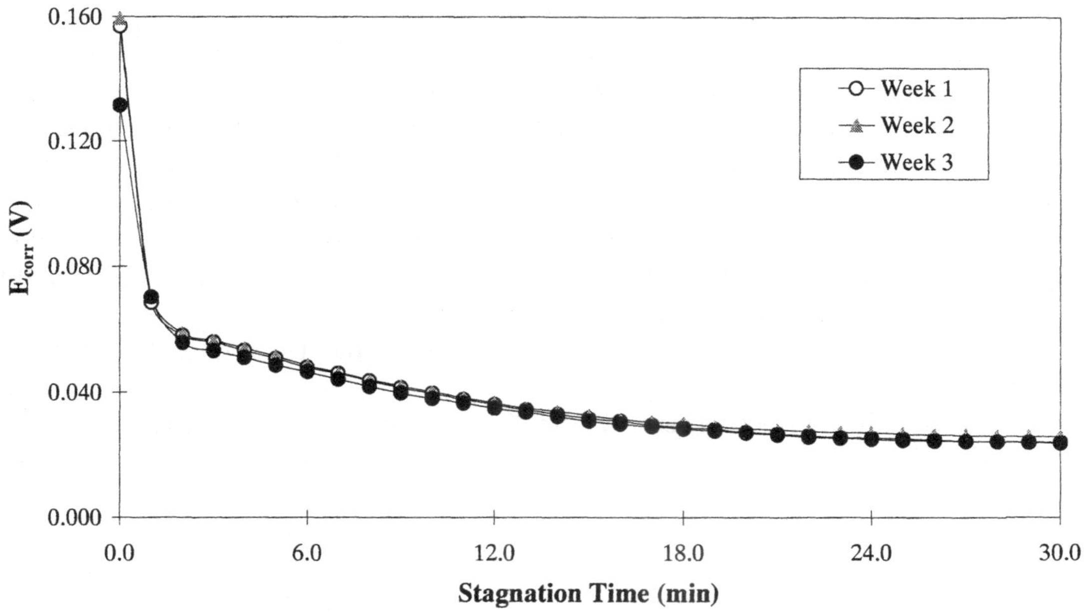

Figure D.20 Comparison of the initial phase of E_{corr} vs. stagnation time profiles obtained during first three weeks of exposure. Pinellas County tap water, CPSF NRC cell.

Measurements of copper concentrations in one-liter samples (withdrawn from the CPSF cell at the end of an eight-hour stagnation) showed that total soluble metal in those samples was largely stable after 10 days of exposure (Figure D.21) and was close to 25 μg/L, which corresponds to the copper concentration of 1250 μg/L if measured directly in the CPSF cell. This is very close to the 1.3 mg/L action level for copper.

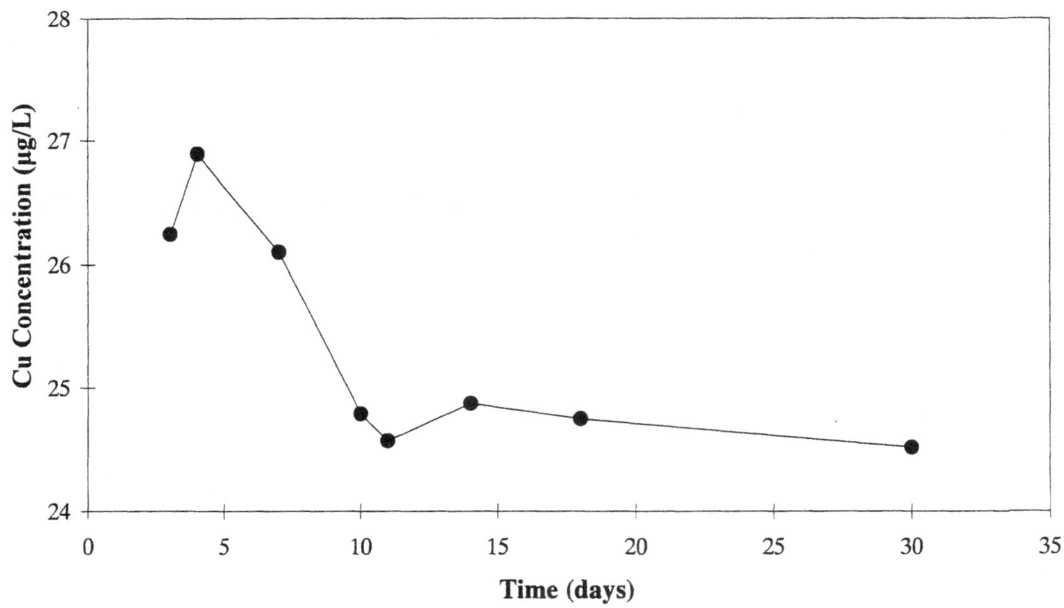

Figure D.21 Concentrations of copper at the end of an 3-hour stagnation period. Pinellas County tap water, one-liter sample, CPSF NRC cell.

Generally, experiments were performed for stagnation times between one minute and 180 minutes. However, on the first day of testing, one experiment with a stagnation time of 480 minutes was performed. The first and last days were the only days on which samples were collected for stagnation times of less than 30 minutes. Direct measurements of copper release in samples with stagnation times of less than 30 minutes showed two main features (Figure D.22): the aging of the surface for almost two months caused only minor changes in the rate of copper accumulation in the CPSF cell; and a substantial concentration of copper was found in the cell for stagnation times as low as one minute. Copper corrosion did continue to increase with time but the comparison of copper concentrations determined for 30 and 480-minute stagnation times showed only a 25% difference. The rate of copper release in the CPSF cell was notably higher in Pinellas County water than in the experiments carried out in the University of Washington laboratory (Figure D.23).

Despite the specific effects seen in Pinellas County water because of the high concentration of chlorine, the correlation between total copper concentrations and E_{corr} were as linear and strong as in all other cases (Figure D.24).

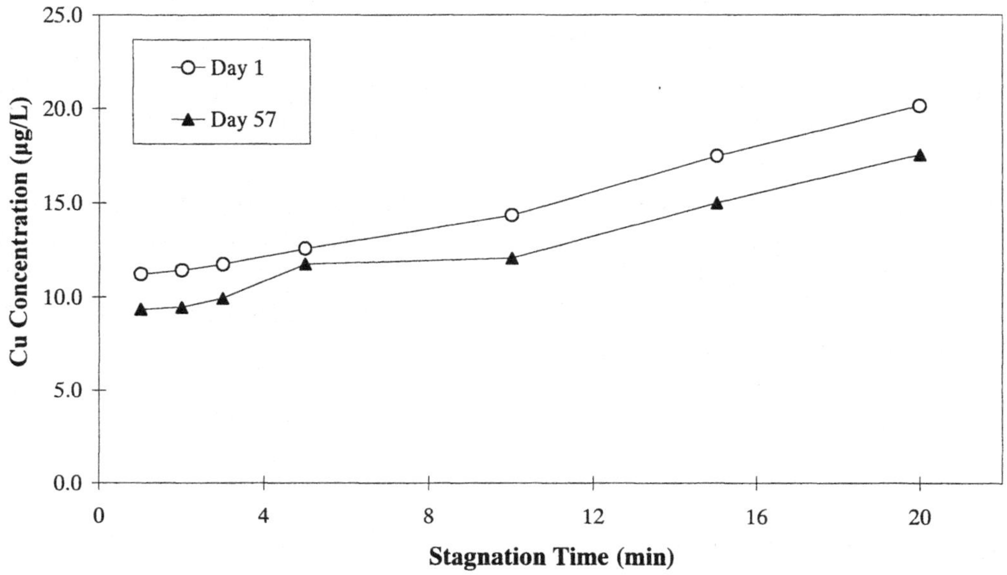

Figure D.22 Comparison of copper accumulation on Day 1 and 57 of experiments. Pinellas County tap water, one-liter sample, CPSF NRC cell.

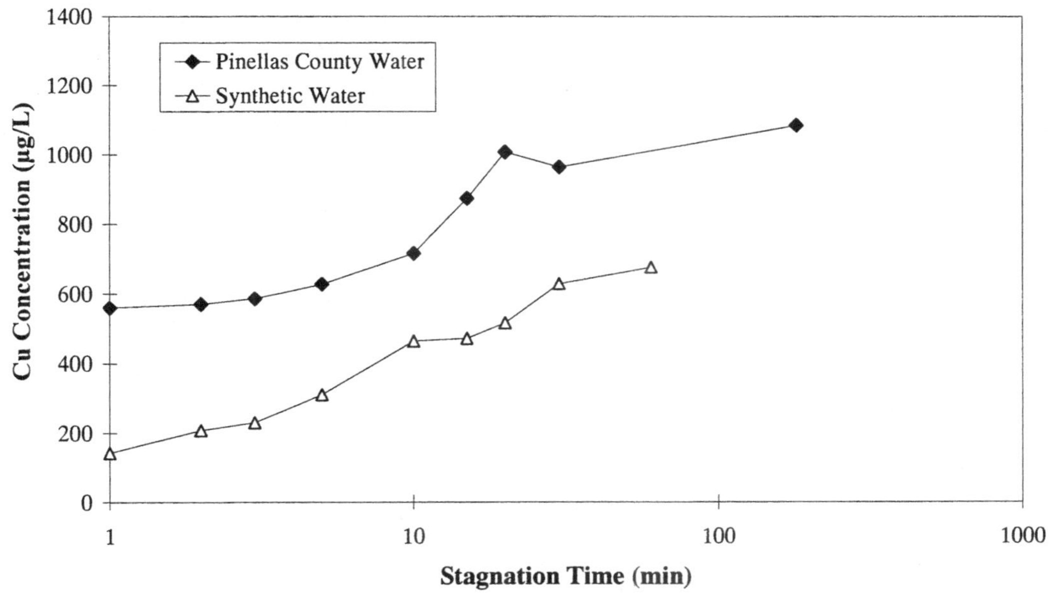

Figure D.23 Comparison of metal release in Pinellas County tap water ([Cl] = 3.7 mg/L) and synthetic water quality ([Cl] = 1.5 mg/L). Metal concentrations shown represent total soluble metal inside the cell. CPSF NRC cell.

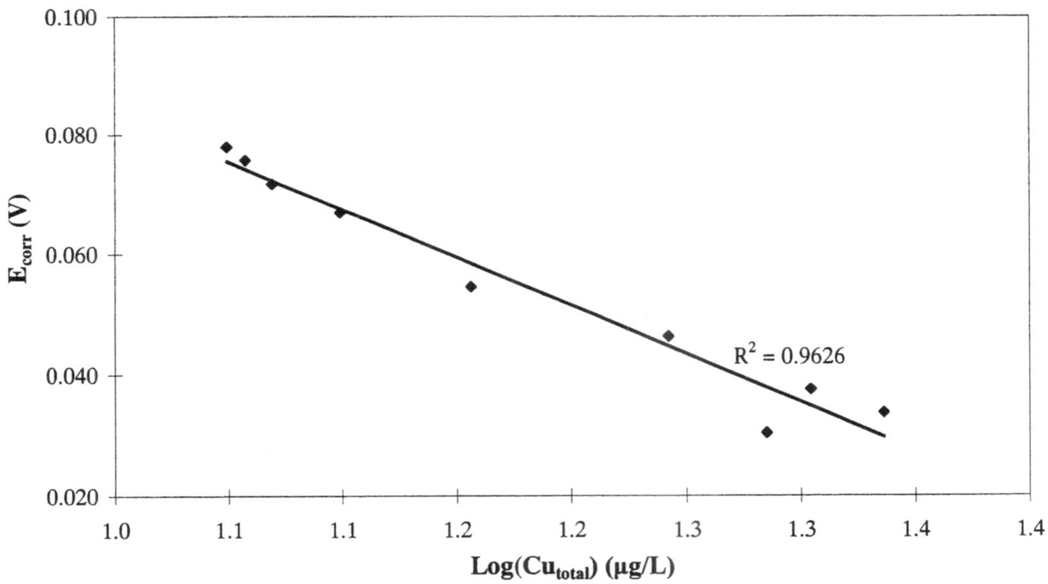

Figure D.24 Correlation between E_{corr} and total copper concentrations. Pinellas County tap water, one-liter sample, CPSF NRC cell.

When the E_{corr} and log(Cu_{total}) correlation for Pinellas County water is compared with the correlation for synthetic drinking water (with a total chlorine concentration 1.5 mg/L), there is an obvious difference (Figure D.25). First, the values of E_{corr} are much higher in Pinellas County water. This is undoubtedly due to a higher chlorine concentration. Second, the slope of the correlation between E_{corr} and log(Cu_{total}) are also different.

Currently, the CPSF method is not exhaustive enough to explain the difference in slope. It could be caused by different chlorine concentrations, alkalinities, different exposure times, or all of the above. Further experiments and numerical modeling are necessary to explore this issue. However, based on the obvious strength of the log(Cu_{total}) vs. E_{corr} correlation observed for this water quality, we can conclude that the CPSF method is likely to succeed in on-line monitoring of corrosion and metal release.

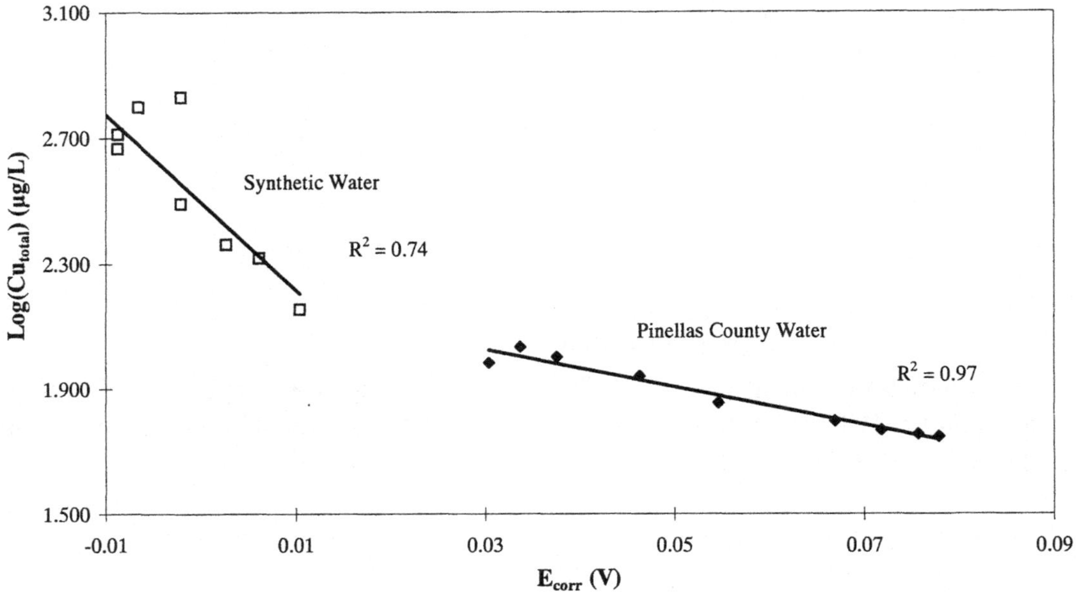

Figure D.25 Correlation between the values of E_{corr} and copper concentrations in water inside the CPSF NRC cell for Pinellas County tap water ([Cl] = 3.7 mg/L) and synthetic water ([Cl] = 1.5 mg/L).

APPENDIX E
PIPE RACK FIELD STUDIES

Introduction

As part of an earlier Awwa Research Foundation project, entitled *Development of a Pipe Loop Protocol for Lead Control*, both the Portland Water Bureau (PWB) and the Fort Worth Water Department (Fort Worth) constructed and operated pipe racks containing either three replicate copper loops soldered with 50-50 lead/tin solder or three lead loops. These pipe racks were originally constructed and operated in 1991-1992. Both utilities continued operation of the pipe racks and collected standing lead and copper levels intermittently until the spring of 2001. This was done in order to assess the value of pipe rack data as a predictor of changes in tap lead and copper levels. In addition, the Portland Water Bureau provided data from a brass block testing system placed in their distribution system.

Objectives

The objectives of the pipe rack field studies conducted at the Portland Water Bureau and the Fort Worth Water Department were to:

- Summarize the lead and copper data collected from pipe rack and brass block systems,
- Compare the pipe rack and brass block lead and copper data with tap monitoring data collected during the same time period, and
- Discuss the variability and sensitivity of these alternative monitoring systems as predictors of changes in tap lead and copper levels over time.

Description of Monitoring Program

PWB constructed one pipe rack containing three replicate loops of lead soldered copper. Fort Worth constructed one pipe rack with three lead soldered copper loops and one pipe rack containing three lead piping loops. Flow through the pipe racks was controlled to simulate typical home plumbing use with various on/off cycles followed by an eight hour standing time. The pipe racks were operated during 1991 and 1992 and standing water quality samples were collected for lead, copper, pH, alkalinity, and chlorine residual levels. Both utilities continued to operate their pipe racks through the spring of 2002, but water quality samples were not collected during the entire time period. Fort Worth collected samples for lead and copper levels intermittently throughout March 2001 and PWB resumed analysis of lead and copper levels from March 1999 through mid-September, 2001.

In addition to the pipe rack, PWB constructed and operated four brass block monitoring devices that were installed in various locations in the distribution system. These blocks were composed of cast red brass blocks which contain approximately 3% lead, to represent the typical material in home faucet fixtures Flow was controlled through the testers to simulate typical home use. Weekly samples were collected for standing lead and copper levels from the brass blocks from March 1999 through September 2001.

The following field studies are descriptions of the pipe rack and brass block monitoring systems, summaries of the data collected, comparisons of the data with regulatory tap monitoring data, and finally, a discussion of the sensitivity of these monitoring systems with respect to assessing changes in lead and copper levels.

PORTLAND WATER BUREAU

Background

The Portland Water Bureau (PWB) serves water to a metropolitan population of approximately 821,000. The primary water source is the Bull Run watershed, a highly protected unfiltered source characterized by low dissolved solids, low alkalinity, and neutral to slightly acidic pH. Treatment consists of disinfection with chloramine and pH adjustment using sodium hydroxide (as NaOH solution) prior to distribution. The pH of the water treatment plant effluent is currently 7.8. The pH of the treated water in the distribution system is typically in the range of 7.0 to 7.9 and the alkalinity is typically 6 to 12 mg/L as $CaCO_3$. The PWB system also has a supplemental groundwater supply that is used during the summer months or emergencies. Groundwater is disinfected with chlorine, followed by an ammonia addition prior to being pumped to the distribution system.

The major sources of lead in the PWB system were originally lead-soldered copper plumbing in homes and service line lead goosenecks. Both lead and copper levels exceeded the action level in the first two rounds of monitoring in 1992. PWB has replaced the lead goosenecks in its system, and from 1993 through 1997 initiated evaluation studies of optimal corrosion control treatment, non-chemical measures to reduce lead and copper, and the health benefits that might result to the community if treatment were installed. These evaluations indicated that "optimal" corrosion control treatment would involve increasing the pH of Bull Run treated water to 9.0-9.5, and increasing alkalinity to at least 20 mg/L (as $CaCO_3$). PWB also determined that implementing this treatment would be of limited effectiveness in reducing the community's blood lead levels. Therefore, PWB has implemented a Lead Hazard Reduction Program (LHRP) which includes:

- Water Treatment and Water Quality Monitoring
- A Home Lead Hazard Reduction Program
- Stakeholder Involvement, Public Education and Community Outreach
- Lead in Water Testing Program

The water treatment component of the LHRP includes limited corrosion control treatment to provide pH adjustment up to 7.5 to meet the copper action level and reduce lead levels in water by an estimated 40 percent. This treatment was implemented in 1997. Portland has conducted lead and copper monitoring at customer taps as directed by the Oregon Health Department since 1992 with two rounds of monitoring completed in 1992, as well as in 1997 through 2001 (Table E.1). Significant reductions in 90th percentile lead and copper levels have resulted from implementation of this treatment approach.

In addition to the regulatory tap monitoring that has been conducted since 1992, PWB has operated alternative monitoring systems that have been used to collect standing lead and copper samples over time. These systems include a pipe rack containing three replicate loops of lead soldered copper, and four brass block monitoring devices that have been installed in various locations in the distribution system.

Table E.1
90th percentile lead and copper levels from regulatory home tap monitoring

Year	Number of homes	Lead (µg/L)	Copper (mg/L)
Spring 1992	126	44	1.8
Fall 1992	125	53	1.3
Spring 1997	130	12	0.65
Fall 1997	129	13	0.68
Spring 1998	129	15	0.67
Fall 1998	127	19	0.67
Spring 1999	126	20	0.64
Fall 1999	126	12	0.60
Spring 2000	125	15	0.66
Fall 2000	124	22	0.71
Spring 2001	144	17	0.61

Description of Monitoring Program

Pipe Rack

The pipe rack was constructed at the Capital Highway Pump Station, in an area of the distribution system that contained a majority of PWB's regulatory home tap monitoring sites, (i.e. homes with lead soldered copper plumbing built between 1983 and 1985). A schematic of the pipe rack is shown in Figure E.1. As part of a previous AwwaRF project, the pipe rack was operated from June 1991 through mid-July 1992 and standing water quality samples were collected weekly for lead, copper, pH, alkalinity, and chlorine residual levels. From June 1992 through February 1999, water continued to flow through the loops in the rack. However water quality samples were not collected during this time period. From March 1999 through mid-September 2001, standing samples were once again collected for the parameters listed above. During the two sampling periods, June 1991 through mid-July 1992, and March 1999 through mid-September 2001, the flow to the pipe rack was controlled in order to simulate typical home plumbing use. The on:off cycles were as shown in Table E.2. After the 8 hour standing time, one-liter samples were collected from each test loop to simulate the regulatory protocol for collection of standing lead and copper levels.

Table E.2
Cycle times for pipe rack (PWB)

Time	Cycle	Time	Cycle
6:30 a.m.	ON	10:30 p.m.	ON
7:30 a.m.	OFF	11:00 p.m.	OFF
3:30 p.m.	ON	1:30 a.m.	ON
4:00 p.m.	OFF	2:00 a.m.	OFF
7:30 p.m.	ON	4:30 a.m.	ON
8:00 p.m.	OFF	5:00 a.m.	OFF

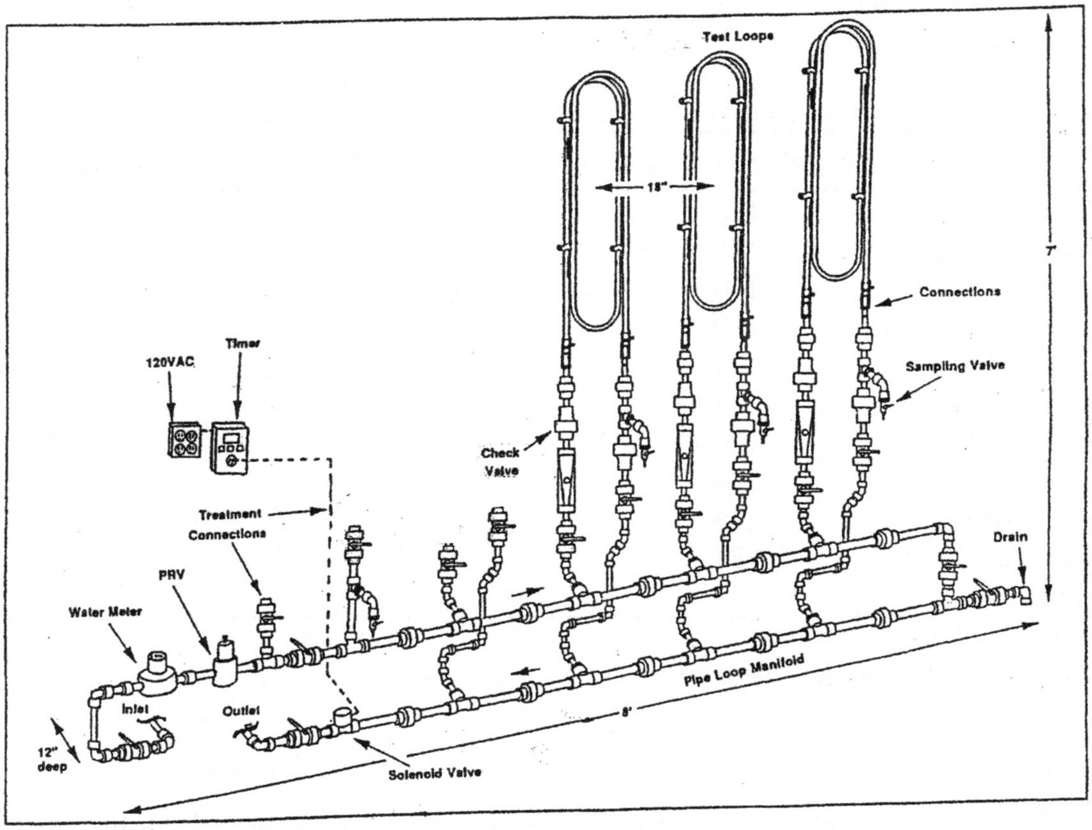

Figure E.1 Pipe Rack Schematic (*Source*: EES and ISWS 1990)

Brass Blocks

The brass block testers consist of 4x4x24-inch cast red brass blocks which contain approximately 3% lead and represent the typical material in home faucet fixtures.

These testers were installed at four locations in the distribution system :

- 112th Avenue Pump Station
- Capitol Highway Pump Station
- Linnton Pump Station
- Reservoir 6 Inlet

Flow was controlled through the testers to simulate typical home use. Weekly samples were collected from the brass blocks between March 1999 and September 2001 and were analyzed for pH, chlorine residual, lead, and copper. Monthly samples were collected for alkalinity.

LCR Tap Monitoring

As mentioned previously, Portland has conducted regulatory lead and copper monitoring at customer taps since 1992. Approximately 125 to 145 sites are sampled during each monitoring round, and all locations are Tier 1 sites, i.e. sites with lead soldered copper piping built between January 1, 1983 and June 30, 1985.

Lead and Copper Results

Pipe Rack

A summary of the lead and copper data collected from each of the replicate loops in the pipe rack is shown in Table E.3 and Figures E.2 and E.3.

Table E.3
Pipe rack summary statistics for lead and copper levels at PWB

	Lead levels (µg/L)					
	November 1991 through July 1992			March 1999 through September 2001		
	Loop A	Loop B	Loop C	Loop A	Loop B	Loop C
Average (x)	55	36	21	2.2	2.2	2.1
Max-Min	1-18	8-250	8-89	1-12	1-11	1-15
25%-75%	43-59	30-45	13-27	1-3	1-2	1-2
90th Percentile	66	48	54	4	4	3
Std. Dev.(s)	25.2	38.3	18.5	1.8	1.9	2.0
CV (s/x)	0.46	1.1	0.88	0.82	0.89	0.95
	Copper levels (mg/L)					
	November 1991 through July 1992			March 1999 through September 2001		
	Loop A	Loop B	Loop C	Loop A	Loop B	Loop C
Average (x)	1.20	1.20	1.14	0.36	0.37	0.36
Max-Min	0.90 – 3.60	0.86 – 4.10	0.91 – 2.10	0.07 – 0.70	0.12 - 0.69	0.12 - 0.70
25%-75%	1.00 – 1.20	1.0 – 1.2	1.0 – 1.2	0.30 - 0.40	0.30 - 0.40	0.30 - 0.40
90th Percentile	1.40	1.40	1.40	0.48	0.50	0.50
Std. Dev. (s)	0.45	0.54	0.23	0.09	0.09	0.09
CV (s/x)	0.37	0.45	0.20	0.25	0.24	0.25

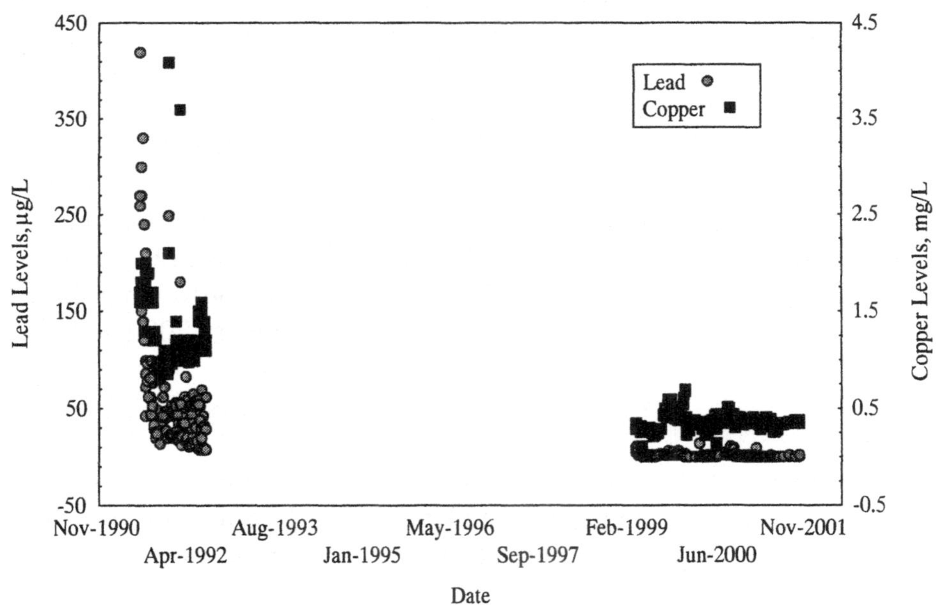

Figure E.2 All lead and copper levels from Portland pipe rack

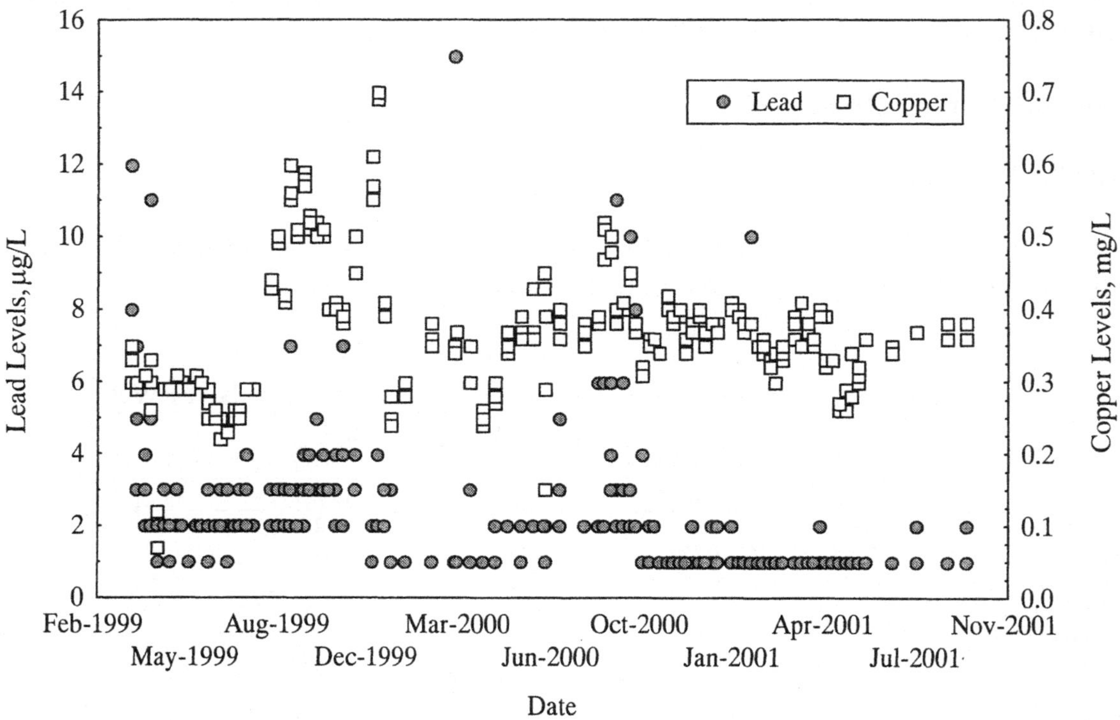

Figure E.3 Lead and copper levels from March 1999 through September 2001 for Portland pipe rack

During the initial operation of the pipe rack in 1991 and 1992, both lead and copper levels displayed a definite downward trend until the end of September 1991, after which the levels appeared to stabilize. Average lead and copper levels after stabilization were 29 - 53 µg/L for lead and 1.2 mg/L for copper for this initial evaluation period.

From March 1999 through September 2001 the lead and copper levels measured from the pipe racks were significantly lower than in 1991 and 1992 (Figures E.2). This is most likely due to the combined effects of pH adjustment, which was implemented in 1997, and continued leaching of lead and copper from these materials as they aged. Average lead and copper levels for this time period were 2 µg/L and 0.36 mg/L respectively. Lead and copper levels measured from the three replicate loops were statistically similar during this time period. Lead and copper levels were stable for the entire period of March 1999 through September 2001, however copper levels appeared to exhibit a narrower range of values after January 2000.

Brass Block

Figures E.4 and E.5 display the lead and copper data from the brass blocks. Average lead levels for the four brass block devices ranged from 7.7 to 14.6 µg/L during the time period from March 1999 through mid-September 2001 (Table E.4). Average copper levels were low, ranging from 0.05 - 0.09 mg/L.

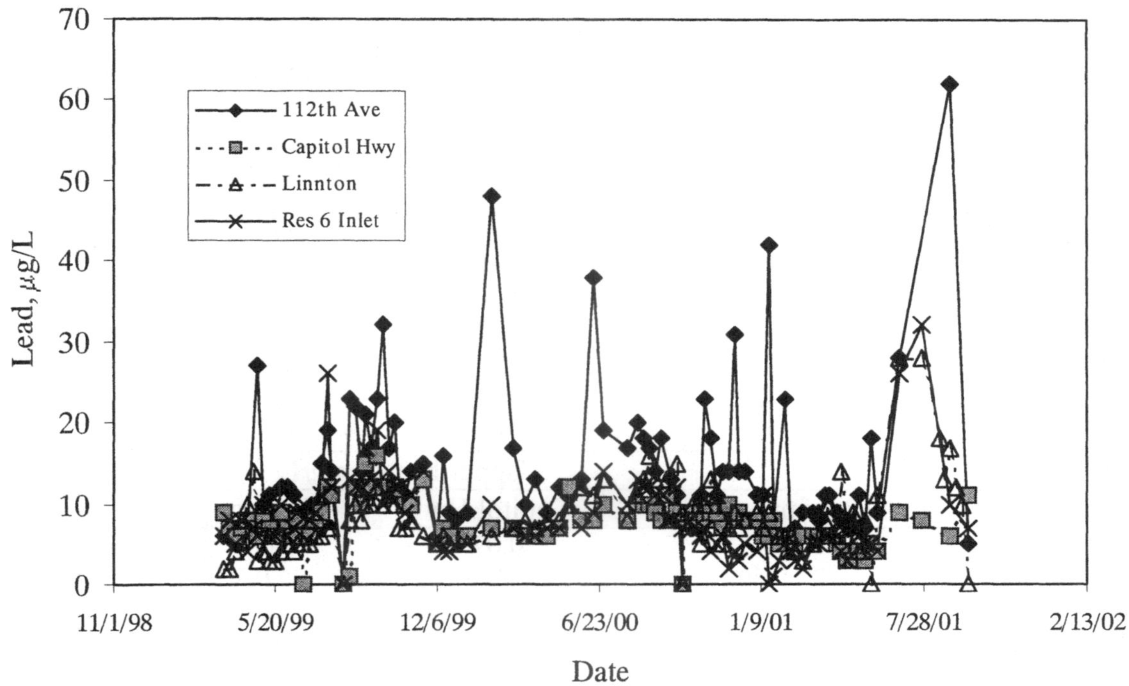

Figure E.4 PWB brass block lead levels

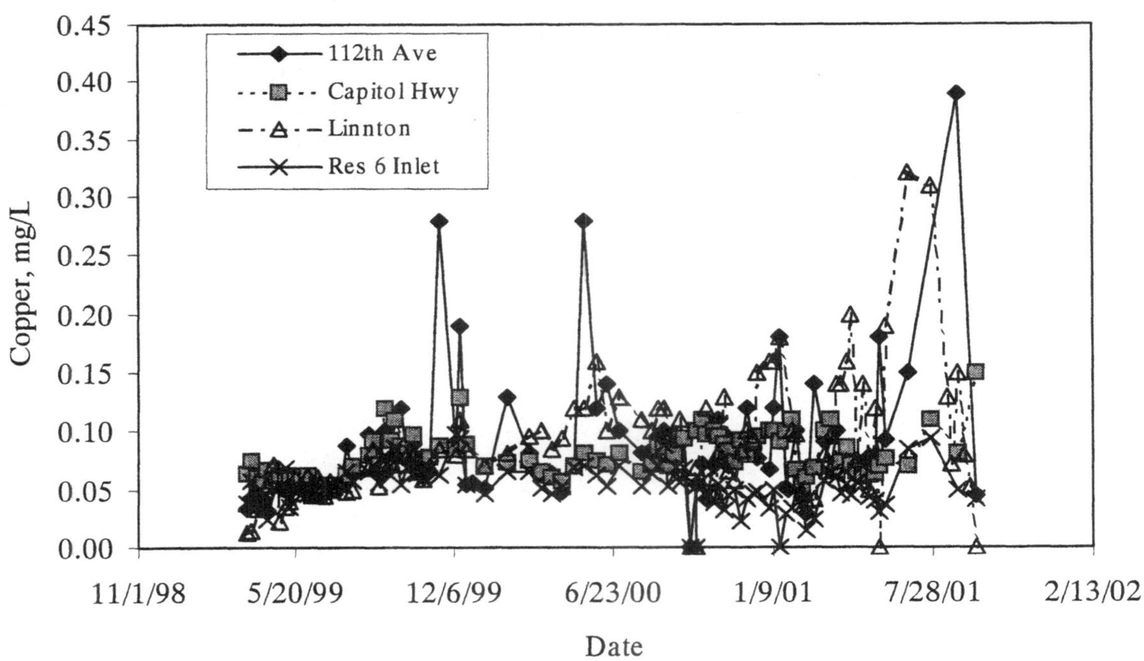

Figure E.5 PWB brass block copper levels

Table E.4
Summary statistics for brass block lead and copper levels at PWB - March 1999 through September 2001

	112 pump station	Capitol Hwy	Linnton	Res. 6 Inlet
Lead levels, μg/L				
	112 pump station	Capitol Hwy	Linnton	Res. 6 Inlet
Average (x)	14.6	7.7	8.4	8.3
Min-Max	5 – 62	1 – 16	1 – 28	0 – 32
25%-75%	9 – 17	6 – 9	5 – 10	5 – 11
90th Percentile	23	12	13	13
Std. Dev.(s)	9.4	2.7	4.56	5.05
CV (s/x)	0.64	0.35	0.54	0.61
N	90	94	95	95
Copper levels, mg/L				
	112 pump station	Capitol Hwy	Linnton	Res. 6 Inlet
Average (x)	0.086	0.079	0.089	0.053
Min-Max	0 - 0.39	0.045 - 0.15	0 - 0.32	0 - 0.10
25%-75%	0.055 - 0.098	0.064 - 0.092	0.054 - 0.11	0.045 - 0.064
90th Percentile	0.14	0.10	0.15	0.07
Std. Dev.(s)	0.055	0.019	0.053	0.018
CV (s/x)	0.64	0.24	0.60	0.34
N	90	91	93	93

LCR Tap Monitoring Data

PWB has conducted 11 rounds of monitoring, two rounds in 1992, and two rounds each year from 1997 through 2001. Tables E.5 and E.6 summarize the results of the lead and copper monitoring for regulatory monitoring rounds from 1997 through 2001. This data is also displayed in Figures E.6 and E.7. Since 1997, the 90th percentile copper levels have been below the action level, and 90th percentile lead levels have ranged from 12 to 22 μg/L. Summary statistics for all of the tap lead and copper levels measured during the time period from Spring 1999 through Spring 2001 are displayed in Table E.7. Over 600 samples were analyzed during this time period, with an average lead level of 7.7 μg/L.

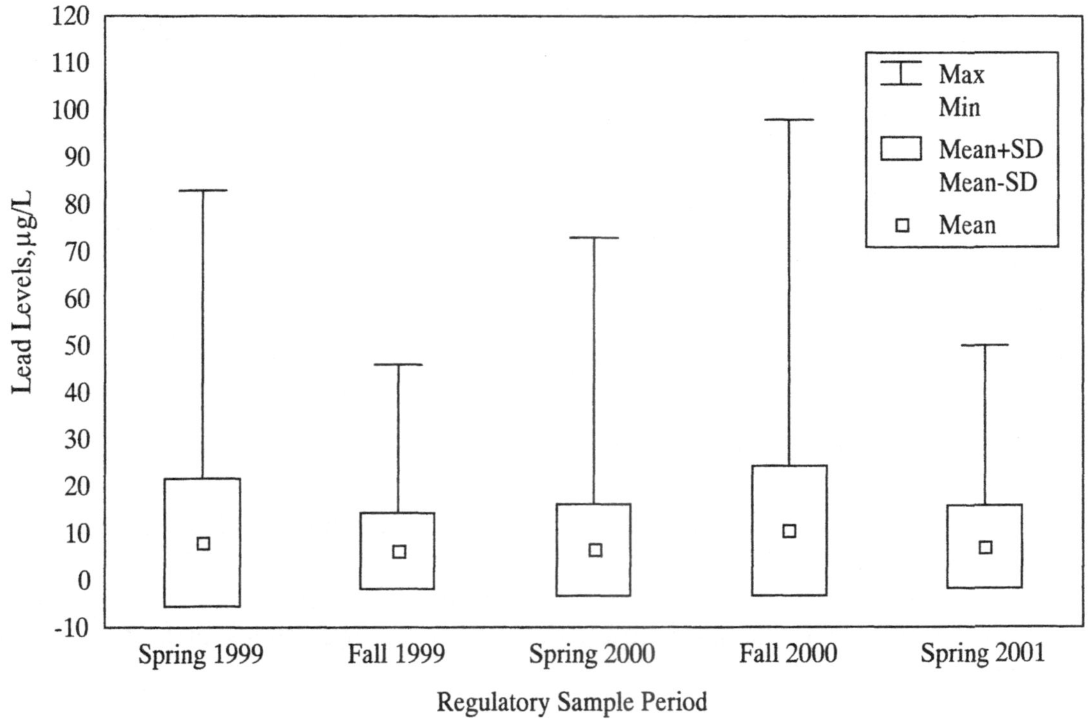

Figure E.6 PWB lead levels from regulatory monitoring

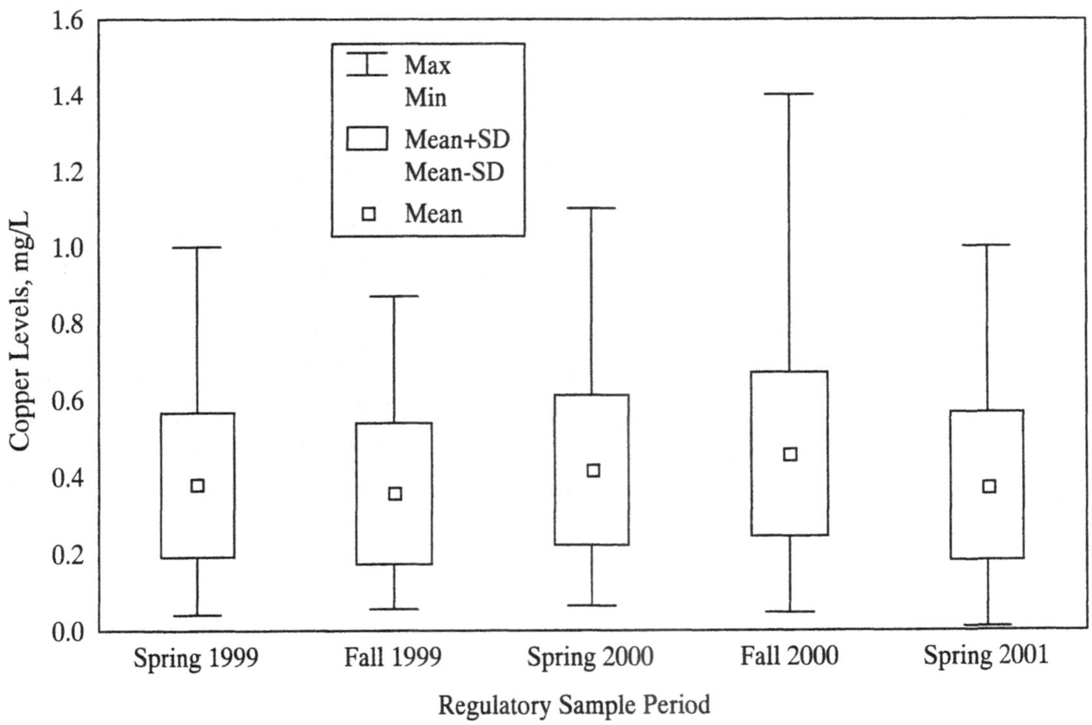

Figure E.7 PWB copper levels from regulatory monitoring

Table E.7
PWB summary statistics for regulatory lead and copper levels– all data from Spring 1999 through Spring 2001

	Lead Levels (μg/L)	Copper Levels (mg/L)
Average (x)	7.7	0.40
Min-Max	1 – 98	0.01 – 1.4
25%-75%	2 – 8	0.25 - 0.51
90th Percentile	17	0.66
Std. Dev.(s)	11.1	0.20
CV (s/x)	1.44	0.38
N	644	644

Lead and Copper Level Comparisons

Comparisons of pipe rack, brass block, and LCR data were completed for data collected from Spring 1999 through Fall 2001.

Table E.8 lists the summary statistics for lead levels collected during this time period. Lead data from the pipe rack was much lower than either the brass block or LCR data, and was also less variable.

Table E.8
PWB lead levels comparisons from Spring 1999 through Spring 2001 (Lead, μg/L)

Range of Values	Pipe Rack	Brass Block	LCR
Average	2.2	8.3	7.7
Min – Max	1 – 15	1 – 98	1 – 98
90th Percentile	4.0	18.0	17.0
Median (50%)	2.0	4.0	4.0
Standard Deviation (s)	1.9	12.1	11.1
CV (s/x)	0.9	1.4	1.5

The LCR tap monitoring data collected from Spring 1999 through Spring 2001 exhibited the highest lead level variability, which would be expected with over 600 samples collected over the time period from approximately 125 sites. Pipe rack lead values were the least variable. The brass block data was statistically similar to the LCR data collected during the same time period. Pipe rack lead levels were statistically different than both LCR lead levels and brass block lead levels. Figure E.8 displays a visual comparison of the lead data from each of the monitoring programs for each regulatory sample period from Spring of 1999 through Spring of 2001. The similarity between the LCR and brass block lead results for the same time period is evident.

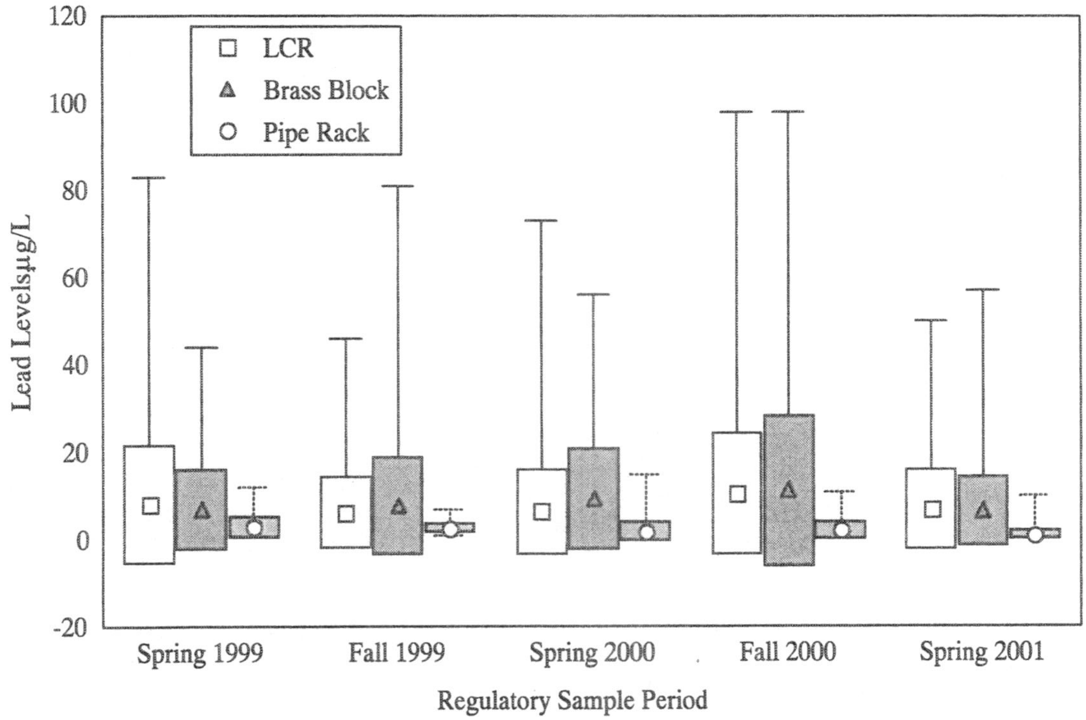

Figure E.8 Comparison of lead levels from sampling programs at PWB – Spring 1999 through Spring 2001

Comparisons of copper levels from the pipe rack, brass block, and LCR data collected from Spring of 1999 through Spring 2001 (Table E.9) indicated that the brass block copper data was much lower than either the pipe rack or LCR data.

Table E.9
PWB copper level comparisons from Spring 1999 through Spring 2001 (mg/L)

Range of Values	Pipe Rack	Brass Block	LCR
Average	0.36	0.10	0.40
Min – Max	0.07 - 0.70	0 - 0.32	0.01 – 1.4
90th Percentile	0.50	0.11	0.66
Median	0.36 - 0.37	0.07	0.38
Standard Deviation (s)	0.09	0.04	0.20
CV (s/x)	0.2	0.5	0.5

Neither the pipe rack nor the brass block copper was statistically similar to the LCR copper data collected during the same time period, however visual observation of the data indicates that the copper levels from the pipe rack were within the range of values measured during regulatory monitoring (Figure E.9). Copper levels from the pipe rack exhibited the lower variability than the brass block or regulatory copper data.

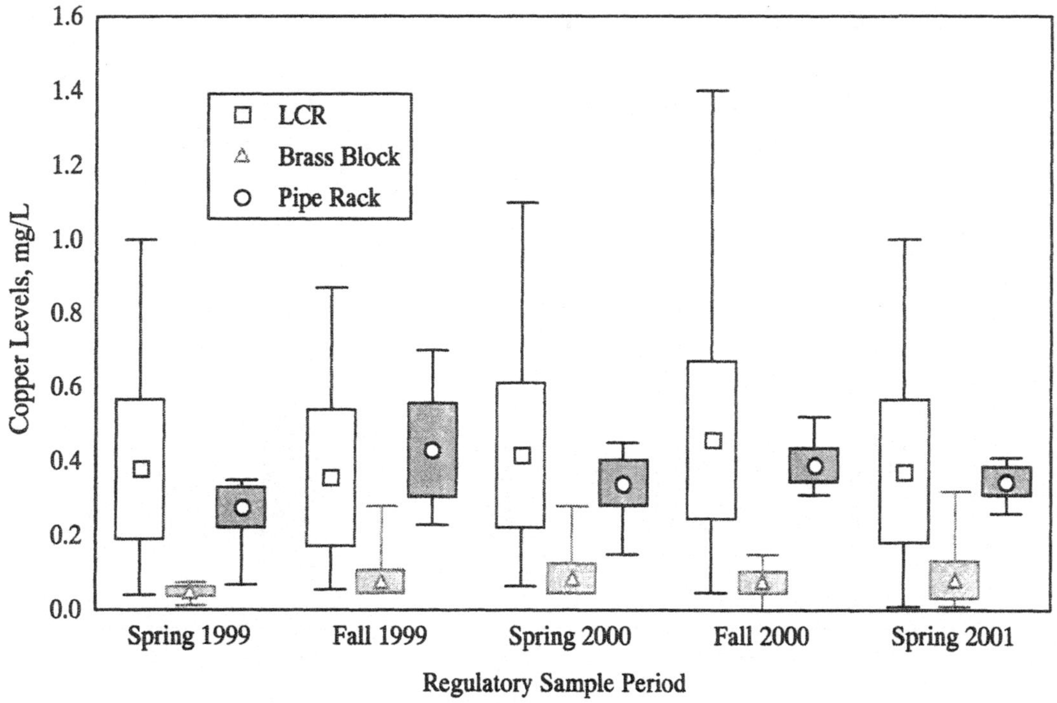

Figure E.9 PWB comparison of copper levels from sampling programs – Spring 1999 through Spring 2001

Conclusions

Based on the evaluation of lead and copper data from the pipe rack, brass blocks, and regulatory monitoring for the time period extending from Spring 1999 through Spring 2001, the following conclusions were drawn:

- Standing lead and copper levels measured from the pipe rack were significantly reduced after several years of operation (more than 9 years).
- Replicate loops in the pipe rack generated standing lead and copper levels that were statistically similar.
- Lead levels from the brass blocks were similar to lead levels from regulatory monitoring sites collected during the same time period.
- The variability of lead levels from the brass block data was similar to variability measured in regulatory monitoring data.
- Lead levels from the pipe rack were significantly lower than regulatory have monitoring data, and exhibited lower variability.
- The pipe rack exhibited copper levels which were similar to regulatory monitoring data.
- Copper levels from the pipe rack exhibited lower variability than copper levels from regulatory monitoring data.
- Copper levels from the brass block were extremely low and were not similar to regulatory copper levels.

FORT WORTH WATER DEPARTMENT

Background

The Fort Worth Water Department (Fort Worth) serves water to approximately 556,700 people in the City of Fort Worth, Texas and provides water to 29 communities on a wholesale basis. The sources of supply include the West Fork of the Trinity River (Lake Worth, Eagle Mountain Lake, and Lake Bridgeport), and two east Texas reservoirs (Cedar Creek and Richland-Chambers). Four conventional water treatment plants provide treated water to the distribution system – North Holly, South Holly, Rolling Hills, and Eagle Mountain. With the exception of Eagle Mountain, the treatment plants are conventional treatment facilities including pre-chlorination, coagulation (ferric sulfate and polymer), sedimentation, and dual media filtration. Free chlorine is used for primary disinfection and chloramines are used for distribution system residual maintenance. Eagle Mountain employs mono media (anthracite) filtration with ozone as the primary disinfectant and chloramines for distribution system residual maintenance. Lime is added at all four plants for pH and alkalinity adjustment. Finished water pH ranges from 7.8 to 8.7 with alkalinity levels of 80 - 126 mg/L as $CaCO_3$. The water system maintained by Fort Worth has been optimized for corrosion control since 1994.

The major source of lead in Fort Worth's system are lead soldered copper piping. However, there are an unknown number of service connections that are constructed of lead. Fort Worth has had a lead service line replacement program in place for several years in which lead services are replaced as needed when replacing old water lines and mains. Regulatory tap monitoring has been conducted since 1992. Table E.10 lists the 90th percentile lead and copper levels measured for each monitoring round.

Table E.10
90th percentile lead and copper levels for Fort Worth

Sample Period	Number of Homes	90th Percentile Lead Levels (μg/L)	90th Percentile Copper Levels (mg/L)
Jan-Jun 1992	100	12	0.71
Jul-Dec 1992	100	10	0.94
Jan-Mar 1997	100	8.7	0.59
Sept-Oct 1997	100	5.8	0.63
Aug 1998	50	4.9	0.32
July 1999	50	5.8	0.38

In addition to the regulatory tap monitoring that has been conducted since 1992, Fort Worth has intermittently operated two pipe racks, one containing three replicate loops of lead soldered copper, and one containing three replicate lead pipe loops.

Description of Monitoring Programs

Pipe Racks

Fort Worth constructed and operated two pipe racks in 1991 as part of a previous Awwa Research Foundation project. The copper pipe-lead solder pipe rack (PR-1) contains three replicate copper loops soldered with 50-50 lead tin solder and the other pipe pack (PR-2) consists of three replicate lead loops. PR-1 is located at the Alta Mesa pump station which serves an area of the distribution system that contains newer homes with lead-soldered copper plumbing. PR-2 is located at the Como Pump Station that serves an older area of Fort Worth where lead service connections are known to exist.

As part of the previous AwwaRF project, the pipe rack was operated from October 1991 through October 1992. Concurrently, standing water quality samples were collected weekly for lead, copper, pH, alkalinity, and chlorine residual levels. The flow to the pipe rack was controlled in order to simulate typical home plumbing use. The on:off cycles were as shown in Table E.11. After the 8 hour standing time, one-liter samples were collected from each test loop to simulate the regulatory protocol for collection of standing lead and copper levels.

Table E.11
Cycle times for pipe rack (Fort Worth)

Como Pipe Rack		Alta Mesa Pipe Rack	
Time	Cycle Mode	Time	Cycle Mode
7:00 a.m.	On	7:30 a.m.	On
8:00 a.m.	Off	8:30 a.m.	Off
4:00 p.m.	On	4:30 p.m.	On
4:30 p.m.	Off	5:00 p.m.	Off
7:00 p.m.	On	7:30 p.m.	On
7:30 p.m.	Off	8:00 p.m.	Off
10:00 p.m.	On	10:30 p.m.	On
10:30 p.m.	Off	11:00 p.m.	Off
1:00 a.m.	On	1:30 a.m.	On
1:30 a.m.	Off	2:00 a.m.	Off
4:00 a.m.	On	4:30 a.m.	On
4:30 a.m.	Off	5:00 a.m.	Off

After the initial AwwaRF monitoring period in 1991 and 1992, Fort Worth continued operation of the pipe racks and began monthly sample collection for lead and copper levels in January 1993. Samples were collected at the Alta Mesa pipe rack through February 2001, with the exception of the last half of 1993 and the last two months of 1995. The Como pipe rack was sampled approximately monthly through March 2001, with the exception of the last half of 1993, June through September 1998, and August 1999 through December 2000.

Historical LCR Tap Monitoring

Fort Worth has conducted six rounds of regulatory monitoring; two rounds in 1992, and two rounds each year from 1997 through 2001. Approximately 100 sites were sampled in the 1992 and 1997 sample rounds, and 50 sites were sampled in 1998 and 1999. In all sample rounds, both lead soldered copper sites and lead service line sites were sampled.

Lead and Copper Results

Pipe Racks

Alta Mesa Pipe Rack. During the initial operation of the pipe rack in 1991 – 1992, lead levels displayed a somewhat flat trend that approached stabilization after 200 days (March 1992). The exceptions were two periods when somewhat higher levels were measured. Copper levels were very erratic for the first 200 days of operation, and then exhibited a narrower range of values. Average lead and copper levels after stabilization for this initial period of operation (1991 – 1992) were 4.2 μg/L for lead and 0.59 mg/L for copper. With continued operation of the pipe rack, lead levels continued to remain stable, and the range of values narrowed again after approximately 1500 days of operation (October 1995). Average lead levels for the period of October 1995 through February 2001 were 5.9 μg/L and all three loops were statistically similar for this time period. Copper levels exhibited a definite downward trend through the end of 1996, after which they stabilized at an average level of 0.2 mg/L for the period from January 1997 through February 2001. Copper levels from Loop A were statistically different than copper levels from Loops B and C, and Loops B and C were similar for this time period. A summary of the lead and copper data collected from the lead soldered copper pipe rack at Alta Mesa is shown in Table E.12 and Figure E.10.

Table E.12
Summary statistics for Alta Mesa pipe rack lead and copper levels
(lead soldered copper loops)

	Lead Levels, μg/L					
	March 1992 - October 1992			October 1995 - February 2001		
	Loop A	Loop B	Loop C	Loop A	Loop B	Loop C
Average (x)	6.3	7.7	6.7	5.9	5.9	5.8
Samples	124	127	124	43	42	42
Max-Min	1 - 128	2 - 140	1 - 130	2 – 29	2 – 23	2 - 21
Median (50%)	5.0	5.0	5.0	5.0	5.0	5.0
90th Percentile	12.4	12.0	10.9	11.0	11.0	10.0
Std. Dev.(s)	11.8	15.5	17.7	4.7	3.7	3.3
CV (s/x)	1.9	2.0	1.9	0.8	0.6	0.6
	Copper Levels, mg/L					
	November 1991 - July 1992			March 1999 - September 2001		
	Loop A	Loop B	Loop C	Loop A	Loop B	Loop C
Average (x)	0.39	0.40	0.39	0.16	0.23	0.21
Samples	138	136	138	42	41	41
Max-Min	0.0 – 1.1	0.0 – 1.2	0.0 - 1.2	0.0 - 0.3	0.0 - 0.4	0.0 - 0.4
Median (50%)	0.3	0.3	0.4	0.1	0.3	0.2
90th Percentile	0.7	0.8	0.7	0.3	0.3	0.3
Std. Dev. (s)	0.25	0.24	0.23	0.07	0.09	0.09
CV (s/x)	0.7	0.6	0.6	0.5	0.4	0.5

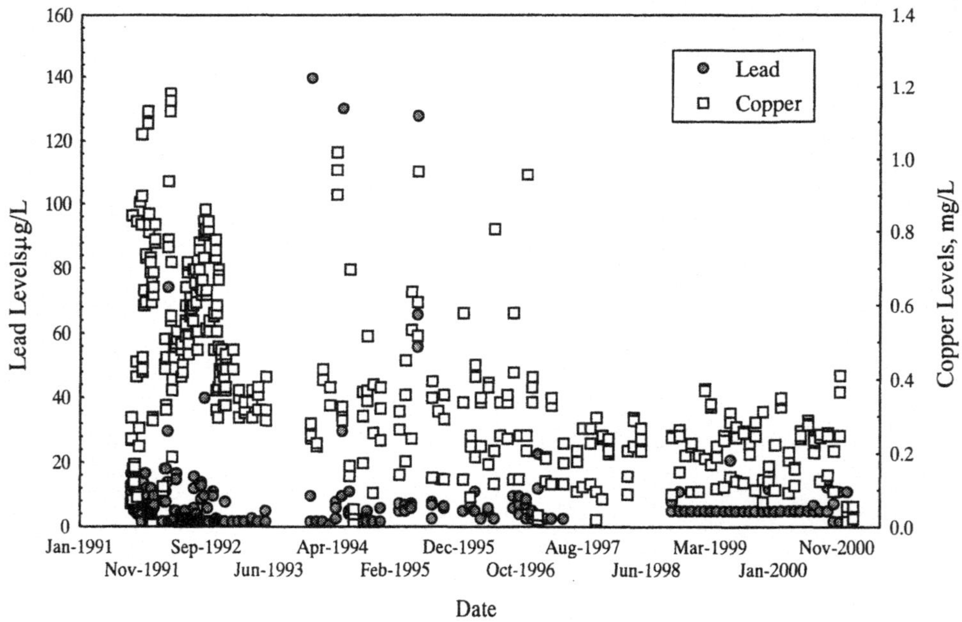

Figure E.10 Lead and copper levels from Fort Worth's Alta Mesa pipe rack (lead soldered copper pipe)

Como Pipe Rack. Lead levels averaged 223 μg/L and did not stabilize during the initial operation of the pipe rack in 1991 – 1992. As would be expected, copper levels were extremely low (0.01 mg/L). From March 1997 through March 2001, lead levels stabilized, with average levels of 126 μg/L. All three loops exhibited statistically similar lead levels for this time period. A summary of the lead and copper data collected from the lead pipe rack at the Como Pump Station is shown in Table E.13 and Figure E.11.

Table E.13
Summary statistics for Como pipe rack lead and copper levels (lead loops)

	Lead Levels, μg/L					
	October 1991 – October 1992			March 1997 - March 2001		
	Loop A	Loop B	Loop C	Loop A	Loop B	Loop C
Average (x)	207.9	197.2	207.6	118.3	122.5	135.4
Samples	112	112	112	22	22	22
Max-Min	10 – 1280	2 – 1910	2 - 3020	10 – 404	6 - 305	5 - 342
Median (50%)	151.5	158.1	138.5	121	99	110
90th Percentile	428.0	421.0	342.0	180.0	279.0	275.0
Std. Dev.(s)	219.4	226.5	330.5	91.4	97.9	106.4
CV (s/x)	1.1	1.1	1.6	0.8	0.8	0.8

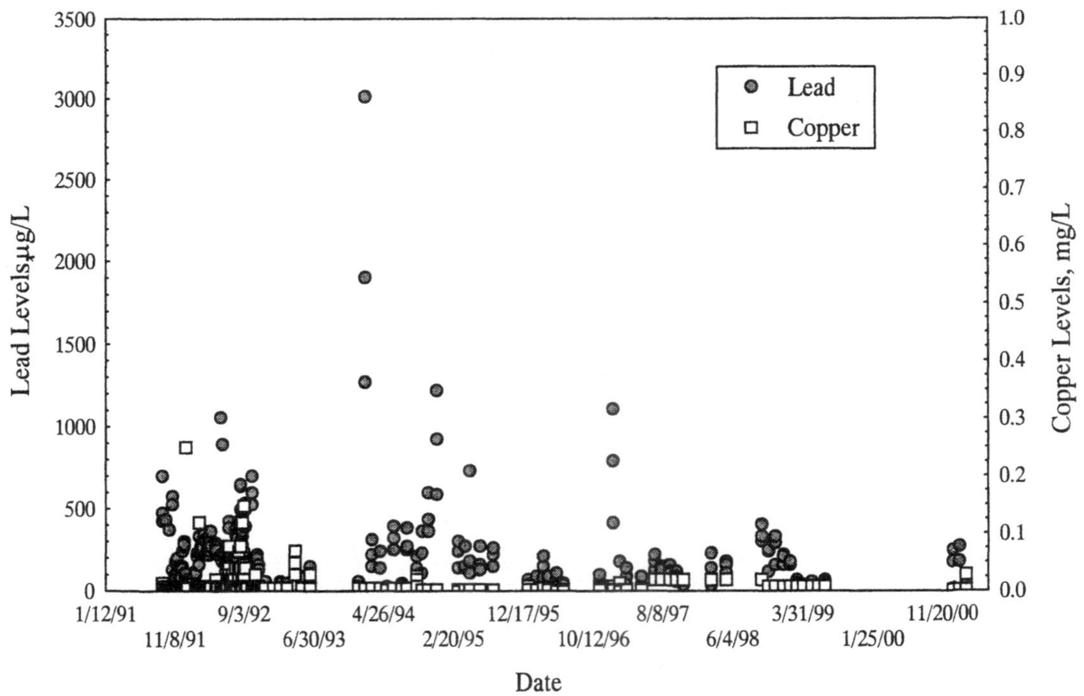

Figure E.11 Lead and copper levels from Fort Worth's Como pipe rack (lead pipe)

Historical Tap Monitoring

Table E.14 summarizes the results of the lead and copper monitoring for each of the six regulatory monitoring rounds, and Figures E.12 and E.13 display lead and copper levels, respectively, for the regulatory sample periods from 1997 through 1999. The 90th percentile levels for both lead and copper have been below the action levels for all six rounds of monitoring. Since Spring 1997, both lead and copper levels have been further reduced and copper levels have become less variable with time.

Table E.14
Results of regulatory monitoring from all sites at Fort Worth

	Lead, μg/L					
	Spring 92	Fall 92	Spring 97	Fall 97	Fall 98	Fall 99
Average (x)	4.4	4.9	3.3	2.8	2.3	3.1
Number of Samples (N)	106	102	100	100	50	50
Min – Max	0 – 53.4	1 – 46.4	0.3 – 56.2	0.3 – 22.3	0.3 – 14.0	0.3 – 31.2
Median (50%)	2	2.6	1.5	1.5	1.3	1.7
90th Percentile	11.9	10.3	8.7	5.8	4.9	5.8
Std. Dev(s)	6.9	6.6	6.2	3.5	2.6	4.9
CV (s/x)	1.6	1.3	1.9	1.3	1.1	1.6
	Copper Levels, mg/L					
	Spring 92	Fall 92	Spring 97	Fall 97	Fall 98	Fall 99
Average (x)	0.32	0.45	0.27	0.28	0.16	0.20
Number of Samples (N)	106	102	100	100	50	50
Min – Max	0 – 1.04	0.01 – 1.41	0 - 0.84	0.01 - 0.86	0 - 0.44	0.01 - 0.48
Median (50%)	0.17	0.28	0.21	0.15	0.15	0.18
90th Percentile	0.71	0.94	0.59	0.63	0.32	0.38
Std. Dev(s)	0.30	0.41	0.23	0.25	0.13	0.15
CV (s/x)	0.9	0.9	0.9	0.9	0.8	0.7

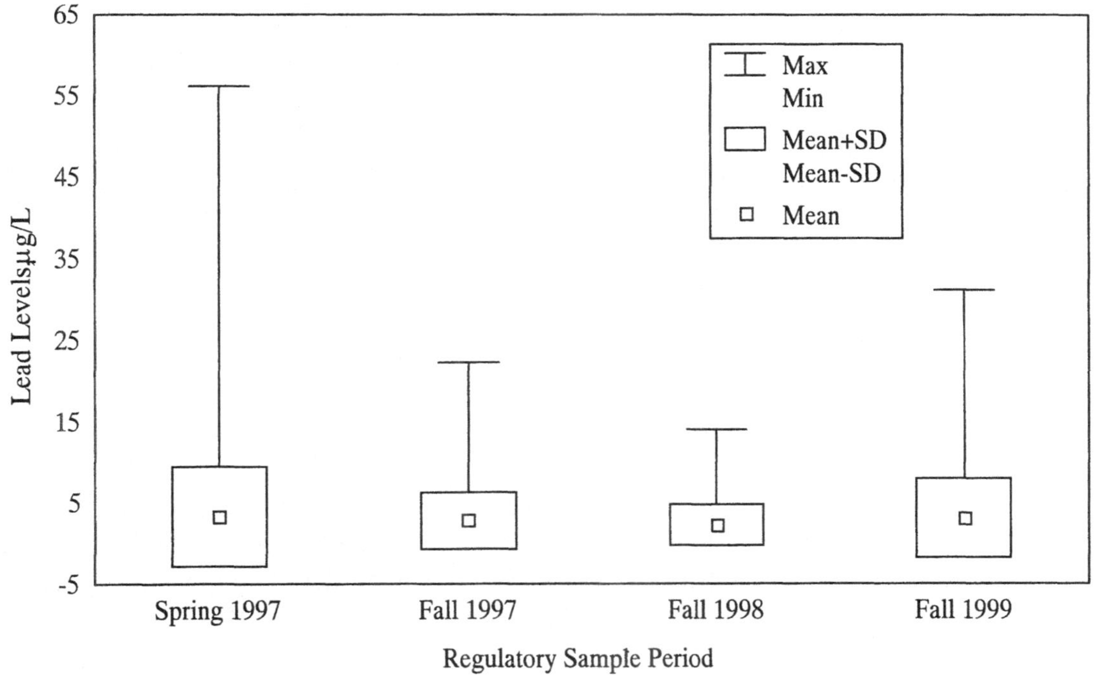

Figure E.12 Lead levels from regulatory monitoring at Fort Worth

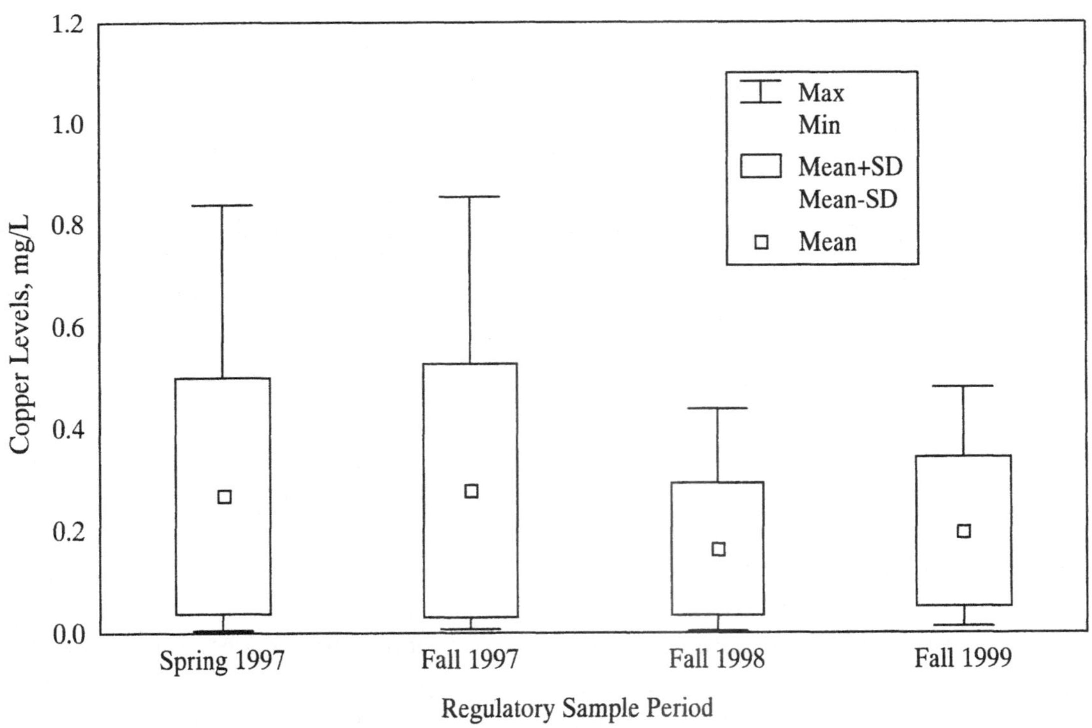

Figure E.13 Copper levels from regulatory monitoring at Fort Worth

Summary statistics for lead levels from lead service line sites sampled during regulatory monitoring are listed in Table E.15.

Table E.15
Lead service site results from regulatory monitoring at Fort Worth, (Lead µg/L)

	Spring 92	Fall 92	Spring 97	Fall 97	Fall 98	Fall 99
Average (x)	5.2	5.9	5.6	4.2	3.9	5.2
Samples	49	50	50	51	21	21
Min – Max	0 – 27.3	1 – 22.3	0.5 – 56.2	0.3 – 22.3	0.3 – 14.0	0.9 – 31.2
Median (50%)	3.0	4.5	3.4	3.0	2.9	3.1
90th Percentile	15.1	12.3	11.6	8.5	6.9	6.6
Std. Dev. (s)	6.2	5.0	8.2	4.4	3.2	7.0
CV (s/x)	1.2	0.8	1.5	1.0	0.8	1.4

Approximately one half of the sites sampled in 1992 contained lead service lines. In the 1998 and 1999 sampling periods, 21 of 50 total sites sampled had lead service lines. Summary statistics for lead and copper levels from sites containing lead soldered copper piping are listed in Table E.16. Lead levels from these sites were lower than the lead service line sites, however the variability in lead levels was similar.

Table E.16
Lead soldered copper site results from regulatory monitoring at Fort Worth

Lead (µg/L)	Spring 92	Fall 92	Spring 97	Fall 97	Fall 98	Fall 99
Average (x)	3.8	3.9	1.0	1.4	1.2	1.5
Samples	57	52	50	49	29	29
Min – Max	0 – 53.4	1 – 46.6	0.3 – 3.4	0.3 – 8.6	0.3 – 4.7	0.3 – 6.8
Median (50%)	1.4	1.6	0.9	0.9	1.0	1.0
90th Percentile	8.8	4.8	2.0	2.8	3.3	4.4
Std. Dev(s)	7.5	7.8	0.7	1.3	1.0	1.5
CV (s/x)	2.0	2.0	0.7	1.0	0.9	1.0
Copper (mg/L)	Spring 92	Fall 92	Spring 97	Fall 97	Fall 98	Fall 99
Average (x)	0.50	0.81	0.45	0.44	0.23	0.28
Samples	57	52	50	49	29	29
Min – Max	0 – 1.0	0.2 – 1.4	0.1 - 0.8	0.1 - 0.9	0.0 - 0.4	0.1 - 0.5
Median (50%)	0.6	0.8	0.5	0.5	0.3	0.3
90th Percentile	0.8	1.0	0.7	0.7	0.4	0.4
Std. Dev(s)	0.27	0.23	0.17	0.20	0.10	0.10
CV (s/x)	0.5	0.3	0.4	0.5	0.4	0.4

Lead and Copper Level Comparisons

Lead and copper levels measured from regulatory monitoring for 1997 through 1999 and from the pipe racks during the same time period were compared. Lead and copper levels from the Alta Mesa pipe rack were compared to regulatory sites containing lead soldered copper piping, and lead levels from the Como pipe rack were compared to regulatory sites with lead service lines.

Lead levels from the Como pipe rack were significantly higher than lead levels from the lead service line regulatory monitoring sites for this period of time (Table E.17 and Figure E.14).

Table E.17
Lead levels from Fall 1999 through Fall 2001 – lead pipe

Range of Values	Como Pipe Rack (µg/L)	Lead Service Sites (µg/L)
Average	156	4.8
Min – Max	5 – 1111	0.3 – 56.2
Median (50%)	120	3.0
90th Percentile	305	9.9
Standard Deviation (s)	178	6.2
CV (s/x)	1.1	1.3

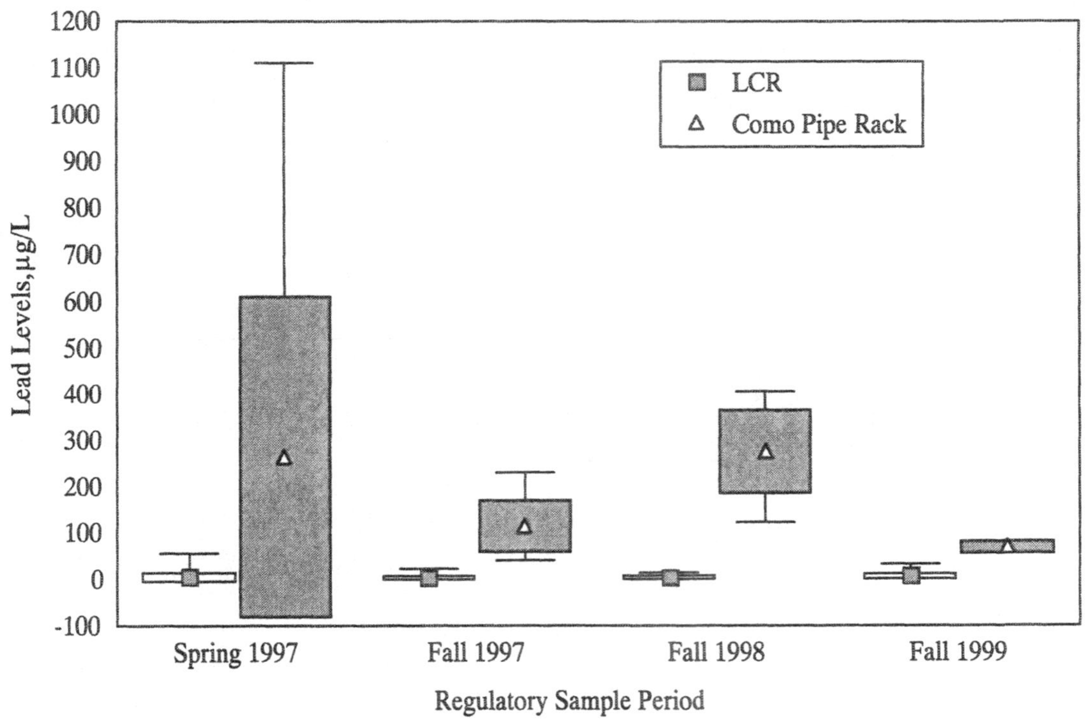

Figure E.14 Comparison of lead levels from lead sites at Fort Worth

Lead levels from the Alta Mesa pipe rack were also higher than regulatory monitoring during this time period, whereas copper levels from the pipe rack were lower (Tables E.18 and E.19 and Figures E.15 and E.16).

Table E.18
Lead levels from Fall 1999 through Fall 2001 - lead soldered copper pipe

Range of values	Alta Mesa pipe rack (µg/L)	LCR – lead soldered copper sites (µg/L)
Average	6.3	1.2
Min – max	3.0 – 29	0.3 – 8.6
Median (50%)	5.0	1.0
90th percentile	12	2.3
Standard deviation (s)	5.1	1.1
CV (s/x)	0.8	0.9

Table E.19
Copper levels from Fall 1999 through Fall 2001 – lead soldered copper pipe

Range of values	Alta Mesa pipe rack (mg/L)	LCR – lead soldered copper sites (mg/L)
Average	0.20	0.37
Min – max	0.02 - 0.38	0.02 – 0.86
Median (50%)	0.21	0.36
90th percentile	0.30	0.62
Standard deviation (s)	0.09	0.19
CV (s/x)	0.4	0.5

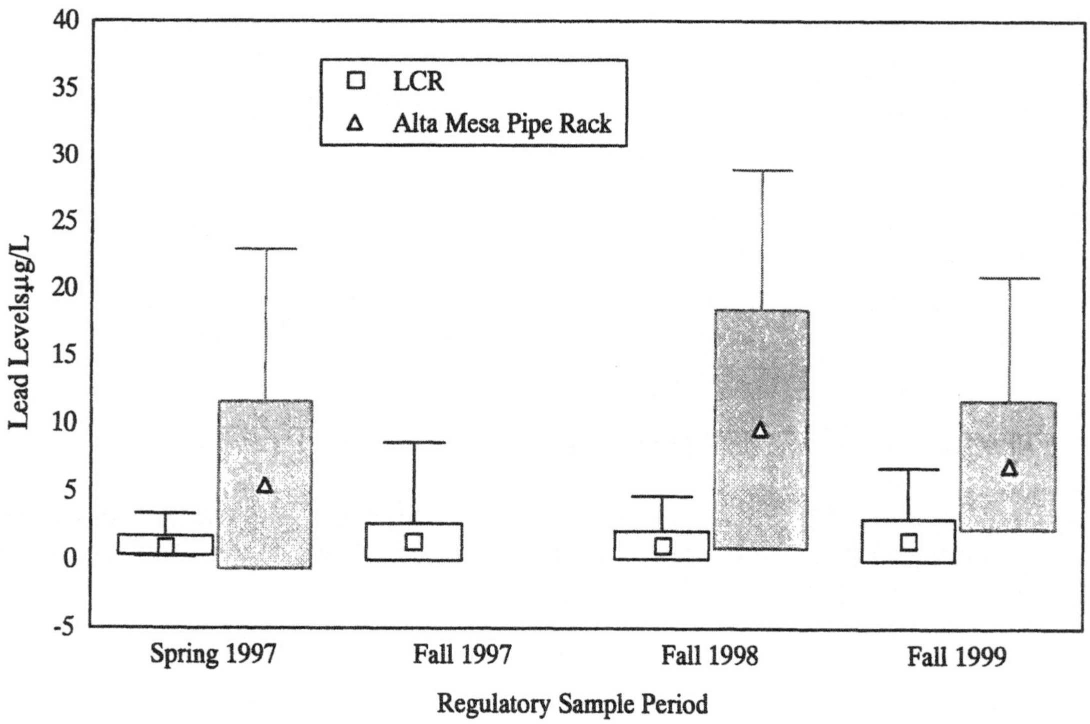

E.15 Comparison of lead levels from lead soldered copper sites at Fort Worth

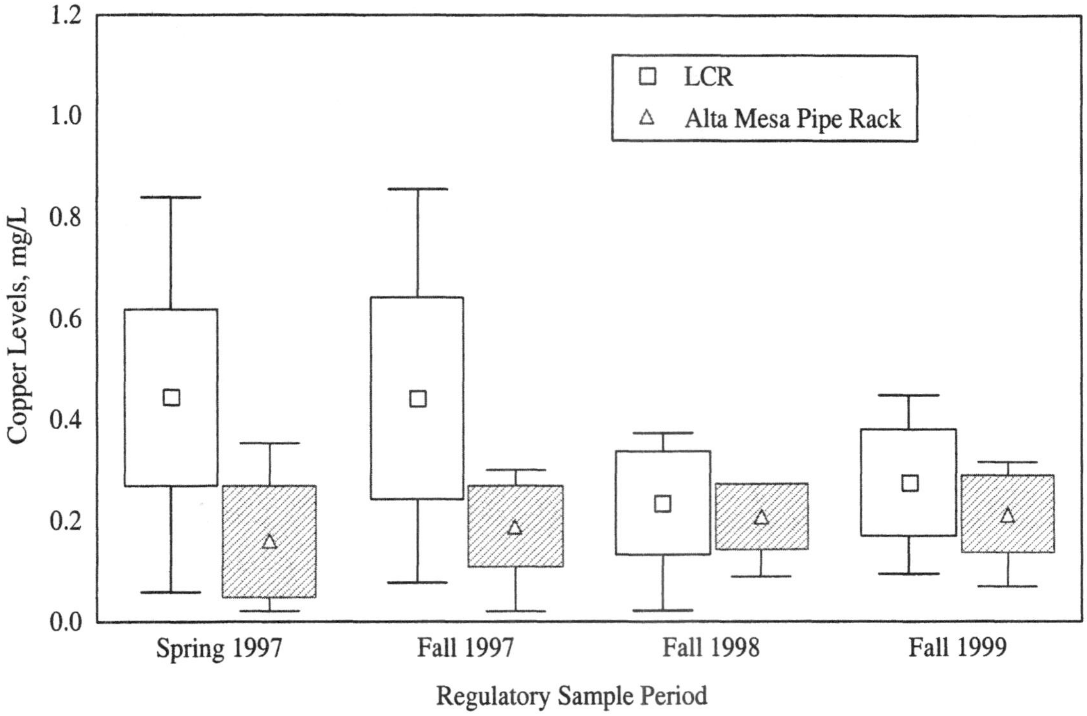

Figure E.16 Comparison of copper levels from lead soldered copper sites at Fort Worth

Neither lead nor copper levels from the Alta Mesa pipe rack were statistically similar to regulatory monitoring lead levels for this time period. However, copper levels measured from the pipe rack were similar to copper levels from regulatory monitoring during the fall 1998 sample period.

Conclusions

Based on the evaluation of lead and copper data from the two pipe racks operated by Fort Worth, the following conclusions can be drawn:

- Standing lead levels measured for the lead soldered pipe rack remained stable with continued long-term operation of the system (longer than 9 years).
- A narrower range of lead levels was observed after approximately 1500 days of operation (October 1995).
- Lead levels were statistically similar for all three replicate loops in both the lead soldered copper pipe rack and the pipe rack containing lead piping.
- Lead levels from the pipe rack containing lead piping were higher than lead levels from the lead soldered copper pipe rack.
- Lead levels from the lead pipe rack did not stabilize until March 1997.
- Copper levels from the lead soldered copper pipe rack declined with continued long-term operation of the system (longer than 9 years). Copper levels stabilized at an average level of 0.2 mg/L. for the period from January 1997 through February 2001.

- Lead levels from both the lead soldered pipe rack (Alta Mesa) and the lead pipe rack (Como) were significantly higher than lead levels from regulatory monitoring sites with similar piping types.
- In general, copper levels from the lead soldered copper pipe rack were lower than copper levels from regulatory monitoring sites containing lead soldered copper piping, however both data sets exhibited low values (below 0.5 mg/L).
- Variability in copper levels from the lead soldered copper pipe rack was similar to variability in copper levels from regulatory monitoring for Fall 1999 through Fall 2001.

REFERENCES

American Public Health Association, American Water Works Association, Water Pollution Control Federation (APHA, AWWA, and WPCF). 1995. *Standard Methods for the Examination of Water and Wastewater*, 18th Edition. APHA, AWWA, and WPCF: Washington, D.C.

American Society for Metals. 1979. *Source Book on Copper and Copper Alloys*. Metals Park, Ohio: American Society for Metals.

American Water Works Association. 1990. *Lead Control Strategies*. Denver, CO.:AWWA.

AwwaRF. 1990. *Lead Control Strategies*. AwwaRF and American Water Works Association. Denver, CO.

AwwaRF and DVGW-Technologiezentrum Wasser. 1996. *Internal Corrosion of Water Distribution Systems*, Second Edition. Denver, CO: AwwaRF and DVGW-Technologiezentrum Wasser.

Baitz, S., and H. Follner. 2002. Nanocrystalline Coatings in Copper Pipes for Cold Water Plumbing Systems. *Chem. Eng. Technol.*, 25 (8): 841-848.

Barceloux, D. G. 1999. Copper. *J. Toxicol.-Clinincal Toxicol.*, 37 (2): 217-230.

Benjamin, M. M. 2002. *Water Chemistry*. McGraw Hill, Boston, MA.

Boffardi, B. P., and A. M. Sherbondy. 1991. Control of Lead Corrosion by Chemical Treatment. *Corrosion*, 47: 966-975.

Boulay, N., and M. Edwards. 2001. Role of Temperature, Chlorine, and Organic Matter in Copper Corrosion By-Product Release in Soft Water. *Water Res.*, 35 (3): 683-690.

Britton, A., and W. N. Richards. 1981. Factors Influencing Plumbosolvency in Scotland. *Jour. Inst. Wat. Eng. Sci.* 35: 349-364.

Broo, A. E., B. Berghult, and T. Hedberg. 1999. Drinking Water Distribution - The Effect of Natural Organic Matter (NOM) on the Corrosion of Iron and Copper. *Water Sci. Technol.*, 40 (9): 17-24.

Clement, M., R. Seux and S. Rabarot. 2000. A Practical Model for Estimating Total Lead Intake from Drinking Water. Water Res., 34 (5): 1533-1542.

Colling, J. H., P. A. E. Whindcup, and C. R. Hayes. 1987. The Measurement of Plumbosolvency Propensity to Guide the Control of Lead in Tap Waters. *J. Inst. Water & Envir. Manag.*, , 1: 263-269.

CRC Handbook of Chemistry and Physics. 1990. CRC Press: Baton Rouge.

Cussler, E. L. 1984. *Diffusion. Mass Transfer in Fluid System*. Cambridge University Press, Cambridge.

de Mora, S. J., R. M. Harrison, and S. J. Wilson. 1987. The Effect of Water Treatment on the Speciation and Concentration of Lead in Domestic Tap Water Derived from a Soft Upland Source. *Water Res.*, 21: 83-94.

Edwards, M, S. Jacobs and R. J. Taylor. 2000. "The Blue Water Phenomenon". *Jour. AWWA*. Vol 90(7): 72 – 82.

Edwards, M., S. Jacobs, and D. Dodrill. 1999. Desktop Guidance for Mitigating Pb and Cu Corrosion By-Products. *Jour. AWWA*, 91 (5): 66-77

Edwards, M. A., T. E. Meyer, J. Rehring, J. F. Ferguson, G. V. Korshin, and S. A. L. Perry. 1996a. *Role of Inorganic Anions, NOM and Water Treatment Processes in Copper Corrosion*. AwwaRF and AWWA: Denver, CO.

Edwards, M., M. R. Schock, and T. E. Meyer. 1996b. Alkalinity, pH, and Copper Corrosion By-Product Release. *Jour. AWWA,* 88 (3): 81-94

Edwards, M. A., J. F. Ferguson, and S. H. Reiber. 1994. The Pitting Corrosion of Copper. *Jour. AWWA*, 86 (7): 74-90.

Edwards, M. A., and J. F. Ferguson. 1993. Accelerated Testing of Copper Corrosion. *Jour. AWWA*, 85 (10): 105-113.

EES and ISWS (Economic and Engineering Services, Inc. and Illinois State Water Survey). 1990. *Lead Control Strategies*. Denver, Colo.: AwwaRF and AWWA.

Ferguson, J. F., O. von Franqué, and M. R. Schock. 1996. Corrosion of Copper in Potable Water systems. In: *Internal Corrosion of Water Distribution Systems*. 1996. Denver, CO: Second Edition. AwwaRF and DVGW-Technologiezentrum Wasser, pp. 231-268.

Frateur, I., C. Deslouis, M. E. Orazem, and B. Tribollet. 1999. Modeling of the Cast Iron/Drinking Water System by Electrochemical Impedance Spectroscopy. *Electrochem. Acta*, 44 (24): 4345-4356.

Gardels, M. C., and T. J. Sorg. 1989. A Laboratory Study of the Leaching of Lead from Water Faucets. *Jour. AWWA*, 81(7): 101-113.

Gregory, R. J. 1990. Galvanic Corrosion of Lead Solder in Copper Pipework. *Inst. Water & Envir. Manag.*, 4: 112-118.

Hidmi, L. and M. Edwards. 1999. Role of Temperature and pH in $Cu(OH)_2$ Solubility. *Environ. Science Technol.*, 33 (15): 2607-2610

Hong, P. K. A., and Y. Y. MacAuley. 1998. Corrosion and Leaching of Copper Tubing Exposed to Chlorinated Drinking Water. *Water Air Soil Pollut.*, 108 (3-4): 457-471.

Hulsmann, A.D. 1990. Particulate Lead in Water Supplies. *J. Inst. Water & Envir. Manag.* 4: 19-25.

Jacobs, S., S. Reiber, and M. Edwards. 1998. Sulfide-Induced Copper Corrosion. *Jour. AWWA.,* 90 (7): 62-73.

Korshin, G. V., J. F. Ferguson, and A. N. Lancaster. 2000. Influence of Natural Organic Matter on the Corrosion of Leaded Brass in Potable Water. Behavior of the Lead Phase. *Corrosion Science*, 42 (1): 53-66.

Koryta, J. 1975. *Ion-Selective Electrodes*. Cambridge University Press, New York, NY.

Kuch, A., and I. Wagner. 1983. Mass Transfer Model to Describe Lead Concentrations in Drinking Water. *Water Research,* 17(10): 1303-1307.

Leban, M., A. Legat, and V. Dolecek. 2001. Electrochemical Noise during Non-Stationary Corrosion Processes. *Mater. Corros.*, 52 (6): 418-425

Lee, R.G., W.C.Becker and D.W.Collins. 1989. Lead at the Tap: Sources and Control. *Jour. AWWA,* 81 (7): 52-62.

Levin 1989. Exposure to Lead in U.S. Drinking Water. Paper presented at the *Twenty-third Annual Conference on Trace Substances in Environmental Health.* May 30 – June 1, Cincinnati, OH.

Lytle, D. A, and M. R. Schock. 2000. Impact of Stagnation Time on Metal Dissolution from Plumbing Materials in Drinking Water. *J. Water Supply Res Technol.-Aqua*, 49 (5): 243-257.

McClave, J. T. and P. G. Benson. 1982. Statistics for Business and Economics. Second Edition. Prentice Hall. Saddle River, N.J. 1028 pages.

Merkel, T. 2003. Untersuchungen zu den Chemischen Reaktionen bei der Flachenkorrosion des Kupfers in Trinkwasserinstallationen. Ph.D. Dissertation, University of Karlsruhe, Karslruhe, Germany.

Merkel, T., H.-J. Grob, W. Werner, T. Dahlke, S. Reicherter, G. Beuchle and S. H. Eberle. 2002. Copper Corrosion By-Product Release in Long-Term Stagnation Experiments. *Water Res.*, 36 (6): 1547-1555.

Merkel, T., H.-J. Groß, G. Beuchle and S. H. Eberle. 2001. Kupferkorrosion. Untersuchungen zur Deckschichtbildung und Korrosionskinetik. *GWA*, 81 (8): 511-519.

Marani, D., G. Macchi, and M. Pagano. 1995. Lead Precipitation in the Presence of Sulfate and Carbonate: Testing of Thermodynamic Predictions. *Water Res,* 29 (4): 1085-1092.

Miller, I. And J. Freund. 1985. Probability and Statistics for Engineers. Third Ed. Prentice-Hall, Inc. Englewood Cliffs, NJ

Neff, C. H., M. R. Schock and J. I. Marden. 1987. *Relationships Between Water Quality and Corrosion of Plumbing Materials in Buildings.* Illinois State Water Survey project report for Grant No. CR80856010. EPA/600/S2-87/036. USEPA.

Newman, J. S. 1972. *Electrochemical Systems.* Prentice-Hall. Englewood Cliffs, N.J.

Paige, J. I., and B. S. Covino Jr. 1992. Leachibility of Lead from Selected Copper-based Alloys. *Corrosion*, 48(12): 1040-1046.

Pankow, J. F. 1991. *Aquatic Chemistry Concepts.* Lewis Publishers, Chelsea, MI.

Rehring, J. P. and M. Edwards. 1996. Copper Corrosion in Potable Water Systems: Impacts of Natural Organic Matter and Water Treatment Processes. *Corrosion*, 52 (4): 307-317.

Reiber, S. 1991. Galvanic Stimulation of Corrosion of Lead-Tin Solder-Sweated Joints. *Journ. AWWA*, 83(7): 83-91.

Reiber, S., R. A. Ryder, and I. Wagner. 1996. Corrosion Assessment Technologies. In *Internal Corrosion of Water Distribution Systems.* Denver, CO: AWWA.

Schecher, W. D. 2001. *Thermochemical Data used in MINEQL+ Version 4.5.* Environmental Research Software, Hallowell, ME.

Schecher, W. D., and D.C. McAvoy. 1998. *MINEQL+. A Chemical Equilibrium Modeling System. Version 4.0 for Windows.* Environmental Research Software, Hallowell, ME.

Schmitt, G., E. Slavcheva, and P. Plagemann. 2001. ECN-Measurements at Copper in Artificial Tap Water - Investigation of Anion Effects. *Mater. Corros.*, 52 (6): 439-444.

Schock, M. R., I. Wagner, and R. J. Oliphant. 1996. Corrosion and Solubility of Lead in Drinking Water. In *Internal Corrosion of Water Distribution Systems.* Denver, CO: AWWA.

Schock, M. R., D. A. Lytle, J. A. Clement. 1995. *Effect of pH, DIC, Orthophosphate and Sulfate on Drinking Water Cuprosolvency.* EOA/600/R-95/085. National Risk Management Research laboratory. Office of Research and Development. USEPA. Cincinnati, OH.

Schock, M. R. 1990. Causes of Temporal Variability of Lead in Domestic Plumbing Systems. *Env. Monitor. Assess.*, 15: 59-82.

Schock, M. R. 1989. Understaning Corrosion Control Strategies for Lead. *Jour. AWWA*, 81(7): 88-100.

Schock, M. R., R. Levin 1988. "The Significance of Sources of Temporal Variability of Lead in Corrosion Evaluation and Monitoring Program Design. " *Proc. AWWA Water Quality Technology Conference*, St. Louis, MO, AWWA.

Schock, M.R., and C.H.Neff. 1988. Traced Metal Contamination from Brass Fittings. *Journ. AWWA*, 80 (11): 47-56.

Schock, M.R., and I. Wagner. 1985. The Corrosion and Solubility of Lead in Drinking Water. Chapter 4 in AwwaRF and DVGW Forschungsstelle, *Internal Corrosion of Water Distribution Systems*. Denver; AwwaRF.

Schock, M. R., and M. C. Gardels. 1983. Plumbosolvency Reduction by High pH and Low Carbonate - Solubility Relationship. *Jour. AWWA*, 75 (2): 87-91.

Snoeyink, V. L., and I. Wagner. 1996. Principles of Corrosion of Water Distribution Systems. In *Internal Corrosion of Water Distribution Systems*. Denver, CO: AWWA.

Swistock, B. R., W. E. Sharpe, and P. D. Robillard. 1993. A Survey of Lead, Nitrate and Radon Contamination of Private Individual Water Systems in Pennsylvania. *Jour. Env. Health*, 55: 6-13.

Taylor, P., and V. J. Lopata. 1984. Stability and Solubility Relationships between Some Solids in the System PbO-CO_2-H_2O. *Can. Jour. Chem.*, 62: 395-402.

USEPA (U.S. Environmental Protection Agency). 1991. *Lead and Copper Rule Guidance Manual. Volume 1. Monitoring.* Denver, Co.: American Water Works Association.

USEPA (U.S. Environmental Protection Agency). 1992. *Lead and Copper Rule Guidance Manual. Volume 2. Corrosion Control and Treatment.* Denver, Co.: American Water Works Association.

Van Der Leer, D., N. P. Weatherill, R. J. Sharp, and C. R. Hayes. 2002. Modeling the Diffusion of Lead into Drinking Water. *Appl. Math. Modelling*, 26 (6): 681-699.

Walker, R., and R. Oliphant. 1982. Contamination of Potable Water from Lead-Tin Solders. *Anti-Corrosion*, July: 13-15.

Webster, B.J., S. E. Werner, D. B. Wells, and P. J. Bremer. 2000. Microbiologically Influenced Corrosion of Copper in Potable Water Systems - pH Effects. *Corrosion*, 56 (9): 942-950

Werner, W. 1992. Untersuchungen zum Mechanismus der Flächenkorrosion. in Trinkwasserleitungen aus Kupfer. Ph.D. Dissertation, University of Karlsruhe, Karlsruhe, Germany.

Zhang, X. H., S. O. Pehkonen, N. Kocherginsky, and G. A. Ellis. 2002. Copper Corrosion in Mildly Alkaline Water with the Disinfectant Monochloramine. *Corrosion Sci.*, 44 (11): 2507-2528.

ACRONYMS AND ABBREVIATIONS

A	fitting parameter (factor) in linear correlation between logarithm of total copper concentration and corrosion potential in the presence of chlorine
A/cm^2	Ampere per square centimeter
AE	atomic emission
AL	action level
Alk.	alkalinity
B	fitting parameter (constant) in linear correlation between logarithm of total copper concentration and corrosion potential in the presence of chlorine
BHCD	Bureau of Housing and Community Development
CCT	Corrosion Control Treatment
cm	centimeter
cm^2/s	square centimeter per second (unit of diffusion coefficient)
$C^0_{O_2}$	initial concentration of dissolved oxygen
CPSF	Corrosion Potential Stagnation/Flow
conc.	concentration
CT	concentration xtime
Cu_{diff}	concentration of copper in the diffusion layer
Cu_p	concentration of copper at which malachite begins to precipitate
Cu_{total}	total copper
CV	coefficient of variation
D	diffusion coefficient
DBP	disinfection by-products
D_{Cu}	diffusion coefficient for copper
D/DBPR	Disinfection/Disinfectant By-product Rule
DEP	Department of Environmental Protection
DHS	Department of Human Services
D_{O_2}	diffusion coefficient for oxygen
DIC	dissolved inorganic carbon
DO	dissolved oxygen
DOC	dissolved organic carbon
DOH	Department of Health
DWSD	Detroit Water and Sewerage Department
E	electrochemical potential
E_0	electrochemical potential of a redox couple
E_{Cu}	standard electrochemical potential of the Cu/Cu^{2+} couple at a current copper concentration
E^0_{Cu}	normal standard electrochemical potential of the Cu/Cu^{2+} couple
EIS	electrochemical impedance spectroscopy

E_{corr}	corrosion potential
E_{O_2}	standard electrochemical potential of the O_2/H_2O couple at current oxygen concentration
$E^0_{O_2}$	normal standard electrochemical potential of the O_2/H_2O couple
F	Faraday constant
g	grams
H	activity of H^+ ion
hr	hour
i_{corr}	corrosion current
i^0_{Cu}	standard exchange current for copper oxidation
i_{Cu}	electrochemical current of copper oxidation
in.	inch
$i^0_{O_2}$	standard exchange current for oxygen reduction
i_{O_2}	electrochemical current of oxygen reduction
ICP	inductively coupled plasma
ISE	ion-selective electrode
K_1	first deprotonation constant
K_2	second deprotonation constant
k_p	rate of formation of malachite
L	liter
l_{cell}	thickness of a CPSF corrosion cell
LCR	Lead and Copper Rule
LHRP	Lead Hazard Reduction Program
LPS	linear polarization scan
L/min	liter per minute
M	molarity
MGD	million gallons per day
mg/L	milligram per liter
MΩ/cm	megaohm per centimeter
μg/L	micrograms per liter
μmhos/cm	micro mhos per centimeter
μm	micrometers
mL	milliliter
MUD	Metropolitan Utilities District
mV	millivolt

NCCWC	North Clackamas County Water Commission
n.h.e.	normal hydrogen electrode
NOM	natural organic matter
nm	nanometers
NRC	Narrow Rectangular Channel
NTNC	non-transient/non-community
$O_{2\,diff}$	concentration of oxygen in the diffusion layer
O&M	operations and maintenance
OLCC	on-line corrosion cell
OLWD	Oak Lodge Water District
PDS	potentiodynamic scan
PVC	polyvinyl chloride
PWD	Philadelphia Water District
PWB	Portland Water Bureau
R^2	regression coefficient
R_p	polarization resistance
s	second
SD	standard deviation
SDWA	Safe Drinking Water Act
SL	Super Long Coil Copper
SPU	Seattle Public Utilities
t	time
t_{relax}	relaxation time
T_{Cua}	anodic Tafel coefficient for the Cu/Cu^{2+} couple
T_{Cuc}	cathodic Tafel coefficient for the Cu/Cu^{2+} couple
TDS	total dissolved solids
THAA	total haloacetic acid
THM	trihalomethane
TNRCC	Texas Natural Resources Conservation Commission
T_{O2a}	anodic Tafel coefficient for the O_2/H_2O couple
TOC	total organic carbon
TTHM	total trihalomethane
USEPA	United States Environmental Protection Agency
V	volt
WQP	water quality parameters
WTP	water treatment plant
XLC	excellence and leadership for communities

α_a	anodic Tafel slope for Cu oxidation
α_c	cathodic Tafel slope for Cu oxidation
$^*\beta_1$	formation constant of the monohydroxo complex
$^*\beta_2$	formation constant of the dihydroxo complex
$^*\beta_3$	formation constant of the trihydroxo complex
$^*\beta_4$	formation constant of the tetrahydroxo complex
β_{CuCO_3}	formation constant of the carbonate complex
$\beta_{Cu(CO_3)_2}$	formation constant of the dicarbonate complex
β_{CuHCO_3}	formation constant of the bicarbonate complex
β_{O_2}	cathodic Tafel coefficient for the O_2/H_2O couple
γ	slope of linear correlation between corrosion potential and logarithm of total copper concentration in the absence of chlorine
δ_{diff}	thickness of diffusion layer
Φ	Nernst factor

IWA Publishing's authorised EU representative for General Product Safety Regulations is Diane D'Arras, 15 rue Duret, 75116 Paris, France, e-mail: safety@iwap.co.uk.

Printed and bound by CPI Group (UK) Ltd, Croydon, CR0 4YY
17/09/2026
02218799-0006